NAOMI KLEIN is an award-winning journalist, author, and film-maker. Her most recent book, *The Shock Doctrine: The Rise of Disaster Capitalism*, was a *New York Times* and number one international bestseller; it is being translated into twenty-seven languages. She writes an internationally syndicated column for *The Nation* and *The Guardian*, is a contributing editor at *Harper's* and reporter for *Rolling Stone*. In 2004, she released *The Take*, a feature documentary about Argentina's occupied factories, co-produced with director Avi Lewis. She is a former Miliband Fellow at the London School of Economics and holds an honorary Doctor of Civil Laws degree from the University of King's College, Nova Scotia.

www.naomiklein.org

From the reviews of *No Logo*:

'Naomi Klein brilliantly charts the protean nature of consumer capitalism, how it absorbs radical challenges to its dominance and turns them into consumer products' Madeleine Bunting, *Guardian*

'The bible for anti-corporate militancy'
Christopher Hirst, *Independent*

'Personable and well-informed, prescient, necessary and ultimately optimistic, *No Logo* paints a vivid picture of spirited, creative rebellion' William Georgiades, *Literary Review*

'Naomi Klein catches the anti-capitalist mood so well it seems unbelievable that *No Logo* was written before the "Battle of Seattle". She expresses brilliantly the rage that so many people feel about what is going on in the world, giving us ammunition against the bosses and the governments' Judith Orr, *Socialist Review*

'A brilliant book' Peter York, *The Times*

guments and gratifyingly cross with not
our own eagerness to buy into it, *No*
etter timed' *Independent on Sunday*

'*No Logo* should be read by anyone who thinks that the Seattle demonstrations were an aberration' *The Economist*

'Athletic, expansive and an antidote to sloppy thinking . . . It's impossible not to notice the prescience of her argument'
Austin Bunn, *Sunday Herald*

'Unerring and serious . . . This is a juicy, salty book'
Ferdinard Mount, *TLS*

'*No Logo* is a comprehensive account of the potential monster that the global economy has created and the actions to thwart it. So brands watch out, there is a loud and strong message here!'
Ali Parker, *Marketing*

'A passionate, well-written and thoroughly researched book'
Jim Dunne, *Sunday Business*

'Just when you thought multi-nationals and crazed consumerism were too big to fight, along comes Naomi Klein with facts, spirit, and news of successful fighters already out there. *No Logo* is an invigorating call to arms for everybody who wants to save money, justice, or the universe' Gloria Steinem

'What corporations fear most are consumers who ask questions. Naomi Klein offers us the arguments to take on the superbrands'
Billy Bragg

'Essential millennial reading' Richard Benson, *Limb by Limb*

From the American and Canadian reviews:

'Klein is a gifted writer; her paragraphs can be as seductive as the ad campaigns she dissects' *New York Times Book Review*

'Naomi Kein's trenchant book is the perfect introduction to and explanation of those stunning events [in Seattle] . . . this book is the very essence of cool' *Globe and Mail*

'To understand how branding drives the global market you couldn't ask for a better guide than Naomi Klein' *Toronto Star*

'A dense, fact-filled publication that makes plain the jargon spouted by all who put profit before basic human needs . . . with its far reaching vision and clear presentation. A well conceived primer on the machinations of the modern consumer world, *No Logo* is required reading for anyone who thinks people should not be treated like machines' *Eye Weekly*

'Nothing short of a complete, user-friendly handbook on the negative effects that '90s überbrand marketing has had on culture, work, and consumer choice . . . an encyclopedic compilation of the decade's fringe and mainstream anti-corporate actions and mind-sets' *Village Voice*

'A powerful and passionate book' *National Post*

'An incredibly important, timely read, and a powerful call to arms' *Georgia Straight*

'*No Logo* finally puts into perspective what the newest generation of fed-up consumers and anti-corporate activists have been trying to verbalize for the past 10 years' *Ottawa Express*

'Generation-X intellectual Naomi Klein could become the next Douglas Coupland with her *No Logo*. She anticipates a revolt against corporate power by younger people seeking brand-free space. Even if the revolt is not in the works yet, her tart writing might inspire one' *Report on Business*

'At once an impressive journalistic analysis and an impassioned rallying cry' *New Brunswick Telegraph*

By the same author:

The Shock Doctrine
Fences and Windows

You might not see things yet
on the surface, but underground,
it's already on fire.

– Indonesian writer Y.B. Mangunwijaya, July 16, 1998

NO SPACE
NO CHOICE
NO JOBS
NO LOGO

NAOMI KLEIN

FOURTH ESTATE • *London*

For Avi

Fourth Estate
An imprint of HarperCollins*Publishers*
1 London Bridge Street
London SE1 9GF

Visit our authors' blog at www.fifthestate.co.uk
Love this book? www.bookarmy.com

This 10th anniversary edition published by Fourth Estate 2010

14

Published in paperback by Harper Perennial, 2005 (reprinted 6 times)
First published in paperback by Flamingo in 2001 (reprinted 35 times)
First published in Great Britain by Flamingo in 2000

A catalogue record for this book is available from the British Library

ISBN 978-0-00-734077-4

Set in Rotis Semi Serif by Archetype

Printed and bound by CPI Group (UK) Ltd, Croydon, CR0 4YY

ACKNOWLEDGMENTS

The four-year process of taking *No Logo* from an idea to a finished book has been exhilarating. It has not, however, been painless and I have relied heavily on the support, understanding and expertise of those around me.

It has been my great honor to have as my editor Louise Dennys, whose intellectual rigor and personal commitment to freedom of expression and human rights have sharpened the arguments in this book and smoothed my rough edges as a writer. She transformed this book in magical ways.

My research assistant, Paula Thiessen, has tracked down many of the most obscure facts and sources. For more than two years she worked tirelessly collecting the statistics that make up this book's many original charts, extracting facts from cagey retail chains and cajoling government agencies around the world to send unpublished reports. She also conducted the book's photo research and has been a calming influence and supportive colleague during what is often lonely work.

My agents at the Westwood Creative Artists, Bruce Westwood and Jennifer Barclay, took on what many would have seen as a risky project, with boundless enthusiasm and determination. They searched the international book world for kindred spirits who would not just publish *No Logo*, but would champion it: Reagan Arthur and Philip Gwyn Jones.

The exceptional team at Knopf Canada has been warm-hearted and cool-headed no matter what the crisis. I am grateful to Michael Mouland, Nikki Barrett, Noelle Zitzer and Susan Burns, as well as to the talented and dedicated team of editors who have strengthened, polished, trimmed and checked this text: Doris Cowan, Alison Reid and Deborah Viets.

I am deeply indebted to John Honderich, publisher of *The Toronto Star*, who gave me a regular column in his newspaper when I was far too young; a space that for almost five years allowed me to develop both the ideas and the contacts that form the foundation of this book. My editors at *The Star*—Carol Goar, Haroon Siddiqui and Mark Richardson—have been enormously supportive through leaves of absence and even wished me well when I left the column to focus my full attention on this project. The writing for *No Logo* began in earnest as a piece for *The Village Voice* on culture jamming and I am indebted to Miles Seligman for his editorial insights. My editor at *Saturday Night*, Paul Tough, has supported me with extended deadlines, research leads, and *No Logo*-themed assignments, including a trip to the Roots Lodge, which helped deepen my understanding of the utopian aspirations of branding.

I received valuable research assistance from Idella Sturino, Stefan Philipa and Maya Roy. Mark Johnston hooked me up in London, Bern Jugunos did the same in Manila and Jeff Ballinger did it in Jakarta. Hundreds of individuals and organizations also cooperated with the research, but a few individuals went far out of their way to ply me with stats and facts: Andrew Jackson, Janice Newson, Carly Stasko, Leah Rumack, Mark Hosler, Dan Mills, Bob Jeffcott, Lynda Yanz, Trim Bissell, Laird Brown, and most of all, Gerard Greenfield. Unsolicited juicy tidbits arrived by post and E-mail from Doug Saunders, Jesse Hirsh, Joey Slinger, Paul Webster and countless other electronic angels. The Toronto Reference Library, the International Labour Organization, the Corporate Watch Web site, the Maquila Solidarity Network, *The Baffler*, SchNEWS, *Adbusters* and the Tao Collective listserves were all invaluable to my research.

I am also grateful to Leo Panitch and Mel Watkins for inviting me to speak at conferences that helped me to workshop the thesis early on, and to my colleagues on the *This Magazine* editorial board for their generosity and encouragement.

Several friends and family members have read the manuscript and offered advice and input: Michele Landsberg, Stephen Lewis, Kyo Maclear, Cathie James, as well as Bonnie, Michael, Anne and Seth Klein. Mark Kingwell has been a dear friend and intellectual mentor. Sara Borins was my first and most enthusiastic reader – of both the proposal and the first draft – and it was the ever-fabulous Sara who insisted that *No Logo* must have a design that matched the spirit of its content. Nancy Friedland, John Montesano, Anne Baines and Rachel Giese stood by me when I was nowhere to be found. My late grandfather, Philip Klein, who worked as an animator for Walt Disney, taught me a valuable lesson early in life: always look for the dirt behind the shine.

My greatest debt is to my husband, Avi Lewis, who for years greeted me every morning with a cup of coffee and a stack of clippings from the business section. Avi has been a partner in this project in every possible way: he stayed up late into the night helping to evolve the ideas in this book; accompanied me on numerous research escapades, from suburban monster malls to Indonesia's export factory zones; and edited the manuscript with centurion attention at multiple stages. For the sake of *No Logo* he allowed our lives to be totally branded by this book, giving me the great freedom and luxury to be fully consumed.

CONTENTS

CONTENTS

NO LOGO

NO LOGO AT TEN

As I write this introduction, thinking about how much branding has changed in ten years, a couple of developments seem worth mentioning off the top. In May of 2009, Absolut Vodka launched a limited-edition line called "Absolut No Label." The company's global public relations manager Kristina Hagbard explains that, "for the first time we dare to face the world completely naked. We launch a bottle with no label and no logo, to manifest the idea that no matter what's on the outside, it's the inside that really matters. . . . We encourage people to think twice about their prejudice, because in an Absolut world, there are no labels."

A few months later, Starbucks tried to avoid being judged by its own label by opening its first unbranded coffee shop in Seattle, called 15th Avenue E Coffee and Tea. This "stealth Starbucks" (as the anomalous outlet immediately became known) was decorated with "one-of-a-kind" fixtures and customers were invited to bring in their own music for the stereo system as well as their own pet social causes—all to help develop what the company called "a community personality." Customers had to look hard to find the small print on the menus: "inspired by Starbucks." Tim Pfeiffer, a Starbucks senior vice president, explained that unlike the ordinary Starbucks outlet that used to occupy the very same piece of retail space, "This one is definitely a little neighborhood coffee shop." After spending two decades blasting its logo onto 16,000 stores worldwide, Starbucks was now trying to escape its own brand.

Clearly the techniques of branding have both thrived and adapted since I published *No Logo*. But in the past ten years I have written very little about developments like these. I realized why while reading William Gibson's 2003 novel *Pattern Recognition*. The book's protagonist, Cayce Pollard, is allergic to brands, particularly Tommy Hilfiger and the Michelin Man. So strong is this "morbid and sometimes violent reactivity to the semiotics of the marketplace" that she has the buttons on her Levi's jeans ground smooth so that there are no corporate markings. When I read those words, I immediately realized that I had a similar affliction. It was not one of those conditions that you are born with but one that develops, over time, due to prolonged overexposure. I didn't used to be allergic to brands. As I confess in the pages of this book, as a child and teenager I was almost obsessively drawn to them. But writing *No Logo* required four years of total immersion in ad culture—four years of watching and rewatching Super Bowl ads, scouring *Advertising Age* for the latest innovations in corporate synergy, reading soul-destroying business books on how to get in touch with your personal brand values, attending corporate seminars on brand management, making excursions to Niketowns, to monster malls, to branded towns. And watching some of the worst movies ever made while taking notes in the dark on product placement.

Some of it was fun. But by the end, it was as if I had passed some kind of threshold and, like Cayce, I developed something close to a brand allergy. Brands lost most of their charm for me, which was handy because once *No Logo* was a bestseller, even drinking a Diet Coke in public could land me in the gossip column of my hometown newspaper.

The aversion extended even to the brand that I had accidentally created: No Logo. From studying brands like Nike and Starbucks, I was well acquainted with the basic tenet of brand management: find your message, trademark and protect it and repeat yourself ad nauseam through as many synergized platforms as possible. I set out to break these rules whenever the opportunity arose. The offers for No Logo spin-off projects (feature film, TV series, clothing line . . .) were rejected. So were the ones from the megabrands and cutting-edge advertising agencies that wanted me to give them seminars on why they were so hated (there was a career to be made, I was learning, in being a kind of anti-corporate dominatrix, making overpaid executives feel good by telling them what bad, bad brands they were). And against all sensible advice, I stuck by the decision not to trademark the title (that means no royalties from a line of Italian No Logo food products, though they did send me some lovely olive oil).

Most important to my marketing detox program, I changed the subject.

Less than a year after *No Logo* came out I put a personal ban on all talk of corporate branding. In interviews and public appearances I would steer discussion away from the latest innovation in viral marketing and Prada's new superstore and toward the growing resistance movement against corporate rule, the one that had captured world attention with the militant protests against the World Trade Organization in Seattle. "But aren't you your own brand?" clever interviewers would ask me endlessly. "Probably," I would respond. "But I try to be a really crap one."

Changing the subject from branding to politics was no great sacrifice because politics was what brought me to marketing in the first place. The first articles I published as a journalist were about the limited job options available to me and my peers—the rise of short-term contracts and McJobs, as well as the ubiquitous use of sweatshop labor to produce the branded gear sold to us. As a token "youth columnist," I also covered how an increasingly voracious marketing culture was encroaching onto previously protected non-corporate spaces—schools, museums, parks—while ideas that my friends and I had considered radical were absorbed almost instantly into the latest marketing campaigns for Nike, Benetton and Apple.

I decided to write *No Logo* when I realized these seemingly disparate trends were connected by a single idea—that corporations should produce brands, not products. This was the era when corporate epiphanies were striking CEOs like lightning bolts from the heavens: Nike isn't a running shoe company, it is about *the idea of transcendence through sports*, Starbucks isn't a coffee shop chain, it's about *the idea of community*. Down on earth these epiphanies meant that many companies that had manufactured their products in their own factories, and had maintained large, stable workforces, embraced the now ubiquitous Nike model: close your factories, produce your products through an intricate web of contractors and subcontractors and pour your resources into the design and marketing required to fully project your big idea. Or they went for the Microsoft model: maintain a tight control center of shareholder/employees who perform the company's "core competency" and outsource everything else to temps, from running the mailroom to writing code. Some called these restructured companies "hollow corporations" because their goal seemed to be to transcend the corporeal world of things so they could be an utterly unencumbered brand. As corporate guru Tom Peters, quoted in this book, put it: "You're a damn fool if you own it!"

The frantic corporate quest to get out of the product business and into the ideas business explained several trends at once. Companies were constantly on the lookout for new meaningful ideas, as well as pristine spaces on which to project them, because creating meaning was their new act of production. And of course jobs were getting crummier: these companies no longer saw producing things as their "core" business.

For me, the appeal of x-raying brands like Nike or Starbucks was that pretty soon you were talking about everything except marketing—from how products are made in the deregulated global supply chain to industrial agriculture and commodity prices. Next thing you knew you were also talking about the nexus of politics and money that locked in these wild-west rules through free-trade deals and at the World Trade Organization, and made following them the pre-condition of receiving much-needed loans from the International Monetary Fund. In short, you were talking about how the world works.

By the time *No Logo* came out, the movement the book documents in its nascent form was already at the gates of the powerful institutions that were spreading corporatism around the world. Tens and then hundreds of thousands of demonstrators were making their case outside trade summits and G8 meetings from Seattle to New Delhi, in several cases stopping new agreements in their tracks. What the corporate media insisted on calling the "anti-globalization movement" was nothing of the sort. At the reformist end it was anti-corporate; at the radical end it was anti-capitalist. But, as I document in this book, what made it unique was its insistent internationalism. All of these developments meant that when I was on book tour, there were many more interesting things to talk about than logos—like where this movement came from, what it wanted and whether there were viable alternatives to the ruthless strain of corporatism that went under the innocuous pseudonym of "globalization." It was a thrilling period and on a personal level, I was deeply relieved not to have to read *Advertising Age* anymore.

In recent years, however, I have found myself doing something I swore I had finished with: re-reading the branding gurus quoted in this book. Guys like

Peters ("Brand! Brand!! Brand!!! That's the message . . . for the late '90s and beyond.") and Scott Bedbury ("a great brand raises the bar—it adds a greater sense of purpose to the experience"). This time, however, it wasn't to try to understand what was happening at the mall but rather at the White House—first under the presidency of George W. Bush and now under Barack Obama, the first U.S. president who is also a superbrand.

There are many acts of destruction for which the Bush years are rightly reviled—the illegal invasions, the defiant defenses of torture, the tanking of the global economy. But the administration's most lasting legacy may well be the way it systematically did to the U.S. government what branding-mad CEOs did to their companies a decade earlier: it hollowed it out, handing over to the private sector many of the most essential functions of government, from protecting borders to responding to disasters to collecting intelligence. This hollowing out was not a side project of the Bush years, it was a central mission, reaching into every field of governance. And though the Bush clan was often ridiculed for its incompetence, the process of auctioning off the state, leaving behind only a shell—or a brand—was approached with tremendous focus and precision. They were good at this. Explaining his administration's mission, Bush's budget director Mitch Daniels said, "The general idea—that the business of government is not to provide services, but to make sure that they are provided—seems self-evident to me."

One company that took over many of those services was Lockheed Martin, the world's largest defense contractor. "Lockheed Martin doesn't run the United States," observed a 2004 *New York Times* exposé. "But it does help run a breathtakingly big part of it. . . . It sorts your mail and totals your taxes. It cuts Social Security checks and counts the United States census. It runs space flights and monitors air traffic. To make all that happen, Lockheed writes more computer code than Microsoft."

No one approached the task of auctioning off the state with more zeal than Bush's much-maligned defense secretary, Donald Rumsfeld. Having spent twenty-odd years in the private sector, heading pharmaceutical and technology companies and sitting on the boards of such blue-chip firms as Sears and Kellogg's, Rumsfeld was steeped in the corporate culture of branding and outsourcing. He entered the Defense Department not with the posture of a public servant but channeling a celebrity CEO—the guy with the guts to downsize and offshore and, most of all, rebrand. For Rumsfeld, his department's brand identity was clear: global dominance. The core competency was combat. For everything else, he said, sounding very much like Bill Gates, "we should seek suppliers who can provide these non-core activities efficiently and effectively." And channeling Tom Peters, he argued, it's time "to stop thinking about things, numbers of things and mass." Addressing the Department of Defense in September 2001, Rumsfeld wanted to know, "Why is DoD one of the last organizations around that still cuts its own checks? When an entire industry exists to run warehouses efficiently, why do we own and operate so many of our own? At bases around the world, why do we pick up our own garbage and mop our own floors, rather than contracting services out, as many businesses do?" Rumsfeld even went after the sacred cow of the military establishment: health care for soldiers. Why were there so many doctors? Rumsfeld wanted to know. "Some of those needs, especially where they may involve general practice or specialties unrelated to combat, might be more efficiently delivered by the private sector."

The laboratory for this radical vision was Iraq under U.S. occupation. From the start Rumsfeld planned the troop deployment like a Wal-Mart vice president looking to shave a few more hours from the payroll. The generals wanted 500,000 troops, he would give them 200,000, with contractors and reservists filling the gaps as needed—a just-in-time invasion. In practice, this strategy meant that as Iraq spiraled out of U.S. control, an ever more elaborate privatized war industry took shape to prop up the bare-bones army. Blackwater, whose original contract was to provide bodyguards for U.S. envoy Paul Bremer, soon took on other functions, including engaging in combat in a battle with the Mahdi Army in 2004. And as the war moved into the jails, with tens of thousands of Iraqis rounded up by U.S. soldiers, private contractors even performed prisoner interrogations, with some facing accusations of torture. The sprawling Green Zone, meanwhile, was run as a corporate city-state, with everything from food to entertainment to pest control handled by Halliburton. Just as companies like Nike and Microsoft had pioneered the hollow corporation, this was, in many ways, a hollow war. And when one of the contractors screwed up—Blackwater operatives opening fire in Baghdad's

Nisour Square in 2007, for instance, leaving seventeen people dead, or Halliburton allegedly supplying contaminated water to soldiers—the Bush administration, like so many hollow brands before, was free to deny responsibility: these were independent contractors, they could argue, there was nothing the government could do but review the contract. Blackwater, which had prided itself on being the Disney of mercenary companies, complete with a line of branded clothing and Blackwater teddy bears, responded to the scandals by—what else?—rebranding. Its new name is Xe Services.

The dream of a hollow state was realized in its purest form at the Department of Homeland Security, a branch of government that, because it was brand new, could be built as an empty shell from the start. Jane Alexander, deputy director of the research wing of the DHS, explained, "We don't make things. If it doesn't come from industry, we are not going to be able to get it." She sounded like former Nike CEO Phil Knight when he explained back in the nineties that, "There is no value in making things anymore." Unlike Nike, however, which tells its contractors what kind of products to make, the Department of Homeland Security didn't even do that. When it decided it needed to build "virtual fences" on the U.S. borders with Mexico and Canada, for instance, Michael P. Jackson, deputy secretary of the department, sent the word out to contractors: "This is an unusual invitation. . . . We're asking you to come back and tell us how to do our business." The department's inspector general explained that Homeland Security "does not have the capacity needed to effectively plan, oversee and execute the [Secure Border Initiative] program."

That same kind of can't-do attitude applied even when the financial system imploded in the fall of 2008 and the U.S. Treasury stepped in with a $700-billion bank bailout. Not only did it fail to attach meaningful strings to the money, but it announced that it did not have the capacity to administer the program. It needed to outsource the rescue of the banks to the very banks that created the disaster and were receiving the bailout funds. A case in point was The Bank of New York Mellon, which received $3 billion. The bank was also awarded the juiciest "master custodian" contract, worth an estimated $20 million, to administer the bailout. As Bank of New York Mellon President Gerald Hassell explained, "It's the ultimate outsourcing—because the Federal

Reserve and the Treasury do not have the mechanics to run the entire program, and we're essentially the general contractor across the entire program."

It was a striking admission. By the end of eight years of self-immolation under Bush, the state still has all the trappings of a government—the impressive buildings, presidential press briefings, policy battles—but it no more did the actual work of governing than the employees at Nike's Beaverton campus actually stitched running shoes. Governing, it seemed, was not its core competency.

The Bush administration's determination to mimic the hollow corporations it admired extended to its handling of the anger its actions inspired around the world. Rather than actually changing or even adjusting its policies, it launched a series of ill-fated campaigns to "rebrand America" for an increasingly hostile world. First came Charlotte Beers, hired as undersecretary of state for public diplomacy and public affairs shortly after the invasion of Afghanistan. Despite the seniority of the post, Beers had no previous diplomatic experience. She had, however, held the top job at both the J. Walter Thompson and Ogilvy & Mather ad agencies, where she built brands for everything from dog food to power drills. When Secretary of State Colin Powell came under criticism for the appointment, he shrugged it off: "There is nothing wrong with getting somebody who knows how to sell something. We are selling a product. We need someone who can rebrand American foreign policy, rebrand diplomacy." Besides, he said, "She got me to buy Uncle Ben's rice."

Only a few months in, the experiment was in disarray. Beers's propaganda materials were greeted with derision. And when she went on a mission to Egypt to improve the perception of the United States among Arab opinion makers, Beers ended up getting lectured on U.S. military bases, blanket support for Israel and wars with unacceptably high levels of civilian casualties. After Beers quietly returned to the private sector, the same thing happened to her successor, Karen Hughes, when she went on several "listening tours," focusing in particular on forging a bond as a "working mother" with Muslim women. In Ankara, she was informed by a group of Turkish women's rights activists that the idea that the United States was an advocate for women's freedoms would remain laughable so long as the occupation of Iraq continued. "This war is really, really bringing your positive efforts to the level of zero," Hidayet Sefkatli Tuksal, an activist with the Capital City Women's Forum, told Hughes. Fatma Nevin Vargun, a Kurdish feminist, added that, "War makes the rights of women completely erased, and poverty comes after war—and women pay the price." Hughes kept a low profile for the remainder of her tenure.

Watching these cringeful attempts to rebrand American during the Bush years, I was convinced that Price Floyd, former director of media relations at the State Department, had it right. After resigning in frustration, he said that the United States was facing mounting anger not because of the failure of its messaging but because of the failure of its policies. "I'd be in meetings with other public-affairs officials at State and the White House," Floyd told *Slate* magazine. "They'd say, 'We need to get our people out there on more media.' I'd say, 'It's not so much the packaging, it's the substance that's giving us trouble.' " Exactly. A powerful, imperialist country is not like a hamburger or a running shoe. You can't get the whole world to change its opinion of it just by getting "out there [to] tell our story," as Charlotte Beers put it. America didn't have a branding problem; surely it had a product problem.

I used to think that, but I may have been wrong. When Barack Obama was sworn in as president, the American brand could scarcely have been more battered—Bush was to his country what New Coke was to Coca-Cola, what cyanide in the bottles had been to Tylenol. Yet Obama, in what was perhaps the most successful rebranding campaign of all time, managed to turn things around. "The election and nomination process is the brand relaunch of the year," declared David Brain, CEO of Edelman Europe, Middle East and Africa, a global public relations giant. Kevin Roberts, global CEO of Saatchi & Saatchi, set out to depict visually what the new president represented. In a full-page graphic commissioned by the stylish *Paper Magazine*, he showed the Statue of Liberty with her legs spread, giving birth to Barack Obama. America, reborn.

So, it seemed that the United States government *could* solve its reputation problems with branding—it's just that it needed a branding campaign and product spokesperson sufficiently hip, young and exciting to compete in today's tough market. The nation found that in Obama, a man who clearly has a natural feel for branding and who has surrounded himself with a team of top-flight marketers. His social networking guru, for instance, is Chris Hughes, one of the young founders of Facebook. His social secretary is Desirée Rogers, a glamorous Harvard MBA and former marketing executive. And David Axelrod, Obama's top advisor, was formerly a partner in ASK Public Strategies, a PR firm which, according to *BusinessWeek* "has quarterbacked campaigns" for everyone from Cablevision to AT&T. Together, the team has marshaled every tool in the modern marketing arsenal to create and sustain the Obama brand: the perfectly calibrated logo (sunrise over stars and stripes); expert viral marketing (Obama ringtones); product placement (Obama ads in sports video games); user-generated content (Obama Girl? Genius!); a thirty-minute infomercial (which could have been cheesy but was universally heralded as "authentic"); and the choice of strategic brand alliances (Oprah for maximum reach, the Kennedy family for gravitas, and no end of hip-hop stars for street cred).

The first time I saw the "Yes We Can" video, the one produced by Black Eyed Peas front man will.i.am, featuring celebrities speaking and singing over a Martin Luther King–esque Obama speech, I thought—finally, a politician with ads as cool as Nike. The ad industry agreed. A few weeks before he won the presidential elections, Obama beat out Nike, Apple, Coors and Zappos to win the Association of National Advertisers' top annual award, Marketer of the Year. It was certainly a shift. In the nineties, brands upstaged politics completely. Now corporate brands were rushing to piggyback on Obama's caché (to wit: Pepsi-Cola's "Choose Change" campaign, Ikea's "Embrace Change '09" and Southwest Airlines' offer of "Yes You Can" tickets).

Indeed everything Obama and his family touches turns to branding gold. J.Crew saw its stock price increase 200 percent in the first six months of Obama's presidency, thanks in part to Michelle's well known fondness for the brand. Obama's much-discussed attachment to his BlackBerry has been similarly good news for Research In Motion. The surest way to sell magazines and newspapers in these difficult times is to have an Obama on the cover, and you only need to call three ounces of vodka and some fruit juice an Obamapolitan or a Barackatini and you can get $15 for it, easy. In February 2009, *Portfolio* magazine put the size of "the Obama economy"—the tourism he generates and the swag he inspires—at $2.5 billion. Not at all bad in an economic crisis. Desirée Rogers, the White House's social secretary, got into trouble with some of her colleagues when she spoke too frankly with *The Wall Street Journal.* "We have the best brand on earth: the Obama brand," she said. "Our possibilities are endless."

The exploration of those possibilities did not end, or even slow, with the election victory. Bush had used his ranch in Crawford, Texas, as a backdrop to perform his best impersonation of the Marlboro Man, forever clearing brush, having cookouts and wearing cowboy boots. Obama has gone much further, turning the White House into a kind of never-ending reality show starring the lovable Obama clan. This too can be traced to the mid-nineties branding craze, when marketers grew tired of the limitations of traditional advertising and began creating three-dimensional "experiences"—branded temples where shoppers could crawl inside the personality of their favorite brands. Desirée Rogers sounds just like those branding gurus when she speaks about the White House as the "crown jewel" of the Obama brand, a physical space where the administration can embody the values of transparency, change and diversity that drew so many voters out on Election Day.

That means much talk of the importance of healthy eating ("Let's hear it for vegetables!" Michelle and a gaggle of school children cheered at the unveiling of the South Lawn garden. "Let's hear it for fruit!")—but also field trips to Five Guys Burgers so no one thinks the Obamas are too preachy. It means corralling A-list celebrities to perform random acts of mentorship—but also staying down-to-earth enough to redecorate the girls' bedrooms at Pottery Barn. And most of all it has meant, in addition to the usual state dinners, an endless procession of multicultural celebrations: the fountain dyed green on Saint Patrick's Day, a seder on Passover, a special gathering for the Mexican holiday Cinco de Mayo. As a brand, the Obama White House's identity is probably closest to Starbucks: hip, progressive, approachable—a small luxury you can feel good about even during tough economic times.

Perhaps there is nothing wrong with this. Why shouldn't a president who wants to change the country benefit from marketing as good as Starbucks and Nike? Every transformative movement in history has used strong graphic design, catchy slogans and, yes, fashion to build its base. Fifteen years ago,

Nike appropriated the imagery of the civil rights movement and the icons of sixties counterculture to inspire cult-like devotion to running shoes. Obama has used our faded memories of those same movements to revive interest in actual politics; surely that's a step up. So the problem is not that Obama is using the same tricks and tools as the superbrands; anyone wanting to move the culture these days pretty much has to do that. The problem is that, as with so many other lifestyle brands before him, his actions do not come close to living up to the hopes he has raised.

Though it's too soon to issue a verdict on the Obama presidency, we do know this: he favors the grand symbolic gesture over deep structural change every time. So he will make a dramatic announcement about closing the notorious Guantánamo Bay prison—while going ahead with an expansion of the lower profile but frighteningly lawless Bagram prison in Afghanistan, and opposing accountability for Bush officials who authorized torture. He will boldly appoint the first Latina to the Supreme Court, while intensifying Bush-era enforcement measures in a new immigration crackdown. He will make investments in green energy, while championing the fantasy of "clean coal" and refusing to tax emissions, the only sure way to substantially reduce the burning of fossil fuels. Similarly, he will slam the unacceptable greed of banking executives, even as he hands the reins of the economy to consummate Wall Street insiders Timothy Geithner and Larry Summers, who have predictably rewarded the speculators and failed to break up the banks. And most importantly, he will claim to be ending the war in Iraq, and will retire the ugly "war on terror" phrase—even as the conflicts guided by that fatal logic escalate in Afghanistan and Pakistan.

This preference for symbols over substance, and this unwillingness to stick to a morally clear if unpopular course, is where Obama decisively parts ways with the transformative political movements from which he has borrowed so much (his pop-art posters from Che, his cadence from King, his "Yes We Can!" slogan from the migrant farmworkers' *Si Se Puede*). These movements made unequivocal demands of existing power structures: for land distribution, higher wages, ambitious social programs. Because of those high-cost demands, these movements had not only committed followers but serious enemies. Obama, in sharp contrast not just to social movements but to transformative presidents like FDR, follows the logic of marketing: create an appealing canvas on which all are invited to project their deepest desires but stay vague enough not to lose anyone but the committed wing nuts (which, granted, constitute a not inconsequential demographic in the United States). *Advertising Age* had it right when it gushed that the Obama brand is "big enough to be anything to anyone yet had an intimate enough feel to inspire advocacy." And then their highest compliment: "Mr. Obama somehow managed to be both Coke and Honest Tea, both the megabrand with the global awareness and distribution network and the dark-horse, upstart niche player."

Another way of putting it is that Obama played the anti-war, anti–Wall Street party crasher to his grassroots base, which imagined itself leading an insurgency against the two-Party monopoly through dogged organization and donations gathered from lemonade stands and loose change found in the crevices of the couch. Meanwhile, he took more money from Wall Street than any other presidential candidate, swallowed the Democratic Party establishment in one gulp after defeating Hillary Clinton, then pursued "bipartisanship" with crazed Republicans once in the White House.

Does Obama's failure to live up to his lofty brand cost him? It didn't at first. An international study by Pew's Global Attitudes Project, conducted five months after he took office, asked people whether they were confident Obama would "do the right thing in world affairs." Despite the fact that there was already plenty of evidence that Obama was continuing many of Bush's core international policies (albeit with a far less arrogant style), the vast majority said they approved of Obama—in Jordan and Egypt, a fourfold increase from the Bush era. In Europe the change in attitude could give you whiplash: Obama had the confidence of 91 percent of French respondents and 86 percent of Britons—compared with 13 percent and 16 percent respectively under Bush. The poll was proof that "Obama's presidency essentially erased the battering the USA's image took during eight years of the Bush administration," according to *USA Today*. David Axelrod puts it like this: "What has happened is that anti-Americanism isn't cool anymore."

That was certainly true, and had very real consequences. Obama's election and the world's corresponding love affair with his rebranded America came at a crucial time. In the two months before the election, the financial crisis rocking world markets was being rightly blamed not just on the contagion of Wall Street's bad bets but on the entire economic model of deregulation and privatization (called "neoliberalism" in most parts of the world) that had been preached from U.S.-dominated institutions like the International Monetary Fund and the World Trade Organization. If the United States were led by someone who didn't happen to be a global superstar, U.S. prestige would have continued to plummet and the rage at the economic model at the heart of the global meltdown would likely have turned into sustained demands for new rules to rein in (and seriously tax) speculative finance.

Those rules were supposed to have been on the agenda when G20 leaders met at the height of the economic crisis in London in April 2009. Instead, the press focused on excited sightings of the fashionable Obama couple, while world leaders agreed to revive the ailing International Monetary Fund—a chief culprit in this mess—with up to a trillion dollars in new financing. In short, Obama didn't just rebrand America, he resuscitated the neoliberal economic project when it was at death's door. No one but Obama, wrongly perceived as a new FDR, could have pulled it off.

Yet re-reading *No Logo* after ten years provides many reminders that success in branding can be fleeting, and that nothing is more fleeting than the quality of being cool. Many of the superbrands and branded celebrities that looked untouchable not so long ago have either faded or are in deep crisis today. Some overstretched. For others, their actual products began to feel rather disappointing next to the thrill of their marketing (a black woman breastfeeding a white child to sell . . . Benetton sweater sets? Really?). And sometimes it was precisely their claims of political enlightenment that tempted activists to contrast their marketing image with their labor practices, with disastrous results for the brands.

The Obama brand could well suffer a similar fate. Of course many people supported Obama for straightforward strategic reasons: they rightly wanted the Republicans out and he was the best candidate. But what will happen when the throngs of Obama faithful realize that they gave their hearts not to a movement that shared their deepest values but to a devoutly corporatist political party, one that puts the profits of drug companies before the need for affordable health care, and Wall Street's addiction to financial bubbles before the needs of millions of people whose homes and jobs could have been saved with a better bailout?

The risk—and it is real—is that the response will be waves of bitter cynicism, particularly among the young people for whom the Obama campaign was their first taste of politics. Most won't switch parties, they'll just do what young people used to do during elections: stay home, tune out. Another, more hopeful possibility is that Obamamania will end up being what the U.S. president's advisors like to call "a teachable moment." Obama is a gifted politician with a deep intelligence and a greater inclination toward social justice than any leader of his party in recent memory. If he cannot change the system in order to keep his election promises, it's because the system itself is utterly broken.

That was the conversation many of us were having in that brief period between the anti-WTO protests in Seattle in November 1999 and the beginning of the so-called War on Terror. Perhaps it was a limitation, but for the movement the media insisted on calling "anti-globalization," it mattered little which political party happened to be in power in our respective countries. We were focused squarely on the rules of game, and how they had been distorted to serve the narrow interests of corporations at every level of governance—from international free-trade agreements to local water privatization deals.

Looking back on this period, what I liked most was the unapologetic wonkery of it all. In the two years after *No Logo* came out, I went to dozens of teach-ins and conferences, some of them attended by thousands of people (tens of thousands in the case of the World Social Forum), that were exclusively devoted to popular education about the inner workings of global finance and trade. No topic was too arcane: the science of genetically modified foods, trade-related intellectual property rights, the fine print of bilateral trade deals, the patenting of seeds, the truth about carbon sinks. I sensed in these rooms a hunger for knowledge that I had never witnessed in any university class. It was as if people understood, all at once, that gathering this knowledge was crucial to the survival not just of democracy but of the planet. Yes, this was complicated, but we embraced that complexity because we were finally looking at systems, not just symbols.

In some parts of the world, particularly Latin America, that wave of resistance only spread and strengthened. In some countries, social movements grew strong enough to join with political parties, winning national elections and beginning to forge a new regional fair-trade regime. But elsewhere, September 11 pretty much blasted the movement out of existence. In the United States, progressive politics rallied around a single cause: "taking back" the White House (as if "we" ever had it in the first place), while outside the United States, the coalitions that had been focused on a global economic model now trained their attention on the wars in Iraq and Afghanistan, on a resurgent "U.S. empire" and on resisting increasingly aggressive attacks on immigrants. What we knew about the sophistication of global corporatism—that all the world's injustice could not be blamed on one right-wing political party, or on one nation, no matter how powerful—seemed to disappear.

If there was ever a time to remember the lessons we learned at the turn of the millennium, it is now. One benefit of the international failure to regulate the financial sector even after its catastrophic collapse is that the economic model that dominates around the world has revealed itself not as "free market" but "crony capitalist"—politicians handing over public wealth to private players in exchange for political support. What used to be politely hidden is all out in the open now. Correspondingly, public rage at corporate greed is at its highest point not just in my lifetime but in my parents' lifetime as well. Many of the points supposedly marginal activists were making in the streets ten years ago are now the accepted wisdom of cable news talk shows and mainstream op-ed pages.

And yet missing from this populist moment is what was beginning to emerge a decade ago: a movement that did not just respond to individual outrages but had a set of proactive demands for a more just and sustainable economic model. In the United States and many parts of Europe, it is far-right parties and even neofascism that are giving the loudest voice to anti-corporatist rage.

Personally, none of this makes me feel betrayed by Barack Obama. Rather I have a familiar ambivalence, the way I used to feel when brands like Nike and Apple started using revolutionary imagery in their transcendental branding campaigns. Sure it was annoying, but after the apolitical eighties, when there was, according to Margaret Thatcher, "no such thing as society," it also seemed like a good sign that these brands believed otherwise. All of their high-priced market research had found a longing in people for something more than shopping—for social change, for public space, for greater equality and diversity. Of course the brands tried to exploit that longing to sell lattes and laptops. Yet it seemed to me that we on the left owed the marketers a debt of gratitude for all this: our ideas weren't as passé as we had been told. And since the brands couldn't fulfill the deep desires they were awakening, social movements had a new impetus to try.

Perhaps Obama should be viewed in much the same way. Once again, the market research has been done for us. What the election and the global embrace of Obama's brand proved decisively is that there is a tremendous appetite for progressive change—that many, many people do not want markets opened at gunpoint, are repelled by torture, believe passionately in civil liberties, want corporations out of politics, see global warming as the fight of our time, and very much want to be part of a political project larger than themselves.

Those kinds of transformative goals are only ever achieved when independent social movements build the numbers and the organizational power to make muscular demands of their elites. Obama won office by capitalizing on our profound nostalgia for those kinds of social movements. But it was only an echo, a memory. The task ahead is to build movements that are—to borrow an old Coke slogan—the real thing. As Studs Terkel, the great oral historian, used to say: "Hope has never trickled down. It has always sprung up."

INTRODUCTION

A WEB OF BRANDS

If I squint, tilt my head, and shut my left eye, all I can see out the window is 1932, straight down to the lake. Brown warehouses, oatmeal-colored smokestacks, faded signs painted on brick walls advertising long-discontinued brands: "Lovely," "Gaywear." This is the old industrial Toronto of garment factories, furriers and wholesale wedding dresses. So far, no one has come up with a way to make a profit out of taking a wrecking ball to these boxes of brick, and in this little eight- or nine-block radius, the modern city has been layered haphazardly on top of the old.

I wrote this book while living in Toronto's ghost of a garment district in a ten-story warehouse. Many other buildings like it have long since been boarded up, glass panes shattered, smokestacks holding their breath; their only remaining capitalist function is to hoist large blinking billboards on their tar-coated roofs, reminding the gridlocked drivers on the lakeshore expressway of the existence of Molson's beer, Hyundai cars and EZ Rock FM.

In the twenties and thirties, Russian and Polish immigrants darted back and forth on these streets, ducking into delis to argue about Trotsky and the leadership of the International Ladies' Garment Workers' Union. These days, old Portuguese men still push racks of dresses and coats down the sidewalk, and next door you can still buy a rhinestone bridal tiara if the need for such an item happens to arise (a Hallowe'en costume, or perhaps a school play...). The real action, however, is down the block amid the stacks of edible jewelry at Sugar Mountain, the retro candy mecca, open until 2 a.m. to service the late-night ironic cravings of the club kids. And a store downstairs continues to do a modest trade in bald naked mannequins, though more often than not it's rented out as the surreal set for a film school project or the tragically hip backdrop of a television interview.

The layering of decades on Spadina Avenue, like so many urban neighborhoods in a similar state of postindustrial limbo, has a wonderful accidental charm to it. The lofts and studios are full of people who know they are playing their part in a piece of urban performance art, but for the most part, they do their best not to draw attention to that fact. If anyone claims too much ownership over "the real Spadina," then everyone else starts feeling like a two-bit prop, and the whole edifice crumbles.

Which is why it was so unfortunate that City Hall saw fit to commission a series of public art installations to "celebrate" the history of Spadina Avenue. First came the steel figures perched atop the lampposts: women hunched over sewing machines and crowds of striking workers waving placards with indecipherable slogans. Then the worst happened: the giant brass thimble arrived – right at the corner of my block. There it was: eleven and a half feet high and eleven feet across. Two giant pastel buttons were plopped on the sidewalk next to it, with wimpy little saplings growing out of the holes. Thank goodness Emma Goldman, the famed anarchist and labor organizer who lived on this street in the late 1930s, wasn't around to witness the transformation of the garment workers' struggle into sweatshop kitsch.

The thimble is only the most overt manifestation of a painful new self-consciousness on the grid. All around me, the old factory buildings are being rezoned and converted into "loft-living" complexes with names like "The Candy Factory." The hand-me-downs of industrialization have already been mined for witty fashion ideas – discarded factory workers' uniforms, Diesel's Labor brand jeans and Caterpillar boots. So of course there is also a booming market for condos in secondhand sweatshops, luxuriously reno-ed, with soaking tubs, slate-lined showers, underground parking, skylit gymnasiums and twenty-four-hour concierges.

So far my landlord, who made his fortune manufacturing and selling London Fog overcoats, has stubbornly refused to sell off our building as condominiums with exceptionally high ceilings. He'll relent eventually, but for now he still has a handful of garment tenants left, whose businesses are too small to move to Asia or Central America and who for whatever reason are unwilling to follow the industry trend toward homeworkers paid by the piece. The rest of the building is rented out to yoga instructors, documentary film producers, graphic designers and writers and artists with live/work spaces. The shmata guys still selling coats in the office next door look terribly dismayed when they see the Marilyn Manson clones stomping down the hall in chains and thigh-high leather boots to the communal washroom, clutching tubes of toothpaste, but what can they do? We are all stuck together here for now, caught between the harsh realities of economic globalization and the all-enduring rock-video aesthetic.

JAKARTA – "Ask her what she makes – what it says on the label. You know – label?" I said, reaching behind my head and twisting up the collar of my shirt. By now these Indonesian workers were used to people like me: foreigners who come to talk to them about the abysmal conditions in the factories where they cut, sew and glue for multinational companies like Nike, the Gap and Liz Claiborne. But these seamstresses looked nothing like the elderly garment workers I meet in the elevator back home. Here they were all young, some of them as young as fifteen; only a few were over twenty-one.

On this particular day in August 1997, the abysmal conditions in question had led to a strike at the Kaho Indah Citra garment factory on the outskirts of Jakarta in the Kawasan Berikat Nusantar industrial zone. The issue for the Kaho workers, who earn the equivalent of US$2 per day, was that they were being forced to work long hours of overtime but weren't being paid at the legal rate for their trouble. After a three-day walkout, management offered a compromise typical of a region with a markedly relaxed relationship to labor legislation: overtime would no longer be compulsory but the compensation would remain illegally low. The 2,000 workers returned to their sewing machines; all except 101 young women who – management decided – were the troublemakers behind the strike. "Until now our case is still not settled," one of these workers told me, bursting with frustration and with no recourse in sight.

I was sympathetic, of course, but, being the Western foreigner, I wanted to know what *brand* of garments they produced at the Kaho factory – if I was to bring their story home, I would have to have my journalistic hook. So here we were, ten of us, crowded into a concrete bunker only slightly bigger than a telephone booth, playing an enthusiastic round of labor charades.

"This company produces long sleeves for cold seasons," one worker offered.

I guessed: "Sweaters?"

"I think not sweaters. If you prepare to go out and you have a cold season you have a..."

I got it: "Coat!"

"But not heavy. Light."

"Jackets!"

"Yes, like jackets, but not jackets – long."

You can understand the confusion: there isn't much need for overcoats on the equator, not in the closet and not in the vocabulary. And yet increasingly, Canadians get through their cold winters not with clothing manufactured by the tenacious seamstresses still on Spadina Avenue but by young Asian women working in hot climates like this one. In 1997, Canada imported $11.7 million of anoraks and ski jackets from Indonesia, up from $4.7 million in 1993.[1] That much I knew already. But I still didn't know what brand of long coats the Kaho workers sewed before they lost their jobs.

"Long, yes. And what's on the label?" I asked again.

There was a bit of hushed consultation, and then, finally, an answer: "London Fog."

A global coincidence, I suppose. I started to tell the Kaho workers that my apartment in Toronto used to be a London Fog coat factory but stopped abruptly when it became clear from their facial expressions that the idea of anyone choosing to live in a garment building was nothing but alarming. In this part of the world, hundreds of workers every year burn to death because their dormitories are located upstairs from firetrap sweatshops.

Sitting cross-legged on the concrete floor of the tiny dorm room, I thought of my neighbors back home: the Ashtanga yoga instructor on two, the commercial animators on four, the aromatherapy candle distributors on eight. It seems the young women in the export processing zone are our roommates of sorts, connected, as is so often the case, by a web of fabrics, shoelaces, franchises, teddy bears and brand names wrapped around the planet. Another logo we had in common was Esprit, also one of the brands manufactured in the zone. As a teenager I worked as a clerk in a store that sold Esprit clothes.

And of course, McDonald's: an outlet had just opened near Kaho, frustrating workers, because this so-called bargain food was squarely out of their price range.

Usually, reports about this global web of logos and products are couched in the euphoric marketing rhetoric of the global village, an incredible place where tribespeople in remotest rain forests tap away on laptop computers, Sicilian grandmothers conduct E-business, and "global teens" share, to borrow a phrase from a Levi's Web site, "a world-wide style culture."[2] Everyone from Coke to McDonald's to Motorola has tailored their marketing strategy around this post-national vision, but it is IBM's long-running "Solutions for a Small Planet" campaign that most eloquently captures the equalizing promise of the logo-linked globe.

It hasn't taken long for the excitement inspired by these manic renditions of globalization to wear thin, revealing the cracks and fissures beneath its high-gloss façade. More and more over the past four years, we in the West have been catching glimpses of another kind of global village, where the economic divide is widening and cultural choices narrowing.

This is a village where some multinationals, far from leveling the global playing field with jobs and technology for all, are in the process of mining the planet's poorest back country for unimaginable profits. This is the village where Bill Gates lives, amassing a fortune of $55 billion while a third of his workforce is classified as temporary workers, and where competitors are either incorporated into the Microsoft monolith or made obsolete by the latest feat in software bundling. This is the village where we are indeed connected to one another through a web of brands, but the underside of that web reveals designer slums like the one I visited outside Jakarta. IBM claims that its technology spans the globe, and so it does, but often its international presence takes the form of cheap Third World labor producing the computer chips and power sources that drive our machines. On the outskirts of Manila, for instance, I met a seventeen-year-old girl who assembles CD-ROM drives for IBM. I told her I was impressed that someone so young could do such high-tech work. "We make computers," she told me, "but we don't know how to operate computers." Ours, it would seem, is not such a small planet after all.

It would be naive to believe that Western consumers haven't profited from these global divisions since the earliest days of colonialism. The Third World, as they say, has always existed for the comfort of the First. What is a relatively new development, however, is the amount of investigative interest there seems to be in the unbranded points of origin of brand-name goods. The travels of Nike sneakers have been traced back to the abusive sweatshops of Vietnam, Barbie's little outfits back to the child laborers of Sumatra, Starbucks' lattes to the sun-scorched coffee fields of Guatemala, and Shell's oil back to the polluted and impoverished villages of the Niger Delta.

The title *No Logo* is not meant to be read as a literal slogan (as in No More Logos!), or a post-logo logo (there is already a No Logo clothing line, or so I'm told). Rather, it is an attempt to capture an anticorporate attitude I see emerging among many young activists. This book is hinged on a simple hypothesis: that as more people discover the brand-name secrets of the global logo web, their outrage will fuel the next big political movement, a vast wave of opposition squarely targeting transnational corporations, particularly those with very high name-brand recognition.

I must stress, however, that this is not a book of predictions, but of first-hand observation. It is an examination of a largely underground system of information, protest and planning, a system already coursing with activity and ideas crossing many national borders and several generations.

Four years ago, when I started to write this book, my hypothesis was mostly based on a hunch. I had been doing some research on university campuses and had begun to notice that many of the students I was meeting were preoccupied with the inroads private corporations were making into their public schools. They were angry that ads were creeping into cafeterias, common rooms, even washrooms; that their schools were diving into exclusive distribution deals with soft-drink companies and computer manufacturers, and that academic studies were starting to look more and more like market research.

They worried that their education was suffering, as institutional priority shifted to those programs most conducive to private-sector partnership. They also had serious ethical concerns about the practices of some of the corporations that their schools were becoming entangled with – not so much their on-campus activities, but their practices far away, in countries like Burma, Indonesia and Nigeria.

It had only been a few years since I left university myself, so I knew this was a rather sudden change in political focus; five years earlier, campus politics was all about issues of discrimination and identity – race, gender and sexuality, "the political correctness wars." Now they were broadening out to include corporate power, labor rights, and a fairly developed analysis of the workings of the global economy. It's true that these students do not make up the majority of their demographic group – in fact, this movement is coming, as all such movements do, from a minority, but it is an increasingly powerful minority. Simply put, anticorporatism is the brand of politics capturing the imagination of the next generation of troublemakers and shit-disturbers, and we need only look to the student radicals of the 1960s and the ID warriors of the eighties and nineties to see the transformative impact such a shift can have.

At around the same time, in my reporting for magazines and newspapers, I also started noticing similar ideas at the center of a wave of recent social and environmental campaigns. Like the campus activists I was meeting, the people leading these campaigns were focused on the effects of aggressive corporate sponsorships and retailing on public space and cultural life, both globally and locally. There were small-town wars being waged all over North America to keep out the "big-box" retailers like Wal-Mart. There was the McLibel Trial in London, a case of two British environmentalists who turned a libel suit McDonald's launched against them into a global cyberplatform that put the ubiquitous food franchise on trial. There was an explosion of protest and activity targeting Shell Oil after the shocking hanging of Nigerian author and anti-Shell activist Ken Saro-Wiwa.

There was also the morning when I woke up and every billboard on my street had been "jammed" with anticorporate slogans by midnight bandits. And the fact that the squeegee kids who slept in the lobby of my building all seemed to be wearing homemade patches on their clothing with a Nike "swoosh" logo and the word "Riot."

There was a common element shared by all these scattered issues and campaigns: in each case, the focus of the attack was a brand-name corporation –

Nike, Shell, Wal-Mart, McDonald's (and others: Microsoft, Disney, Starbucks, Monsanto and so on). Before I began writing this book, I didn't know if these pockets of anticorporate resistance had anything in common besides their name-brand focus, but I wanted to find out. This personal quest has taken me to a London courtroom for the handing down of the verdict in the McLibel Trial; to Ken Saro-Wiwa's friends and family; to anti-sweatshop protests outside Nike Towns in New York and San Francisco; and to union meetings in the food courts of glitzy malls. It took me on the road with an "alternative" billboard salesman and on the prowl with "adbusters" out to "jam" the meaning of those billboards with their own messages. And it brought me, too, to several impromptu street parties whose organizers are determined to briefly liberate public space from its captivity by ads, cars and cops. It took me to clandestine encounters with computer hackers threatening to cripple the systems of American corporations found to be violating human rights in China.

Most memorably, it led me to factories and union squats in Southeast Asia, and to the outskirts of Manila where Filipino workers are making labor history by bringing the first unions to the export processing zones that produce the most recognizable brand-name consumer items on the planet.

Over the course of this journey, I came across an American student group that focuses on multinationals in Burma, pressuring them to pull out because of the regime's violations of human rights. In their communiqués, the student activists identify themselves as "Spiders" and the image strikes me as a fitting one for this Web-age global activism. Logos, by the force of ubiquity, have become the closest thing we have to an international language, recognized and understood in many more places than English. Activists are now free to swing off this web of logos like spy/spiders – trading information about labor practices, chemical spills, animal cruelty and unethical marketing around the world.

I have become convinced that it is in these logo-forged global links that global citizens will eventually find sustainable solutions for this sold planet. I don't claim that this book will articulate the full agenda of a global movement that is still in its infancy. My concern has been to track the early stages of resistance and to ask some basic questions. What conditions have set the stage for this backlash? Successful multinational corporations are increasingly finding themselves under attack, whether it's a cream pie in Bill Gates's face or the incessant parodying of the Nike swoosh – what are the forces pushing more and more people to become suspicious of or even downright enraged at multinational corporations, the very engines of our global growth? Perhaps more pertinently, what is liberating so many people – particularly young people – to act on that rage and suspicion?

These questions may seem obvious, and certainly some obvious answers are kicking around. That corporations have grown so big they have superseded government. That unlike governments, they are accountable only to their shareholders; that we lack the mechanisms to make them answer to a broader public. There have been several exhaustive books chronicling the ascendancy of what has come to be called "corporate rule," many of which have proved invaluable to my own understanding of global economics (see Reading List, page 479).

This book is not, however, another account of the power of the select group of corporate Goliaths that have gathered to form our de facto global government. Rather, the book is an attempt to analyze and document the forces opposing corporate rule, and to lay out the particular set of cultural and economic conditions that made the emergence of that opposition inevitable. Part I, "No Space," examines the surrender of culture and education to marketing. Part II, "No Choice," reports on how the promise of a vastly increased array of cultural choice was betrayed by the forces of mergers, predatory franchising, synergy and corporate censorship. And Part III, "No Jobs," examines the labor market trends that are creating increasingly tenuous relationships to employment for many workers, including self-employment, McJobs and outsourcing, as well as part-time and temp labor. It is the collision of and the interplay among these forces, the assault on the three social pillars of employment, civil liberties and civic space, that is giving rise to the anticorporate activism chronicled in the last section of the book, Part IV, "No Logo," an activism that is sowing the seeds of a genuine alternative to corporate rule.

NO SPACE

Two faces of branded comfort. *Top:* Aunt Jemima from Quaker Oats' early packaging, humanizes production for a population fearful of industrialization. *Bottom:* Martha Stewart, one of the new breed of branded humans.

CHAPTER ONE

NEW BRANDED WORLD

As a private person, I have a passion for landscape, and I have never seen one improved by a billboard. Where every prospect pleases, man is at his vilest when he erects a billboard. When I retire from Madison Avenue, I am going to start a secret society of masked vigilantes who will travel around the world on silent motor bicycles, chopping down posters at the dark of the moon. How many juries will convict us when we are caught in these acts of beneficent citizenship?

— David Ogilvy, founder of the Ogilvy & Mather advertising agency, in *Confessions of an Advertising Man*, 1963

The astronomical growth in the wealth and cultural influence of multinational corporations over the last fifteen years can arguably be traced back to a single, seemingly innocuous idea developed by management theorists in the mid-1980s: that successful corporations must primarily produce brands, as opposed to products.

Until that time, although it was understood in the corporate world that bolstering one's brand name was important, the primary concern of every solid manufacturer was the production of goods. This idea was the very gospel of the machine age. An editorial that appeared in *Fortune* magazine in 1938, for instance, argued that the reason the American economy had yet to recover from the Depression was that America had lost sight of the importance of making *things*:

> This is the proposition that the basic and irreversible function of an industrial economy is *the making of things*; that the more things it makes the

> bigger will be the income, whether dollar or real; and hence that the key to those lost recuperative powers lies... in the factory where the lathes and the drills and the fires and the hammers are. It is in the factory and on the land and under the land that purchasing power *originates* [italics theirs].[1]

And for the longest time, the making of things remained, at least in principle, the heart of all industrialized economies. But by the eighties, pushed along by that decade's recession, some of the most powerful manufacturers in the world had begun to falter. A consensus emerged that corporations were bloated, oversized; they owned too much, employed too many people, and were weighed down with *too many things.* The very process of producing – running one's own factories, being responsible for tens of thousands of full-time, permanent employees – began to look less like the route to success and more like a clunky liability.

At around this same time a new kind of corporation began to rival the traditional all-American manufacturers for market share; these were the Nikes and Microsofts, and later, the Tommy Hilfigers and Intels. These pioneers made the bold claim that producing goods was only an incidental part of their operations, and that thanks to recent victories in trade liberalization and labor-law reform, they were able to have their products made for them by contractors, many of them overseas. What these companies produced primarily were not things, they said, but *images* of their brands. Their real work lay not in manufacturing but in marketing. This formula, needless to say, has proved enormously profitable, and its success has companies competing in a race toward weightlessness: whoever owns the least, has the fewest employees on the payroll and produces the most powerful images, as opposed to products, wins the race.

And so the wave of mergers in the corporate world over the last few years is a deceptive phenomenon: it only *looks* as if the giants, by joining forces, are getting bigger and bigger. The true key to understanding these shifts is to realize that in several crucial ways – not their profits, of course – these merged companies are actually shrinking. Their apparent bigness is simply the most effective route toward their real goal: divestment of the world of things.

Since many of today's best-known manufacturers no longer produce products and advertise them, but rather buy products and "brand" them, these companies are forever on the prowl for creative new ways to build and strengthen their brand images. Manufacturing products may require drills, furnaces, hammers and the like, but creating a brand calls for a completely different set of tools and materials. It requires an endless parade of brand extensions, continuously renewed imagery for marketing and, most of all, fresh new spaces to disseminate the brand's idea of itself. In this section of the book, I'll look at how, in ways both insidious and overt, this corporate obsession with brand identity is waging a war on public and individual space: on public institutions such as schools, on youthful identities, on the concept of nationality and on the possibilities for unmarketed space.

The Beginning of the Brand

It's helpful to go back briefly and look at where the idea of branding first began. Though the words are often used interchangeably, branding and advertising are not the same process. Advertising any given product is only one part of branding's grand plan, as are sponsorship and logo licensing. Think of the brand as the core meaning of the modern corporation, and of the advertisement as one vehicle used to convey that meaning to the world.

The first mass-marketing campaigns, starting in the second half of the nineteenth century, had more to do with advertising than with branding as we understand it today. Faced with a range of recently invented products – the radio, phonograph, car, light bulb and so on – advertisers had more pressing tasks than creating a brand identity for any given corporation; first, they had to change the way people lived their lives. Ads had to inform consumers about the existence of some new invention, then convince them that their lives would be better if they used, for example, cars instead of wagons, telephones instead of mail and electric light instead of oil lamps. Many of these new products bore brand names – some of which are still around today – but these were almost incidental. These products were themselves news; that was almost advertisement enough.

The first brand-based products appeared at around the same time as the invention-based ads, largely because of another relatively recent innovation:

the factory. When goods began to be produced in factories, not only were entirely new products being introduced but old products – even basic staples – were appearing in strikingly new forms. What made early branding efforts different from more straightforward salesmanship was that the market was now being flooded with uniform mass-produced products that were virtually indistinguishable from one another. Competitive branding became a necessity of the machine age – within a context of manufactured sameness, image-based difference had to be manufactured along with the product.

So the role of advertising changed from delivering product news bulletins to building an image around a particular brand-name version of a product. The first task of branding was to bestow proper names on generic goods such as sugar, flour, soap and cereal, which had previously been scooped out of barrels by local shopkeepers. In the 1880s, corporate logos were introduced to mass-produced products like Campbell's Soup, H.J. Heinz pickles and Quaker Oats cereal. As design historians and theorists Ellen Lupton and J. Abbott Miller note, logos were tailored to evoke familiarity and folksiness (see Aunt Jemima, page 2), in an effort to counteract the new and unsettling anonymity of packaged goods. "Familiar personalities such as Dr. Brown, Uncle Ben, Aunt Jemima, and Old Grand-Dad came to replace the shopkeeper, who was traditionally responsible for measuring bulk foods for customers and acting as an advocate for products... a nationwide vocabulary of brand names replaced the small local shopkeeper as the interface between consumer and product."[2] After the product names and characters had been established, advertising gave them a venue to speak directly to would-be consumers. The corporate "personality," uniquely named, packaged and advertised, had arrived.

For the most part, the ad campaigns at the end of the nineteenth century and the start of the twentieth used a set of rigid, pseudoscientific formulas: rivals were never mentioned, ad copy used declarative statements only and headlines had to be large, with lots of white space – according to one turn-of-the-century adman, "an advertisement should be big enough to make an impression but not any bigger than the thing advertised."

But there were those in the industry who understood that advertising wasn't just scientific; it was also spiritual. Brands could conjure a feeling –

think of Aunt Jemima's comforting presence—but not only that, entire corporations could themselves embody a meaning of their own. In the early twenties, legendary adman Bruce Barton turned General Motors into a metaphor for the American family, "something personal, warm and human," while GE was not so much the name of the faceless General Electric Company as, in Barton's words, "the initials of a friend." In 1923 Barton said that the role of advertising was to help corporations find their soul. The son of a preacher, he drew on his religious upbringing for uplifting messages: "I like to think of advertising as something big, something splendid, something which goes deep down into an institution and gets hold of the soul of it.... Institutions have souls, just as men and nations have souls," he told GM president Pierre du Pont.[3] General Motors ads began to tell stories about the people who drove its cars—the preacher, the pharmacist or the country doctor who, thanks to his trusty GM, arrived "at the bedside of a dying child" just in time "to bring it back to life."

By the end of the 1940s, there was a burgeoning awareness that a brand wasn't just a mascot or a catchphrase or a picture printed on the label of a company's product; the company as a whole could have a brand identity or a "corporate consciousness," as this ephemeral quality was termed at the time. As this idea evolved, the adman ceased to see himself as a pitchman and instead saw himself as "the philosopher-king of commercial culture,"[4] in the words of ad critic Randall Rothberg. The search for the true meaning of brands—or the "brand essence," as it is often called—gradually took the agencies away from individual products and their attributes and toward a psychological/anthropological examination of what brands mean to the culture and to people's lives. This was seen to be of crucial importance, since corporations may manufacture products, but what consumers buy are brands.

It took several decades for the manufacturing world to adjust to this shift. It clung to the idea that its core business was still production and that branding was an important add-on. Then came the brand equity mania of the eighties, the defining moment of which arrived in 1988 when Philip Morris purchased Kraft for $12.6 billion—six times what the company was worth on paper. The price difference, apparently, was the cost of the word

"Kraft." Of course Wall Street was aware that decades of marketing and brand bolstering added value to a company over and above its assets and total annual sales. But with the Kraft purchase, a huge dollar value had been assigned to something that had previously been abstract and unquantifiable – a brand name. This was spectacular news for the ad world, which was now able to make the claim that advertising spending was more than just a sales strategy: it was an investment in cold hard equity. The more you spend, the more your company is worth. Not surprisingly, this led to a considerable increase in spending on advertising. More important, it sparked a renewed interest in puffing up brand identities, a project that involved far more than a few billboards and TV spots. It was about pushing the envelope in sponsorship deals, dreaming up new areas in which to "extend" the brand, as well as perpetually probing the zeitgeist to ensure that the "essence" selected for one's brand would resonate karmically with its target market. For reasons that will be explored in the rest of this chapter, this radical shift in corporate philosophy has sent manufacturers on a cultural feeding frenzy as they seize upon every corner of unmarketed landscape in search of the oxygen needed to inflate their brands. In the process, virtually nothing has been left unbranded. That's quite an impressive feat, considering that as recently as 1993 Wall Street had pronounced the brand dead, or as good as dead.

The Brand's Death (Rumors of Which Had Been Greatly Exaggerated)

The evolution of the brand had one scary episode when it seemed to face extinction. To understand this brush with death, we must first come to terms with advertising's own special law of gravity, which holds that if you aren't rocketing upward you will soon come crashing down.

The marketing world is always reaching a new zenith, breaking through last year's world record and planning to do it again next year with increasing numbers of ads and aggressive new formulae for reaching consumers. The advertising industry's astronomical rate of growth is neatly reflected in year-to-year figures measuring total ad spending in the U.S. (see Table 1.1 on page 11), which have gone up so steadily that by 1998 the figure was set to reach $196.5 billion, while global ad spending is estimated at $435 billion.[5] According to the 1998 United Nations Human Development Report, the growth in global ad spending "now outpaces the growth of the world economy by one-third."

This pattern is a by-product of the firmly held belief that brands need continuous and constantly increasing advertising in order to stay in the same place. According to this law of diminishing returns, the more advertising there is out there (and there always is more, because of this law), the more aggressively brands must market to stand out. And of course, no one is more keenly aware of advertising's ubiquity than the advertisers themselves, who view commercial inundation as a clear and persuasive call for more – and more intrusive – advertising. With so much competition, the agencies argue, clients must spend more than ever to make sure their pitch screeches so loud it can be heard over all the others. David Lubars, a senior ad executive in the Omnicom Group, explains the industry's guiding principle with more candor than most. Consumers, he says, "are like roaches – you spray them and spray them and they get immune after a while."[6]

So, if consumers are like roaches, then marketers must forever be dreaming up new concoctions for industrial-strength Raid. And nineties marketers, being on a more advanced rung of the sponsorship spiral, have dutifully come up with clever and intrusive new selling techniques to do just that. Recent highlights include these innovations: Gordon's gin experimented with filling British movie theaters with the scent of juniper berries; Calvin Klein stuck "CK Be" perfume strips on the backs of Ticketmaster concert envelopes; and in some Scandinavian countries you can get "free" long-distance calls with ads cutting into your telephone conversations. And there's plenty more, stretching across ever more expansive surfaces and cramming into the smallest of crevices: sticker ads on pieces of fruit promoting ABC sitcoms, Levi's ads in public washrooms, corporate logos on boxes of Girl Guide cookies, ads for pop albums on takeout food containers, and ads for Batman movies projected on sidewalks or into the night sky. There are already ads on benches in national parks as well as on library cards in public libraries, and in December 1998 NASA announced plans to solicit ads on its space stations. Pepsi's ongoing threat to project its logo onto the moon's surface hasn't yet materialized, but Mattel did paint an entire street in Salford, England, "a shriekingly bright bubblegum hue" of pink – houses, porches, trees, road, sidewalk, dogs

Come to
Marlboro Country.

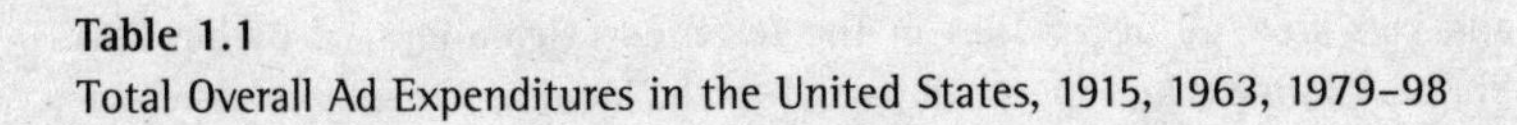

Table 1.1

Total Overall Ad Expenditures in the United States, 1915, 1963, 1979–98

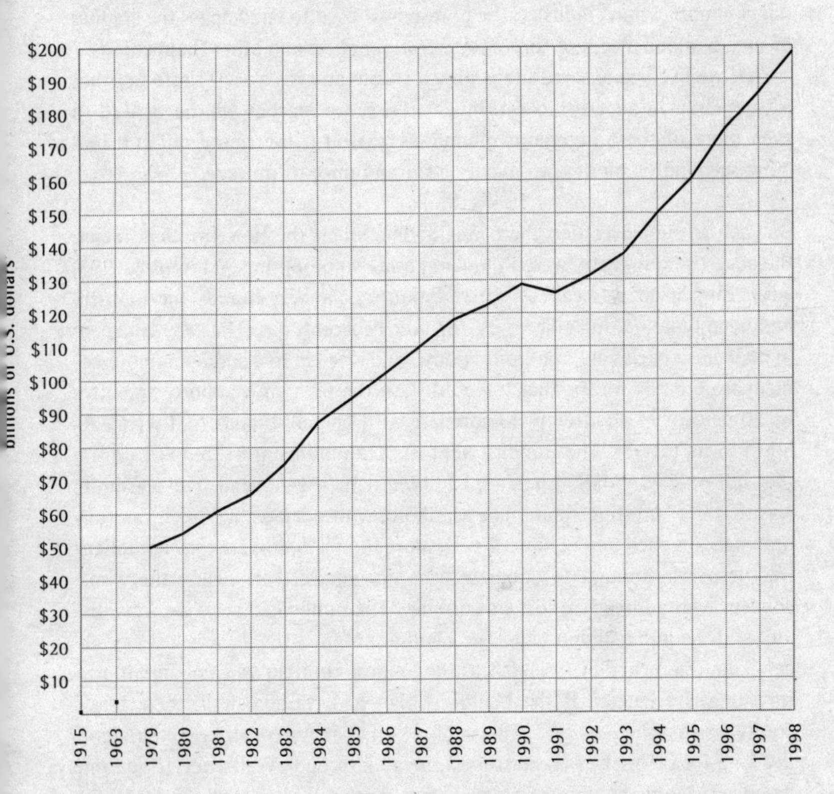

Source: Figures extracted from various articles: *The Economist*, November 14, 1981; *PR Newswire*, May 23, 1983; *Business Week*, August 15, 1983; *Advertising Age*, July 23, 1984; *Ad Age*, May 6, 1985; *Ad Age*, December 16, 1985; *The Record*, January 25, 1986; *Ad Age*, May 12, 1986; *Ad Age*, June 30, 1986; *Ad Age*, August 17, 1987; *Ad Age*, December 14, 1987; *Ad Age*, May 15, 1989; *Marketing*, June 30, 1997; *Ad Age*, December 15, 1997; figures for 1979, 1981 and 1982 are estimates; 1998 figure is a projection based on *Ad Age*, December 15, 1997; all amounts include total of measured and unmeasured ad expenditures in the U.S.

and cars were all accessories in the televised celebrations of Barbie Pink Month.[7] Barbie is but one small part of the ballooning $30 billion "experiential communication" industry, the phrase now used to encompass the staging of such branded pieces of corporate performance art and other "happenings."[8]

That we live a sponsored life is now a truism and it's a pretty safe bet that as spending on advertising continues to rise, we roaches will be treated to even more of these ingenious gimmicks, making it ever more difficult and more seemingly pointless to muster even an ounce of outrage.

But as mentioned earlier, there was a time when the new frontiers facing the advertising industry weren't looking quite so promising. On April 2, 1993, advertising itself was called into question by the very brands the industry had been building, in some cases, for over two centuries. That day is known in marketing circles as "Marlboro Friday," and it refers to a sudden announcement from Philip Morris that it would slash the price of Marlboro cigarettes by 20 percent in an attempt to compete with bargain brands that were eating into its market. The pundits went nuts, announcing in frenzied unison that not only was Marlboro dead, all brand names were dead. The reasoning was that if a "prestige" brand like Marlboro, whose image had been carefully groomed, preened and enhanced with more than a billion advertising dollars, was desperate enough to compete with no-names, then clearly the whole concept of branding had lost its currency. The public had seen the advertising, and the public didn't care. The Marlboro Man, after all, was not any old campaign; launched in 1954, it was the longest-running ad campaign in history. It was a legend. If the Marlboro Man had crashed, well, then, brand equity had crashed as well. The implication that Americans were suddenly thinking for themselves en masse reverberated through Wall Street. The same day Philip Morris announced its price cut, stock prices nose-dived for all the household brands: Heinz, Quaker Oats, Coca-Cola, PepsiCo, Procter and Gamble and RJR Nabisco. Philip Morris's own stock took the worst beating.

Bob Stanojev, national director of consumer products marketing for Ernst and Young, explained the logic behind Wall Street's panic: "If one or two powerhouse consumer products companies start to cut prices for good, there's going to be an avalanche. Welcome to the value generation."[9]

Yes, it was one of those moments of overstated instant consensus, but it was not entirely without cause. Marlboro had always sold itself on the strength of its iconic image marketing, not on anything so prosaic as its price. As we now know, the Marlboro Man survived the price wars without sustaining too much damage. At the time, however, Wall Street saw Philip Morris's decision as symbolic of a sea change. The price cut was an admission that Marlboro's name was no longer sufficient to sustain the flagship position, which in a context where image is equity meant that Marlboro had blinked. And when Marlboro – one of the quintessential global brands – blinks, it raises questions about branding that reach beyond Wall Street, and way beyond Philip Morris.

The panic of Marlboro Friday was not a reaction to a single incident. Rather, it was the culmination of years of escalating anxiety in the face of some rather dramatic shifts in consumer habits that were seen to be eroding the market share of household-name brands, from Tide to Kraft. Bargain-conscious shoppers, hit hard by the recession, were starting to pay more attention to price than to the prestige bestowed on their products by the yuppie ad campaigns of the 1980s. The public was suffering from a bad case of what is known in the industry as "brand blindness."[10]

Study after study showed that baby boomers, blind to the alluring images of advertising and deaf to the empty promises of celebrity spokespersons, were breaking their lifelong brand loyalties and choosing to feed their families with private-label brands from the supermarket – claiming, heretically, that they couldn't tell the difference. From the beginning of the recession to 1993, Loblaw's President's Choice line, Wal-Mart's Great Value and Marks and Spencer's St. Michael prepared foods had nearly doubled their market share in North America and Europe.[11] The computer market, meanwhile, was flooded by inexpensive clones, causing IBM to slash its prices and otherwise impale itself. It appeared to be a return to the proverbial shopkeeper dishing out generic goods from the barrel in a prebranded era.

The bargain craze of the early nineties shook the name brands to their core. Suddenly it seemed smarter to put resources into price reductions and other incentives than into fabulously expensive ad campaigns. This ambivalence began to be reflected in the amounts companies were willing to pay for so-called brand-enhancing advertising. Then, in 1991, it happened: overall advertising spending actually went down by 5.5 percent for the top 100 brands. It was the first interruption in the steady increase of U.S. ad expenditures since a tiny dip of 0.6 percent in 1970, and the largest drop in four decades.[12]

It's not that top corporations weren't flogging their products, it's just that to attract those suddenly fickle customers, many decided to put their money into promotions such as giveaways, contests, in-store displays and (like Marlboro) price reductions. In 1983, American brands spent 70 percent of their total marketing budgets on advertising, and 30 percent on these other forms of promotion. By 1993, the ratio had flipped: only 25 percent went to ads, with the remaining 75 percent going to promotions.

Predictably, the ad agencies panicked when they saw their prestige clients abandoning them for the bargain bins and they did what they could to convince big spenders like Procter and Gamble and Philip Morris that the proper route out of the brand crisis wasn't less brand marketing but more. At the annual meeting of the U.S. Association of National Advertisers in 1988, Graham H. Phillips, the U.S. chairman of Ogilvy & Mather, berated the assembled executives for stooping to participate in "a commodity marketplace" rather than an image-based one. "I doubt that many of you would welcome a commodity marketplace in which one competed solely on price, promotion and trade deals, all of which can easily be duplicated by competition, leading to ever-decreasing profits, decay and eventual bankruptcy." Others spoke of the importance of maintaining "conceptual value-added," which in effect means adding nothing but marketing. Stooping to compete on the basis of real value, the agencies ominously warned, would spell not just the death of the brand, but corporate death as well.

Around the same time as Marlboro Friday, the ad industry felt so under siege that market researcher Jack Myers published *Adbashing: Surviving the Attacks on Advertising*, a book-length call to arms against everyone from supermarket cashiers handing out coupons for canned peas to legislators contemplating a new tax on ads. "We, as an industry, must recognize that adbashing is a threat to capitalism, to a free press, to our basic forms of entertainment, and to the future of our children," he wrote.[13]

Despite these fighting words, most market watchers remained convinced that the heyday of the value-added brand had come and gone. The eighties had gone in for brands and hoity-toity designer labels, reasoned David Scotland, European director of Hiram Walker. The nineties would clearly be all about value. "A few years ago," he observed, "it might have been considered smart to wear a shirt with a designer's logo embroidered on the pocket; frankly, it now seems a bit naff."[14]

And from the other side of the Atlantic, Cincinnati journalist Shelly Reese came to the same conclusion about our no-name future, writing that "Americans with Calvin Klein splashed across their hip pocket aren't pushing grocery carts full of Perrier down the aisles anymore. Instead they're sporting togs with labels like Kmart's Jaclyn Smith and maneuvering carts full of Kroger Co.'s Big K soda. Welcome to the private label decade."[15]

Scotland and Reese, if they remember their bold pronouncements, are probably feeling just a little bit silly right now. Their embroidered "pocket" logos sound positively subdued by today's logomaniacal standards, and sales of name-brand bottled water have been increasing at an annual rate of 9 percent, turning it into a $3.4 billion industry by 1997. From today's logo-quilted perch, it's almost unfathomable that a mere six years ago, death sentences for the brand seemed not only plausible but self-evident.

So just how did we get from obituaries for Tide to today's battalions of volunteer billboards for Tommy Hilfiger, Nike and Calvin Klein? Who slipped the steroids into the brand's comeback?

The Brands Bounce Back

There were some brands that were watching from the sidelines as Wall Street declared the death of the brand. Funny, they must have thought, we don't feel dead.

Just as the admen had predicted at the beginning of the recession, the companies that exited the downturn running were the ones who opted for marketing over value every time: Nike, Apple, the Body Shop, Calvin Klein, Disney, Levi's and Starbucks. Not only were these brands doing just fine, thank you very much, but the act of branding was becoming a larger and larger focus of their businesses. For these companies, the ostensible product was mere filler for the real production: the brand. They integrated the idea of branding into the very fabric of their companies. Their corporate cultures were so tight and cloistered that to outsiders they appeared to be a cross between fraternity house, religious cult and sanitarium. Everything was an ad for the brand: bizarre lexicons for describing employees (partners, baristas, team players, crew members), company chants, superstar CEOs, fanatical attention to design consistency, a propensity for monument-building, and New Age mission statements. Unlike classic household brand names, such as Tide and Marlboro, these logos weren't losing their currency, they were in the midst of breaking every barrier in the marketing world – becoming cultural accessories and lifestyle philosophers. These companies didn't wear their image like a cheap shirt – their image was so integrated with their business that other people wore it as *their* shirt. And when the brands crashed, these companies didn't even notice – they were branded to the bone.

So the real legacy of Marlboro Friday is that it simultaneously brought the two most significant developments in nineties marketing and consumerism into sharp focus: the deeply unhip big-box bargain stores that provide the essentials of life and monopolize a disproportionate share of the market (Wal-Mart *et al.*) and the extra-premium "attitude" brands that provide the essentials of lifestyle and monopolize ever-expanding stretches of cultural space (Nike *et al.*). The way these two tiers of consumerism developed would have a profound impact on the economy in the years to come. When overall ad expenditures took a nosedive in 1991, Nike and Reebok were busy playing advertising chicken, with each company increasing its budget to outspend the other. (See Table 1.2 on page 19.) In 1991 alone, Reebok upped its ad spending by 71.9 percent, while Nike pumped an extra 24.6 percent into its already soaring ad budget, bringing the company's total spending on marketing to a staggering $250 million annually. Far from worrying about competing on price, the sneaker pimps were designing ever more intricate and pseudoscientific air pockets, and driving up prices by signing star athletes to colossal sponsorship deals. The fetish strategy seemed to be working fine: in the six years prior to 1993, Nike had gone from a $750 million company to a $4 billion one and Phil Knight's Beaverton, Oregon, company emerged from the recession with profits 900 percent higher than when it began.

Benetton and Calvin Klein, meanwhile, were also upping their spending on lifestyle marketing, using ads to associate their lines with risqué art and progressive politics. Clothes barely appeared in these high-concept advertisements, let alone prices. Even more abstract was Absolut Vodka, which for some years now had been developing a marketing strategy in which its product disappeared and its brand was nothing but a blank bottle-shaped space that could be filled with whatever content a particular audience most wanted from its brands: intellectual in *Harper's*, futuristic in *Wired*, alternative in *Spin*, loud and proud in *Out* and "Absolut Centerfold" in *Playboy*. The brand reinvented itself as a cultural sponge, soaking up and morphing to its surroundings. (See Table 1.3, Appendix, page 471 and Absolut image, page 32.)

Saturn, too, came out of nowhere in October 1990 when GM launched a car built not out of steel and rubber but out of New Age spirituality and seventies feminism. After the car had been on the market a few years, the company held a "homecoming" weekend for Saturn owners, during which they could visit the auto plant and have a cookout with the people who made their cars. As the Saturn ads boasted at the time, "44,000 people spent their vacations with us, at a car plant." It was as if Aunt Jemima had come to life and invited you over to her house for dinner.

In 1993, the year the Marlboro Man was temporarily hobbled by "brand-blind" consumers, Microsoft made its striking debut on *Advertising Age*'s list of the top 200 ad spenders—the very same year that Apple computer increased its marketing budget by 30 percent after already making branding history with its Orwellian takeoff ad launch during the 1984 Super Bowl (see image on page 86). Like Saturn, both companies were selling a hip new relationship to the machine that left Big Blue IBM looking as clunky and menacing as the now-dead Cold War.

And then there were the companies that had always understood that they were selling brands before product. Coke, Pepsi, McDonald's, Burger King and Disney weren't fazed by the brand crisis, opting instead to escalate the brand war, especially since they had their eyes firmly fixed on global expansion. (See Table 1.4, Appendix, page 471.) They were joined in this project by a wave of sophisticated producer/retailers who hit full stride in the late

CITGO
F1
BOSTON
MARATHON
2:11:39
Eta.
82

Table 1.2
Nike and Reebok Ad Spending, 1985–97

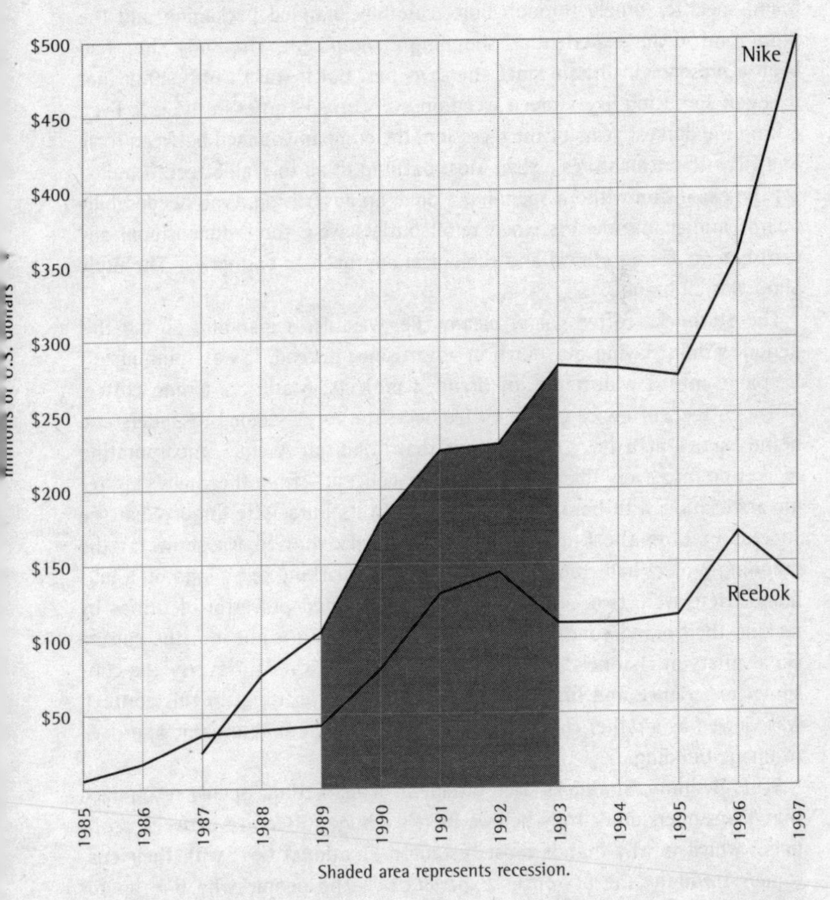

Shaded area represents recession.

Source: "100 Leading National Advertisers," *Advertising Age*, 1985–98; 1985–87 figures for Reebok are from Security Exchange Commission filings. Reebok's 1988 figure is an estimate from *Ad Age*, June 20, 1988, page 3. Nike 1987 figure from "Sneaker Attack," *Ad Age*, June 20, 1988.

eighties and early nineties. The Gap, Ikea and the Body Shop were spreading like wildfire during this period, masterfully transforming the generic into the brand-specific, largely through bold, carefully branded packaging and the promotion of an "experiential" shopping environment. The Body Shop had been a presence in Britain since the seventies, but it wasn't until 1988 that it began sprouting like a green weed on every street corner in the U.S. Even during the darkest years of the recession, the company opened between forty and fifty American stores a year. Most baffling of all to Wall Street, it pulled off the expansion without spending a dime on advertising. Who needed billboards and magazine ads when retail outlets were three-dimensional advertisements for an ethical and ecological approach to cosmetics? The Body Shop was all brand.

The Starbucks coffee chain, meanwhile, was also expanding during this period without laying out much in advertising; instead, it was spinning off its name into a wide range of branded projects: Starbucks airline coffee, office coffee, coffee ice cream, coffee beer. Starbucks seemed to understand brand names at a level even deeper than Madison Avenue, incorporating marketing into every fiber of its corporate concept – from the chain's strategic association with books, blues and jazz to its Euro-latte lingo. What the success of both the Body Shop and Starbucks showed was how far the branding project had come in moving beyond splashing one's logo on a billboard. Here were two companies that had fostered powerful identities by making their brand concept into a virus and sending it out into the culture via a variety of channels: cultural sponsorship, political controversy, the consumer experience and brand extensions. Direct advertising, in this context, was viewed as a rather clumsy intrusion into a much more organic approach to image building.

Scott Bedbury, Starbucks' vice president of marketing, openly recognized that "consumers don't truly believe there's a huge difference between products," which is why brands must "establish emotional ties" with their customers through "the Starbucks Experience."[16] The people who line up for Starbucks, writes CEO Howard Shultz, aren't just there for the coffee. "It's the romance of the coffee experience, the feeling of warmth and community people get in Starbucks stores."[17]

Interestingly, before moving to Starbucks, Bedbury was head of marketing at Nike, where he oversaw the launch of the "Just Do It!" slogan, among other watershed branding moments. In the following passage, he explains the common techniques used to infuse the two very different brands with meaning:

> Nike, for example, is leveraging the deep emotional connection that people have with sports and fitness. With Starbucks, we see how coffee has woven itself into the fabric of people's lives, and that's our opportunity for emotional leverage.... A great brand raises the bar – it adds a greater sense of purpose to the experience, whether it's the challenge to do your best in sports and fitness or the affirmation that the cup of coffee you're drinking really matters.[18]

This was the secret, it seemed, of all the success stories of the late eighties and early nineties. The lesson of Marlboro Friday was that there never really was a brand crisis – only brands that had crises of confidence. The brands would be okay, Wall Street concluded, so long as they believed fervently in the principles of branding and never, ever blinked. Overnight, "Brands, not products!" became the rallying cry for a marketing renaissance led by a new breed of companies that saw themselves as "meaning brokers" instead of product producers. What was changing was the idea of what – in both advertising and branding – was being sold. The old paradigm had it that all marketing was selling a product. In the new model, however, the product always takes a back seat to the real product, the brand, and the selling of the brand acquired an extra component that can only be described as spiritual. Advertising is about hawking product. Branding, in its truest and most advanced incarnations, is about corporate transcendence.

It may sound flaky, but that's precisely the point. On Marlboro Friday, a line was drawn in the sand between the lowly price slashers and the high-concept brand builders. The brand builders conquered and a new consensus was born: the products that will flourish in the future will be the ones presented not as "commodities" but as concepts: the brand as experience, as lifestyle.

Ever since, a select group of corporations has been attempting to free itself from the corporeal world of commodities, manufacturing and products to exist on another plane. Anyone can manufacture a product, they reason (and as the success of private-label brands during the recession proved, anyone did). Such menial tasks, therefore, can and should be farmed out to contractors and subcontractors whose only concern is filling the order on time and under budget (ideally in the Third World, where labor is dirt cheap, laws are lax and tax breaks come by the bushel). Headquarters, meanwhile, is free to focus on the real business at hand – creating a corporate mythology powerful enough to infuse meaning into these raw objects just by signing its name.

The corporate world has always had a deep New Age streak, fed – it has become clear – by a profound need that could not be met simply by trading widgets for cash. But when branding captured the corporate imagination, New Age vision quests took center stage. As Nike CEO Phil Knight explains, "For years we thought of ourselves as a production-oriented company, meaning we put all our emphasis on designing and manufacturing the product. But now we understand that the most important thing we do is market the product. We've come around to saying that Nike is a marketing-oriented company, and the product is our most important marketing tool."[19] This project has since been taken to an even more advanced level with the emergence of on-line corporate giants such as Amazon.com. It is on-line that the purest brands are being built: liberated from the real-world burdens of stores and product manufacturing, these brands are free to soar, less as the disseminators of goods or services than as collective hallucinations.

Tom Peters, who has long coddled the inner flake in many a hard-nosed CEO, latched on to the branding craze as the secret to financial success, separating the transcendental logos and the earthbound products into two distinct categories of companies. "The top half – Coca-Cola, Microsoft, Disney, and so on – are pure 'players' in brainware. The bottom half [Ford and GM] are still lumpy-object purveyors, though automobiles are much 'smarter' than they used to be," Peters writes in *The Circle of Innovation* (1997), an ode to the power of marketing over production.[20]

When Levi's began to lose market share in the late nineties, the trend was widely attributed to the company's failure – despite lavish ad spending – to transcend its products and become a free-standing meaning. "Maybe one of Levi's problems is that it has no Cola," speculated Jennifer Steinhauer in *The New York Times*. "It has no denim-toned house paint. Levi makes what is essentially a commodity: blue jeans. Its ads may evoke rugged outdoorsmanship, but Levi hasn't promoted any particular life style to sell other products."[21]

In this high-stakes new context, the cutting-edge ad agencies no longer sold companies on individual campaigns but on their ability to act as "brand stewards": identifying, articulating and protecting the corporate soul. Not surprisingly, this spelled good news for the U.S. advertising industry, which in 1994 saw a spending increase of 8.6 percent over the previous year. In one year, the ad industry went from a near crisis to another "best year yet."[22] And that was only the beginning of triumphs to come. By 1997, corporate advertising, defined as "ads that position a corporation, its values, its personality and character" were up 18 percent from the year before.[23]

With this wave of brand mania has come a new breed of businessman, one who will proudly inform you that Brand X is not a product but a way of life, an attitude, a set of values, a look, an idea. And it sounds really great – way better than that Brand X is a screwdriver, or a hamburger chain, or a pair of jeans, or even a very successful line of running shoes. Nike, Phil Knight announced in the late eighties, is "a sports company"; its mission is not to sell shoes but to "enhance people's lives through sports and fitness" and to keep "the magic of sports alive."[24] Company president-cum-sneaker-shaman Tom Clark explains that "the inspiration of sports allows us to rebirth ourselves constantly."[25]

Reports of such "brand vision" epiphanies began surfacing from all corners. "Polaroid's problem," diagnosed the chairman of its advertising agency, John Hegarty, "was that they kept thinking of themselves as a camera. But the '[brand] vision' process taught us something: Polaroid is not a camera – it's a social lubricant."[26] IBM isn't selling computers, it's selling business "solutions." Swatch is not about watches, it is about the idea of time. At Diesel Jeans, owner Renzo Rosso told *Paper* magazine, "We don't sell a product, we sell a style of life. I think we have created a movement.... The Diesel concept is everything. It's the way to live, it's the way to wear, it's the way to do something." And as Body Shop founder Anita Roddick explained to me, her stores aren't about what they sell, they are the conveyers of a grand idea – a political philosophy about women, the environment and ethical business. "I just use the company that I surprisingly created as a success – it shouldn't have been like this, it wasn't meant to be like this – to stand on the products to shout out on these issues," Roddick says.

The famous late graphic designer Tibor Kalman summed up the shifting role of the brand this way: "The original notion of the brand was quality, but now brand is a stylistic badge of courage."[27]

The idea of selling the courageous message of a brand, as opposed to a product, intoxicated these CEOs, providing as it did an opportunity for seemingly limitless expansion. After all, if a brand was not a product, it could be anything! And nobody embraced branding theory with more evangelical zeal than Richard Branson, whose Virgin Group has branded joint ventures in everything from music to bridal gowns to airlines to cola to financial services. Branson refers derisively to the "stilted Anglo-Saxon view of consumers," which holds that a name should be associated with a product like sneakers or soft drinks, and opts instead for "the Asian 'trick'" of the *keiretsus* (a Japanese term meaning a network of linked corporations). The idea, he explains, is to "build brands not around products but around reputation. The great Asian names imply quality, price and innovation rather than a specific item. I call these 'attribute' brands: They do not relate directly to one product – such as a Mars bar or a Coca-Cola – but instead to a set of values."[28]

Tommy Hilfiger, meanwhile, is less in the business of manufacturing clothes than he is in the business of signing his name. The company is run entirely through licensing agreements, with Hilfiger commissioning all its products from a group of other companies: Jockey International makes Hilfiger underwear, Pepe Jeans London makes Hilfiger jeans, Oxford Industries make Tommy shirts, the Stride Rite Corporation makes its footwear. What does Tommy Hilfiger manufacture? Nothing at all.

So passé had products become in the age of lifestyle branding that by the late nineties, newer companies like Lush cosmetics and Old Navy clothing began playing with the idea of old-style commodities as a source of retro marketing imagery. The Lush chain serves up its face masks and moisturizers out of refrigerated stainless-steel bowls, spooned into plastic containers with grocery-store labels. Old Navy showcases its shrink-wrapped T-shirts and sweatshirts in deli-style chrome refrigerators, as if they were meat or cheese. When you are a pure, concept-driven brand, the aesthetics of raw product can prove as "authentic" as loft living.

And lest the branding business be dismissed as the playground of trendy consumer items such as sneakers, jeans and New Age beverages, think again. Caterpillar, best known for building tractors and busting unions, has barreled into the branding business, launching the Cat accessories line: boots, backpacks, hats and anything else calling out for a postindustrial *je ne sais quoi*. Intel Corp., which makes computer parts no one sees and few understand, transformed its processors into a fetish brand with TV ads featuring line workers in funky metallic space suits dancing to "Shake Your Groove Thing." The Intel mascots proved so popular that the company has sold hundreds of thousands of bean-filled dolls modeled on the shimmery dancing technicians. Little wonder, then, that when asked about the company's decision to diversify its products, the senior vice president for sales and marketing, Paul S. Otellini, replied that Intel is "like Coke. One brand, many different products."[29]

And if Caterpillar and Intel can brand, surely anyone can.

There is, in fact, a new strain in marketing theory that holds that even the lowliest natural resources, barely processed, can develop brand identities, thus giving way to hefty premium-price markups. In an essay appropriately titled "How to Brand Sand," advertising executives Sam I. Hill, Jack McGrath and Sandeep Dayal team up to tell the corporate world that with the right marketing plan, nobody has to stay stuck in the stuff business. "Based on extensive research, we would argue that you can indeed brand not only sand, but also wheat, beef, brick, metals, concrete, chemicals, corn grits and an endless variety of commodities traditionally considered immune to the process."[30]

Over the past six years, spooked by the near-death experience of Marlboro Friday, global corporations have leaped on the brand-wagon with what can only be described as a religious fervor. Never again would the corporate world stoop to praying at the altar of the commodity market. From now on they would worship only graven media images. Or to quote Tom Peters, the brand man himself: “Brand! Brand!! Brand!!! That’s the message…for the late ’90s and beyond.”[31]

CHAPTER TWO

THE BRAND EXPANDS

How the Logo Grabbed Center Stage

Since the crocodile is the symbol of Lacoste, we thought they might be interested in sponsoring our crocodiles.

– Silvino Gomes, commercial director of the Lisbon Zoo, on the institution's creative corporate sponsorship program, March 1998

I was in Grade 4 when skintight designer jeans were the be-all and end-all, and my friends and I spent a lot of time checking out each other's butt for logos. "Nothing comes between me and my Calvins," Brooke Shields assured us, and as we lay back on our beds Ophelia-style and yanked up the zippers on our Jordache jeans with wire hangers, we knew she was telling no word of a lie. At around the same time, Romi, our school's own pint-sized Farrah Fawcett, used to make her rounds up and down the rows of desks turning back the collars on our sweaters and polo shirts. It wasn't enough for her to see an alligator or a leaping horseman – it could have been a knockoff. She wanted to see the label behind the logo. We were only eight years old but the reign of logo terror had begun.

About nine years later, I had a job folding sweaters at an Esprit clothing store in Montreal. Mothers would come in with their six-year-old daughters and ask to see only the shirts that said "Esprit" in the company's trademark bold block lettering. "She won't wear anything without a name," the moms would confide apologetically as we chatted by the change rooms. It's no secret that branding has become far more ubiquitous and intrusive by now. Labels like Baby Gap and Gap Newborn imprint brand awareness on toddlers and turn babies into mini-billboards. My friend Monica tells me that her seven-year-old son marks his homework not with check marks but with little red Nike swooshes.

Until the early seventies, logos on clothes were generally hidden from view, discreetly placed on the inside of the collar. Small designer emblems did appear on the outside of shirts in the first half of the century, but such sporty attire was pretty much restricted to the golf courses and tennis courts of the rich. In the late seventies, when the fashion world rebelled against Aquarian flamboyance, the country-club wear of the fifties became mass style for newly conservative parents and their preppy kids. Ralph Lauren's Polo horseman and Izod Lacoste's alligator escaped from the golf course and scurried into the streets, dragging the logo decisively onto the outside of the shirt. These logos served the same social function as keeping the clothing's price tag on: everyone knew precisely what premium the wearer was willing to pay for style. By the mid-eighties, Lacoste and Ralph Lauren were joined by Calvin Klein, Esprit and, in Canada, Roots; gradually, the logo was transformed from an ostentatious affectation to an active fashion accessory. Most significantly, the logo itself was growing in size, ballooning from a three-quarter-inch emblem into a chest-sized marquee. This process of logo inflation is still progressing, and none is more bloated than Tommy Hilfiger, who has managed to pioneer a clothing style that transforms its faithful adherents into walking, talking, life-sized Tommy dolls, mummified in fully branded Tommy worlds.

This scaling-up of the logo's role has been so dramatic that it has become a change in substance. Over the past decade and a half, logos have grown so dominant that they have essentially transformed the clothing on which they appear into empty carriers for the brands they represent. The metaphorical alligator, in other words, has risen up and swallowed the literal shirt.

This trajectory mirrors the larger transformation our culture has undergone since Marlboro Friday, sparked by a stampede of manufacturers looking to replace their cumbersome product-production apparatus with transcendent brand names and to infuse their brands with deep, meaningful messages. By the mid-nineties, companies like Nike, Polo and Tommy Hilfiger were ready to take branding to the next level: no longer simply branding their own products, but branding the outside culture as well – by sponsoring cultural events, they could go out into the world and claim bits of it as brand-name outposts. For these companies, branding was not just a matter of adding value to a product. It was about thirstily soaking up cultural ideas and iconography that their brands could reflect by projecting these ideas and images back on the culture as "extensions" of their brands. Culture, in other words, would add value to their brands. For example, Onute Miller, senior brand manager for Tequila Sauza, explains that her company sponsored a risqué photography exhibit by George Holz because "art was a natural synergy with our product."[1]

Branding's current state of cultural expansionism is about much more than traditional corporate sponsorships: the classic arrangement in which a company donates money to an event in exchange for seeing its logo on a banner or in a program. Rather, this is the Tommy Hilfiger approach of full-frontal branding, applied now to cityscapes, music, art, films, community events, magazines, sports and schools. This ambitious project makes the logo the central focus of everything it touches – not an add-on or a happy association, but the main attraction.

Advertising and sponsorship have always been about using imagery to equate products with positive cultural or social experiences. What makes nineties-style branding different is that it increasingly seeks to take these associations out of the representational realm and make them a lived reality. So the goal is not merely to have child actors drinking Coke in a TV commercial, but for students to brainstorm concepts for Coke's next ad campaign in English class. It transcends logo-festooned Roots clothing designed to conjure memories of summer camp and reaches out to build an actual Roots country lodge that becomes a 3-D manifestation of the Roots brand concept. Disney transcends its sports network ESPN, a channel for guys who like to sit around in sports bars screaming at the TV, and launches a line of ESPN Sports Bars, complete with giant-screen TVs. The branding process reaches beyond heavily marketed Swatch watches and launches "Internet time," a new venture for the Swatch Group, which divides the day into one thousand "Swatch beats." The Swiss company is now attempting to convince the on-line world to abandon the traditional clock and switch to its time-zone-free, branded time.

The effect, if not always the original intent, of advanced branding is to nudge the hosting culture into the background and make the brand the star. It is not to sponsor culture but to *be* the culture. And why shouldn't it be? If brands are not products but ideas, attitudes, values and experiences, why can't they be culture too? As we will see later in the chapter, this project has been so successful that the lines between corporate sponsors and sponsored culture have entirely disappeared. But this conflation has not been a one-way process, with passive artists allowing themselves to be shoved into the background by aggressive multinational corporations. Rather, many artists, media personalities, film directors and sports stars have been racing to meet the corporations halfway in the branding game. Michael Jordan, Puff Daddy, Martha Stewart, Austin Powers, Brandy and *Star Wars* now mirror the corporate structure of corporations like Nike and the Gap, and they are just as captivated by the prospect of developing and leveraging their own branding potential as the product-based manufacturers. So what was once a process of selling culture to a sponsor for a price has been supplanted by the logic of "co-branding" – a fluid partnership between celebrity people and celebrity brands.

The project of transforming culture into little more than a collection of brand-extensions-in-waiting would not have been possible without the deregulation and privatization policies of the past three decades. In Canada under Brian Mulroney, in the U.S. under Ronald Reagan and in Britain under Margaret Thatcher (and in many other parts of the world as well), corporate taxes were dramatically lowered, a move that eroded the tax base and gradually starved out the public sector. (See Table 2.1 on page 33.) As government spending dwindled, schools, museums and broadcasters were desperate to make up their budget shortfalls and thus ripe for partnerships with private corporations. It also didn't hurt that the political climate during this time ensured that there was almost no vocabulary to speak passionately about the value of a non-commercialized public sphere. This was the time of the Big Government bogeyman and deficit hysteria, when any political move that was not overtly designed to increase the freedom of corporations was vilified as an endorsement of national bankruptcy. It was against this backdrop that, in rapid order, sponsorship went from being a rare occurrence (in the 1970s) to an exploding growth industry (by the mid-eighties), picking up momentum in 1984 at the Los Angeles Olympics (see Table 2.2 on page 33).

At first, these arrangements seemed win-win: the cultural or educational institution in question received much-needed funds and the sponsoring corporation was compensated with some modest form of public acknowledgment and a tax break. And, in fact, many of these new public-private arrangements were just that simple, successfully retaining a balance between the cultural event or institution's independence and the sponsor's desire for credit, often helping to foster a revival of arts accessible to the general public. Successes like these are frequently overlooked by critics of commercialization, among whom there is an unfortunate tendency to tar all sponsorship with the same brush, as if any contact with a corporate logo infects the natural integrity of an otherwise pristine public event or cause. Writing in *The Commercialization of American Culture*, advertising critic Matthew McAllister labels corporate sponsorship "control behind a philanthropic façade."[2] He writes:

> While elevating the corporate, sponsorship simultaneously devalues what it sponsors.... The sporting event, the play, the concert and the public television program become subordinate to promotion because, in the sponsor's mind and in the symbolism of the event, they exist to promote. It is not Art for Art's Sake as much as Art for Ad's Sake. In the public's eye, art is yanked from its own separate and theoretically autonomous domain and squarely placed in the commercial.... Every time the commercial intrudes on the cultural, the integrity of the public sphere is weakened because of the obvious encroachment of corporate promotion.[3]

This picture of our culture's lost innocence is mostly romantic fiction. Though there have always been artists who have fought fiercely to protect the integrity of their work, neither the arts, sports nor the media have ever, even theoretically, been the protected sovereign states that McAllister imagines. Cultural products are the all-time favorite playthings of the powerful, tossed from wealthy statesmen such as Gaius Cilnius Maecenas, who set up the poet Horace in a writing estate in 33 B.C., and from rulers like Francis I and the

Absolut Vodka/Keith Haring, *Absolut Haring* (detail), 1986.

Table 2.1
Corporate Tax as a Percentage of Total Federal Revenue in the U.S., 1952, 1975 and 1998

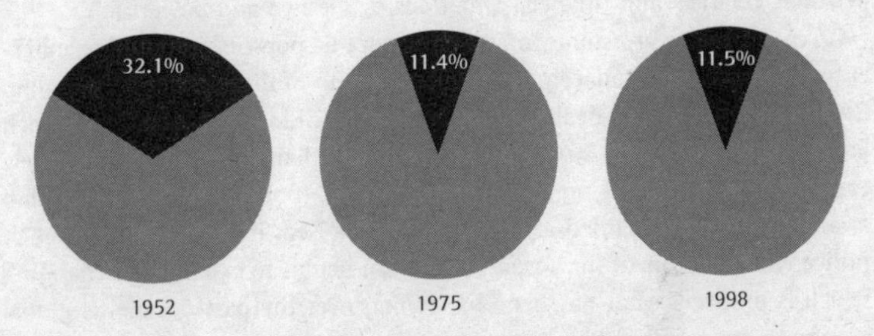

Source: *Time*, March 20, 1987; U.S. Office of Management and Budget; *Revenue Statistics 1965-1996* (1997 edition), OECD; 1999 Federal Budget. (For Canadian figures see Table 2.1a, Appendix, page 472.)

Table 2.2
Increase in U.S. Corporate Sponsorship Spending since 1985

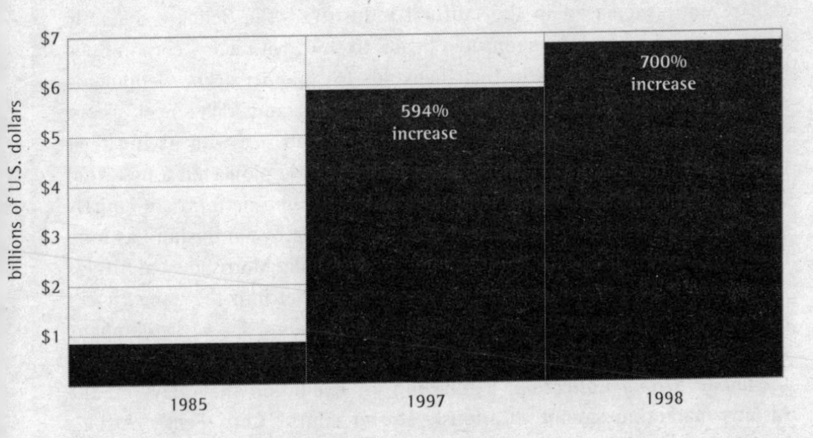

Source: *IEG Sponsorship Report*, December 22, 1997, and December 21, 1998.

Medici family, whose love of the arts bolstered the status of Renaissance painters in the sixteenth century. Though the degree of meddling varies, our culture was built on compromises between notions of public good and the personal, political and financial ambitions of the rich and powerful.

Of course there are some forms of corporate sponsorship that are inherently insidious – the tobacco industry's corralling of the arts springs to mind. But not all sponsorship deals should be so easily dismissed. Not only are such broad strokes unfair to worthy projects but, perhaps more important, they can prevent us from seeing changes in the field. If all corporate sponsorship arrangements are regarded as equally compromised, it becomes easy not to notice when the role of the corporate sponsor begins to expand and change – which is precisely what has been happening over the past decade as global corporate sponsorship has ballooned from a $7-billion-a-year industry in 1991 to a $19.2 billion one in 1999.

When sponsorship took off as a stand-in for public funds in the mid-eighties, many corporations that had been experimenting with the practice ceased to see sponsorship as a hybrid of philanthropy and image promotion and began to treat it more purely as a marketing tool, and a highly effective one at that. As its promotional value grew – and as dependency on sponsorship revenue increased in the cultural industries – the delicate dynamic between sponsors and the sponsored began to shift, with many corporations becoming more ambitious in their demands for grander acknowledgments and control, even buying events outright. Molson and Miller beer, as we will see further on in this chapter, are no longer satisfied with having their logos on banners at rock concerts. Instead, they have pioneered a new kind of sponsored concert in which the blue-chip stars who perform are entirely upstaged by their hosting brand. And while corporate sponsorship has long been a mainstay in museums and galleries, when Philip Morris–owned Altoids mints decided in January 1999 that it wanted to get into the game, it cut out the middleman. Rather than sponsoring an existing show, the company spent $250,000 to buy works by twenty emerging artists and launch its own *Curiously Strong Collection*, a traveling art exhibition that plays on the Altoids marketing slogan, "Curiously strong mints." Chris Peddy, Altoids brand manager, said, "We decided to take it to the next level."[4]

These companies are part of a larger phenomenon explained by Lesa Ukman, executive editor of the *International Events Group Sponsorship Report*, the industry's bible: "From MasterCard and Dannon to Phoenix Home Life and LaSalle Bank, companies are buying properties and creating their own events. This is not because they want to get into the business. It's because proposals sponsors receive don't fit their requirements or because they've had negative experiences buying into someone else's gig."[5] There is a certain logic to this progression: first, a select group of manufacturers transcend their connection to their earthbound products, then, with marketing elevated as the pinnacle of their businesses, they attempt to alter marketing's social status as a commercial interruption and replace it with seamless integration.

The most insidious effect of this shift is that after a few years of Molson concerts, Pepsi-sponsored papal visits, Izod zoos and Nike after-school basketball programs, everything from small community events to large religious gatherings are believed to "need a sponsor" to get off the ground; August 1999, for instance, saw the first-ever private wedding with corporate sponsorship. This is what Leslie Savan, author of *The Sponsored Life*, describes as symptom number one of the sponsored mindset: we become collectively convinced not that corporations are hitching a ride on our cultural and communal activities, but that creativity and congregation would be impossible without their generosity.

The Branding of the Cityscape

The expansive trajectory of branding revealed itself to Londoners in a 1997 holiday season morality play. It began when the Regent Street Association found itself without enough money to replace the dimming Christmas lights that normally adorned the street during the season. Yves Saint Laurent stepped in and generously offered to split the cost of new decorations in exchange for seeing its logo up in lights. But when the time came to hang the Christmas lights, it seemed that the YSL logos were much larger than the agreed-upon size. Every few steps, shoppers were reminded by illuminated signs 5.5 meters high just who had brought them Christmas. The logos were eventually replaced with smaller ones, but the lesson remained: the role of the sponsor, like that of advertising in general, has a tendency to expand.

While yesterday's corporate sponsors may have been satisfied merely propping up community events, the meaning-seeking brand builders will never accept this role for long. Branding is, at its core, a deeply competitive undertaking in which brands are up against not only their immediate rivals (Nike vs. Reebok, Coke vs. Pepsi, McDonald's vs. Burger King, for example) but all other brands in the mediascape, including the events and people they are sponsoring. This is perhaps branding's cruelest irony: most manufacturers and retailers begin by seeking out authentic scenes, important causes and cherished public events so that these things will infuse their brands with meaning. Such gestures are frequently motivated by genuine admiration and generosity. Too often, however, the expansive nature of the branding process ends up causing the event to be usurped, creating the quintessential lose-lose situation. Not only do fans begin to feel a sense of alienation from (if not outright resentment toward) once-cherished cultural events, but the sponsors lose what they need most: a feeling of authenticity with which to associate their brands.

That's certainly what happened to Michael Chesney, the hip-hop adman who painted Canadian billboards into the branding era. He loved Toronto's Queen Street West – the funky clothing stores, the artists on all the patios, and, most of all, the graffiti art that figured large on the walls in that part of town. For Chesney, it was a short step from the public's growing interest in the cultural value of graffiti to the commercial takeover of that pocket of marginal space – a space used and reused by the disenfranchised for political and cultural expression in every city in the world.

From the start, Chesney considered himself a distant relative of the graffiti kids – though less a cousin than a rich uncle. The way he saw it, as a commercial artist and billboard salesman he was also a creature of the streets, because even if he was painting for corporate clients, he, like the graffiti artists, left his mark on walls. It was in this context that Chesney pioneered the advertising practice of the "building takeover." In the late eighties, Chesney's company Murad began painting directly onto building walls, letting the size of each structure dictate the dimensions of the ad. The idea harked back to 1920s Coca-Cola murals on corner grocery stores and to early-industrial urban factories and department stores that painted their names and logos in giant block lettering on their buildings' façades. The walls Chesney rented to Coke, Warner Brothers and Calvin Klein were a little bit bigger, however, reaching their pinnacle at a colossal 20,000-square-foot billboard overlooking one of Toronto's busiest intersections. Gradually, the ads wrapped around the corners of the buildings so that they covered not just one wall, but all of them: the ad as edifice.

In the summer of 1996, when Levi Strauss chose Toronto to test-market its new SilverTab jeans line, Chesney put on his most daring show yet: he called it "The Queen Street Takeover." Between 1996 and 1997, Levi's increased its spending on billboard advertisements by a startling 301 percent – and Toronto saw much of that windfall.[6] For one year, as the centerpiece of the most expensive outdoor ad campaign in Canadian history, Chesney painted his beloved strip silver. He bought up the façades of almost every building on the busiest stretch of Queen and turned them into Levi's billboards, upping the ante of the ad extravaganza even further with 3-D extensions, mirrors and neon. It was Murad's greatest triumph, but the takeover presented some problems for Michael Chesney. When I spent a day with him at the tail end of the SilverTab bonanza, he could barely walk down Queen Street without running into somebody who was furious about the invasion. After ducking a few bullets, he told me a story of bumping into an acquaintance: "she said, 'You took over Queen Street.' She was really almost crying and I just, my heart sank, and she was really bummed out. But, hey, what can you do? It's the future, it's not Queen anymore."

Nearly every major city has seen some variation of the 3-D ad takeover, if not on entire buildings, then on buses, streetcars or taxis. It is sometimes difficult, however, to express dissatisfaction with this brand expansion – after all, most of these venues and vehicles have been carrying some form of advertising for decades. But somewhere along the line, the order flipped. Now buses, streetcars and taxis, with the help of digital imaging and large pieces of adhesive vinyl, have become ads on wheels, shepherding passengers around in giant chocolate bars and gum wrappers, just as Hilfiger and Polo turned clothing into wearable brand billboards.

If this creeping ad expansion seems a mere matter of semantics when applied to taxis and T-shirts, its implications are much more serious when looked at in the context of another marketing trend: the branding of entire neighborhoods and cities. In March 1999, Los Angeles mayor Richard Riordan unveiled a plan to revitalize poor inner-city areas, many of them still scarred from the 1992 riots after the Rodney King verdict: corporations would adopt a run-down part of town and brand its redevelopment. For the time being, the sponsors of Genesis L.A., as the project is called – among them Bank-America and Wells Fargo & Co. – only have the option of seeing these sites named after them, much like a sponsored sports arena. But if the initiative follows the expansive branding trajectory seen elsewhere, the sponsoring companies could well wield more politically powerful roles in these communities soon.

The idea of a fully privatized, branded town or neighborhood is not nearly as far-fetched today as it was only a few years ago, as the inhabitants of Disney's town Celebration, Florida, can attest – and as the citizens of Cashmere, Washington, have quickly learned. A sleepy town of 2,500 people, Cashmere has as its major industry the Liberty Orchard candy factory, which has been making Aplets and Cotlets chewy sweets since it was founded in 1918. It was all very quaint until Liberty Orchard announced in September 1997 that it would leave for greener pastures unless the town agreed to transform itself into a 3-D tourist attraction for the Aplets and Cotlets all-American brand, complete with signs along the highway and a downtown turned into a corporate gift shop. *The Wall Street Journal* reported the company's ransom demands:

> They want all road signs and official correspondence by the city to say "Cashmere, Home of Aplets and Cotlets." They have asked that one of the two main streets in town be changed to Cotlets Avenue, and the other one be renamed Aplets Avenue. The candymaker also wants the Mayor and Council to sell City Hall to them, build new parking lots and possibly go to the bond market to start a tourism campaign on behalf of the worldwide headquarters of a company that says its story is "America in a nutshell."[7]

The Branding of Media

Although there is a clear trajectory in all of these stories, there is little point, at this stage in our sponsored history, in pining for either a mythic brand-free past or some utopian commercial-free future. Branding becomes troubling – as it did in the cases just discussed – when the balance tips dramatically in favor of the sponsoring brand, stripping the hosting culture of its inherent value and treating it as little more than a promotional tool. It is possible, however, for a more balanced relationship to unfold – one in which both sponsor and sponsored hold on to their power and in which clear boundaries are drawn and protected. As a working journalist, I know that critical, independent – even anti-corporate – coverage does appear in corporate-owned media, sandwiched, no less, between the car and tobacco ads. Are these articles tainted by this impure context? No doubt. But if balance (as opposed to purity) is the goal, then maybe print media, where the first mass-market advertising campaigns began, can hold some important lessons for how to cope with the expansionist agenda of branding.

> *I appeal to every producer not to release "sponsored" moving pictures. . . . Believe me, if you jam advertising down their throats and pack their eyes and ears with it, you will build up a resentment that will in time damn your business.*
>
> – Carl Laemmle of Universal Pictures, 1931

It is common knowledge that many advertisers rail at controversial content, pull their ads when they are criticized even slightly and perpetually angle for so-called value-addeds – plugs for their wares in shopping guides and fashion spreads. For example, S.C. Johnson & Co. stipulates that its ads in women's magazines "should not be opposite extremely controversial features or material antithetical to the nature/copy of the advertised product" while De Beers diamonds demands that their ads be far from any "hard news or anti/love-romance themed editorial."[8] And up until 1997, when Chrysler placed an ad it demanded that it be "alerted in advance of any and all editorial content that encompasses sexual, political, social issues or any editorial that might be construed as provocative or offensive."[9] But the advertisers don't always get their way: controversial stories make it to print and to air, even ones critical of major advertisers. At its most daring and uncompromised, the news media can provide workable models for the protection of the public interest even under heavy corporate pressure, though these battles are often won behind closed doors. On the other hand, at their worst, these same media show how deeply distorting the effects of branding can be on our public discourse – particularly since journalism, like every other part of our culture, is under constantly increasing pressure to merge with the brands.

Part of this stepped-up pressure is coming from the explosion of sponsored media projects: magazines, Web sites and television programs that invite corporate sponsors to become involved at the development stage of a venture. That's the role Heineken played in the British music and youth culture show *Hotel Babylon*, which aired on ITV. In an embarrassing incident in January 1996, a memo from a Heineken executive was leaked to the press that berated the producers for insufficiently "Heineken-izing" the as-yet-unaired program. Specifically, Justus Kos objected to male audience members drinking wine as opposed to "masculine drinks like beer, whisky," noted that "more evidence of beer is not just requested but needed" and complained that the show's host "shouldn't stand in the way of the beer columns when introducing guests." Most inflammatory of all was the executive's complaint that there was "too high a proportion of negroes in the audience."[10] After the controversy made its way into the press, Heineken CEO Karel Vuursteen issued a public apology.

Another sponsor scandal erupted during the 1998 Winter Olympics in Nagano, Japan, when CBS investigative journalist Roberta Baskin saw her CBS Sports department colleagues reporting on the games in jackets adorned with bold Nike logos. Nike was the official sponsor of the network's Olympic coverage and it provided news and sports reporters with the swooshed gear because, according to Nike spokesman Lee Weinstein, it "helps us build awareness about our products." Baskin was "dismayed and embarrassed" that CBS reporters seemed to be endorsing Nike products, not only because it represented a further dissolution of the line between editorial and advertising, but because two years earlier, Baskin had broken a news story about physical abuse of workers at a Nike shoe factory in Vietnam. She accused the station of refusing to allow her to pursue a follow-up and of yanking the original story from a scheduled rerun because of its sponsorship deal with Nike. CBS News president Andrew Heyward strenuously denied bowing to sponsor pressure, calling Baskin's allegations "truly preposterous." He did pull the Nike jackets off the news reporters midway through the games, though the sports department kept theirs on.

In some ways, these stories are simply pumped-up versions of the same old tug-of-war between editorial and advertising that journalists have faced for a century and a quarter. Increasingly, however, corporations aren't just asking editors and producers to become their de facto ad agencies by dreaming up ways to plug their wares in articles and photo shoots, they are also asking magazines to become their actual ad agencies, by helping them to create the ads that run in their magazines. More and more magazines are turning their offices into market-research firms and their readers into focus groups in an effort to provide the most cherished "value-added" they can offer their clients: highly detailed demographic information about their readership, amassed through extensive surveys and questionnaires.

In many cases, the magazines then use the readership information to design closely targeted advertisements for their clients. *Details* magazine, for instance, designed a twenty-four-page comic/advertisement strip in October 1997, with products like Hugo Boss cologne and Lee jeans woven into a story line about the daily adventures of a professional in-line skater. On the page following each product's extreme cameo, the company's real ad appeared.

The irony of these branding experiments, of course, is that they only seem to make brands more resentful of the media that host them. Inevitably, the lifestyle brands begin to ask why they need to attach themselves to someone else's media project in the first place. Why, even after proving they can integrate into the most stylish and trendiest of magazines, should they be kept at arm's length or, worse, branded with the word "Advertisement," like the health warnings on packs of cigarettes? So, with lifestyle magazines looking more and more like catalogs for designers, designer catalogs have begun to look more and more like magazines: Abercrombie & Fitch, J. Crew, Harry Rosen and Diesel have all shifted to a storybook format, where characters frolic along sketchily drawn plotlines.

The merger between media and catalog reached a new high with the launch of the teen TV drama *Dawson's Creek* in January of 1998. Not only did the characters all wear J. Crew clothes, not only did the windswept, nautical set make them look as if they had stepped off the pages of a J. Crew catalog, and not only did the characters spout dialogue like "He looks like he stepped out of a J. Crew catalog," but the cast was also featured on the cover of the January J. Crew catalog. Inside the new "freestyle magalog," the young actors are pictured in rowboats and on docks – looking as if they just stepped off the set of a *Dawson's Creek* episode.

To see the birthplace of this kind of brand ambition, you have to go on-line, where there was never really any pretense of a wall existing between editorial and advertisement. On the Web, marketing language reached its nirvana: the ad-free ad. For the most part, the on-line versions of media outlets feature straightforward banner ads similar to their paper or broadcast versions, but many media outlets have also used the Net to blur the line between editorial and advertising much more aggressively than they could in the non-virtual world. For instance, on the *Teen People* site, readers can click and order cosmetics and clothing as they read about them. On the *Entertainment Weekly* site, visitors can click and order the books and CDs being reviewed. In Canada, *The Globe and Mail* has attracted the ire of independent booksellers for the on-line version of its book review section, ChaptersGLOBE.com. After reading *Globe* reviews, readers can click to order books directly from the Chapters chain – a reviewer/retailer partnership that formed "Canada's largest online bookstore." *The New York Times*' on-line partnership with Barnes and Noble has caused similar controversies in the U.S.

These sites are relatively tame examples of the branding-content integration taking place on the Net, however. Sites are increasingly created by "content developers," whose role is to produce editorial that will make an ad-cozy home for the developers' brand-name clients. One such on-line venture is Parent Soup, invented by content developer "iVillage" for Fisher-Price, Starbucks, Procter and Gamble and Polaroid. It calls itself a "parents' community" and attempts to imitate a user-driven newsgroup, but when parents go to Parent Soup to get peer advice, they receive such branded wisdom as: the way to improve your child's self-esteem is by taking Polaroids of her. No need to bully or buy off editors—just publish do-it-yourself content, with ads pre-integrated.

Absolut Vodka's 1997 Absolut Kelly Internet site provided an early preview of the direction in which branded media are headed. The distiller had long since solicited original, brand-centered creations from visual artists, fashion designers and novelists to use in its advertisements—but this was different. On Absolut Kelly, only the name of the site advertised the product; the rest was an illustrated excerpt from *Wired* magazine editor Kevin Kelly's book *Out of Control*. This, it seemed, was what the brand managers had aspired to all along: for their brands to become quietly integrated into the heart of the culture. Sure, manufacturers will launch noisy interruptions if they are locked on the wrong side of the commerce/culture divide, but what they really want is for their brand to earn the right to be accepted, not just as advertising art but simply as art. Off-line, Absolut is still a major advertiser in *Wired*, but on-line, it is Absolut that is the host, and a *Wired* editor the supporting act.

Rather than merely bankrolling someone else's content, all over the Net, corporations are experimenting with the much-coveted role of being "content providers": Gap's site offers travel tips, Volkswagen provides free music samples, Pepsi urges visitors to download video games, and Starbucks offers an on-line version of its magazine, *Joe*. Every brand with a Web site has its own virtual, branded media outlet—a beachhead from which to expand into other non-virtual media. What has become clear is that corporations aren't just selling their products on-line, they're selling a new model for the media's relationship with corporate sponsors and backers. The Internet, because of its anarchic nature, has created the space for this model to be realized swiftly, but the results are clearly made for off-line export. For instance, about a year after the launch of Absolut Kelly, the company reached full editorial integration in *Saturday Night* magazine when the final page of a nine-page excerpt from Mordecai Richler's novel *Barney's Version* was wrapped around the silhouette of an Absolut bottle. This was not an ad, it was part of the story, yet at the bottom of the page were the words

"Absolut Mordecai."[11]

Although magazines and individual television shows are beginning to see the branded light, it is a network, MTV, that is the model for fully branded media integration. MTV started out sponsored, as a joint venture between Warner Communications and American Express. From the beginning, MTV has not been just a marketing machine for the products it advertises around the clock (whether those products are skin cleansers or the albums it moves with its music videos); it has also been a twenty-four-hour advertisement for MTV itself: the first truly branded network. Though there have been dozens of imitators since, the original genius of MTV, as every marketer will tell you, is that viewers didn't watch individual shows, they simply watched MTV. "As far as we were concerned, MTV was the star," says Tom Freston, network founder.[12] And so advertisers didn't want to just advertise on MTV, they wanted to co-brand with the station in ways that are still unimaginable on most other networks: giveaways, contests, movies, concerts, awards ceremonies, clothing, countdowns, listings, credit cards and more.

The model of the medium-as-brand that MTV perfected has since been adopted by almost every other major media outlet, whether magazines, film studios, television networks or individual shows. The hip-hop magazine *Vibe* has extended into television, fashion shows and music seminars. Fox Sports has announced that it wants its new line of men's clothing to be on par with Nike: "We are hoping to take the attitude and lifestyle of Fox Sports off the TV and onto men's backs, creating a nation of walking billboards," said David Hill, CEO of Fox Broadcasting.

The rush to branding has been most dramatic in the film industry. At the same time that brand-name product placement in films has become an indispensable marketing vehicle for companies like Nike, Macintosh and Starbucks, films themselves are increasingly being conceptualized as "branded media properties." Newly merged entertainment conglomerates are always looking for threads to sew together their disparate holdings in cross-promotional webs and, for the most part, that thread is the celebrity generated by Hollywood blockbusters. Films create stars to cross-promote in books, magazines and TV, and they also provide prime vehicles for sports, television and music stars to "extend" their own brands.

I'll explore the cultural legacy of this type of synergy-driven production in Chapter 9, but there is a more immediate impact as well, one that has much to do with the phenomenon of disappearing unmarketed cultural "space" with which this section is concerned. With brand managers envisioning themselves as sensitive culture makers, and culture makers adopting the hard-nosed business tactics of brand builders, a dramatic change in mindset has occurred. Whatever desire might exist to protect a television show from too much sponsor interference, an emerging musical genre from crass commercialism or a magazine from overt advertiser control has been trampled by the manic branding imperative: to disseminate one's own brand "meaning" through whatever means necessary, often in partnership with other powerful brands. In this context, the *Dawson's Creek* brand actively benefits from its exposure in the J. Crew catalog, the Kelly brand grows stronger from its association with the Absolut brand, the *People* magazine brand draws cachet from a close association with Tommy Hilfiger, and the *Phantom Menace* tie-ins with Pizza Hut, Kentucky Fried Chicken and Pepsi are invaluable *Star Wars* brand promotion. When brand awareness is the goal shared by all, repetition and visibility are the only true measures of success. The journey to this point of full integration between ad and art, brand and culture, has taken most of this century to achieve, but the point of no return, when it arrived, was unmistakable: April 1998, the launch of the Gap Khakis campaign.

The Branding of Music

In 1993, the Gap launched its "Who wore khakis?" ads, featuring old photographs of such counterculture figures as James Dean and Jack Kerouac in beige pants. The campaign was in the cookie-cutter co-optation formula: take a cool artist, associate that mystique with your brand, hope it wears off and makes you cool too. It sparked the usual debates about the mass marketing of rebellion, just as William Burroughs's presence in a Nike ad did at around the same time.

Fast forward to 1998. The Gap launches its breakthrough Khakis Swing ads: a simple, exuberant miniature music video set to "Jump, Jive 'n' Wail" – and a great video at that. The question of whether these ads were "co-opting" the artistic integrity of the music was entirely meaningless. The Gap's commercials didn't capitalize on the retro swing revival – a solid argument can be made that they *caused* the swing revival. A few months later, when singer-songwriter Rufus Wainwright appeared in a Christmas-themed Gap ad, his sales soared, so much so that his record company began promoting him as "the guy in the Gap ads." Macy Gray, the new R&B "It Girl," also got her big break in a Baby Gap ad. And rather than the Gap Khaki ads looking like rip-offs of MTV videos, it seemed that overnight, every video on MTV – from Brandy to Britney Spears and the Backstreet Boys – looked like a Gap ad; the company has pioneered its own aesthetic, which spilled out into music, other advertisements, even films like *The Matrix*. After five years of intense lifestyle branding, the Gap, it has become clear, is as much in the culture-creation business as the artists in its ads.

For their part, many artists now treat companies like the Gap less as deep-pocketed pariahs trying to feed off their cachet than as just another medium they can exploit in order to promote their own brands, alongside radio, video and magazines. "We have to be everywhere. We can't afford to be too precious in our marketing," explains Ron Shapiro, executive vice president of Atlantic Records. Besides, a major ad campaign from Nike or the Gap penetrates more nooks and crannies of the culture than a video in heavy rotation on MTV or a cover article in *Rolling Stone*. Which is why piggybacking on these campaign blitzes – Fat Boy Slim in Nike ads, Brandy in Cover Girl commercials, Lil' Kim rapping for Candies – has become, *Business Week* announced with much glee, "today's top 40 radio."[13]

Of course the branding of music is not a story of innocence lost. Musicians have been singing ad jingles and signing sponsorship deals since radio's early days, as well as having their songs played on commercial radio stations and signing deals with multinational record companies. Throughout the eighties – music's decade of the straight-up shill – rock stars like Eric Clapton sang in beer ads, and the pop stars, appropriately enough, crooned for pop: George Michael, Robert Plant, Whitney Houston, Run-DMC, Madonna, Robert Palmer, David Bowie, Tina Turner, Lionel Richie and Ray Charles all did Pepsi or Coke ads, while sixties anthems like the Beatles' "Revolution" became background music for Nike commercials.

During this same period, the Rolling Stones made music history by ushering in the era of the sponsored rock tour – and fittingly, sixteen years later, it is still the Stones who are leading the charge into the latest innovation in corporate rock: the band as brand extension. In 1981, Jovan – a distinctly un-rock-and-roll perfume company – sponsored a Rolling Stones stadium tour, the first arrangement of its kind, though tame by today's standards. Though the company got its logos on a few ads and banners, there was a clear distinction between the band that had chosen to "sell out" and the corporation that had paid a huge sum to associate itself with the inherent rebelliousness of rock. This subordinate status might have been fine for a company out merely to move products, but when designer Tommy Hilfiger decided that the energy of rock and rap would become his "brand essence," he was looking for an integrated experience, one more in tune with his own transcendent identity quest. The results were evident in the Stones' Tommy-sponsored Bridges to Babylon tour in 1997. Not only did Hilfiger have a contract to clothe Mick Jagger, he also had the same arrangement with the Stones' opening act, Sheryl Crow – on stage, both modeled items from Tommy's newly launched "Rock 'n' Roll Collection."

It wasn't until January 1999, however – when Hilfiger launched the ad campaign for the Stones' No Security Tour – that full brand-culture integration was achieved. In the ads, young, glowing Tommy models were pictured in full-page frame "watching" a Rolling Stones concert taking place on the opposite page. The photographs of the band members were a quarter of the size of those of the models. In some of the ads, the Stones were nowhere to be found and the Tommy models alone were seen posing with their own guitars. In all cases, the ads featured a hybrid logo of the Stones' famous red tongue over Tommy's trademarked red-white-and-blue flag. The tagline was "Tommy Hilfiger Presents the Rolling Stones No Security Tour" – though there were no dates or locations for any tour stops, only the addresses of flagship Tommy stores.

In other words, this wasn't rock sponsorship, it was "live-action advertising," as media consultant Michael J. Wolf describes the ads.[14] It's clear from the campaign's design that Hilfiger isn't interested in buying a piece of someone else's act, even if they are the Rolling Stones. The act is a background set, powerfully showcasing the true rock-and-roll essence of the Tommy brand; just one piece of Hilfiger's larger project of carving out a place in the music world, not as a sponsor but as a player – much as Nike has achieved in the sports world.

The Hilfiger/Stones branding is only the highest-profile example of the new relationship between bands and sponsors that is sweeping the music industry. For instance, it was a short step for Volkswagen – after using cutting-edge electronic music in its ads for the new Beetle – to launch DriversFest '99, a VW branded music festival in Long Island, New York. DriversFest competes for ticket sales with the Mentos Freshmaker Tour, a two-year-old traveling music festival owned and branded by a breath-mint manufacturer – on the Mentos Web site, visitors are invited to vote for which bands they want to play the venue. As with the Absolut Kelly Web site and the Altoids' Curiously Strong art exhibition, these are not sponsored events: the brand is the event's infrastructure; the artists are its filler, a reversal in the power dynamic that makes any discussion of the need to protect unmarketed artistic space appear hopelessly naive.

This emerging dynamic is clearest in the branded festivals being developed by the large beer companies. Instead of merely playing in beer ads, as they likely would have in the eighties, acts like Hole, Soundgarden, David Bowie and the Chemical Brothers now play beer-company gigs. Molson Breweries, which owns 50 percent of Canada's only national concert promoter, Universal Concerts, already has its name promoted almost every time a rock or pop star gets up on stage in Canada – either through its Molson Canadian Rocks promotional arm or its myriad venues: Molson Stage, Molson Park, Molson Amphitheatre. For the first decade or so, this was a fine arrangement, but by the mid-nineties, Molson was tired of being upstaged. Rock stars had an annoying tendency to hog the spotlight and, worse, sometimes they even insulted their sponsors from the stage.

Clearly fed up, in 1996 Molson held its first Blind Date Concert. The concept, which has since been exported to the U.S. by sister company Miller Beer, is simple: hold a contest in which winners get to attend an exclusive concert staged by Molson and Miller in a small club – much smaller than the venues where one would otherwise see these megastars. And here's the clincher: keep the name of the band secret until it steps on stage. Anticipation mounts about the concert (helped along by national ad campaigns building up said anticipation), but the name on everyone's lips isn't David Bowie, the Rolling Stones, Soundgarden, INXS or any of the other bands that have played the Dates, it's Molson and Miller. No one, after all, knows who is going to play, but they know who is putting on the show. With Blind Date, Molson and Miller invented a way to equate their brands with extremely popular musicians, while still maintaining their competitive edge over the stars. "In a funny way," says Universal Concerts' Steve Herman, "the beer is bigger than the band."[15]

The rock stars, turned into high-priced hired guns at Molson's bar mitzvah party, continued to find sad little ways to rebel. Almost every musician who played a Blind Date acted out: Courtney Love told a reporter, "God bless Molson.... I douche with it."[16] The Sex Pistols' Johnny Lydon screamed "Thank you for the money" from the stage, and Soundgarden's Chris Cornell told the crowd, "Yeah, we're here because of some fucking beer company... Labatt's." But the tantrums were all incidental to the main event, in which Molson and Miller were the real rock stars and it didn't really matter how those petulant rent-a-bands behaved.

Jack Rooney, Miller's vice president of marketing, explains that his $200 million promotion budget goes toward devising creative new ways to distinguish the Miller brand from the plethora of other brands in the marketplace. "We're competing not just against Coors and Corona," he says, "but Coke, Nike and Microsoft."[17] Only he isn't telling the whole story. In *Advertising Age*'s annual "Top Marketing 100" list of 1997's best brands there was a new arrival: the Spice Girls (fittingly enough, since Posh Spice did once tell a reporter, "We wanted to be a 'household name'. Like Ajax."[18]) And the Spice Girls ranked number six in *Forbes* magazine's inaugural "Celebrity Power 100," in May 1999, a new ranking based not on fame or fortune but on stars' brand "franchise." The list was a watershed moment in corporate history, marking the fact that, as Michael J. Wolf says, "Brands and stars have become the same thing."[19]

But when brands and stars are the same thing, they are also, at times, competitors in the high-stakes tussle for brand awareness, a fact more consumer companies have become ready to admit. Canadian clothing company Club Monaco, for instance, has never used celebrities in its campaigns. "We've thought about it," says vice president Christine Ralphs, "but whenever we go there, it always becomes more about the personality than the brand, and for us, we're just not willing to share that."[20]

There is good reason to be protective: though more and more clothing and candy companies seem intent on turning musicians into their opening acts, bands and their record labels are launching their own challenges to this demoted status. After seeing the enormous profits that the Gap and Tommy Hilfiger have made through their association with the music world, record labels are barreling into the branding business themselves. Not only are they placing highly sophisticated cross-branding apparatus behind working musicians, but bands are increasingly being conceived – and test-marketed – as brands first: the Spice Girls, the Backstreet Boys, N' Sync, All Saints and so on. Prefab bands aren't new to the music industry, and neither are bands with their own merchandising lines, but the phenomenon has never dominated pop culture as it has at the end of the nineties, and musicians have never before competed so aggressively with consumer brands. Sean "Puffy" Combs has leveraged his celebrity as a rapper and record producer into a magazine, several restaurants, a clothing label and a line of frozen foods. And Raekwon, of the rap group Wu-Tang Clan, explains that "the music, movies, the clothing, it is all part of the pie we're making. In the year 2005 we might have Wu-Tang furniture for sale at Nordstrom."[21] Whether it's the Gap or Wu-Tang Clan, the only remaining relevant question in the sponsorship debate seems to be, Where do you have the guts to draw the borders around your brand?

Nike and the Branding of Sports

Inevitably, any discussion about branded celebrity leads to the same place: Michael Jordan, the man who occupies the number-one spot on all of those ranking lists, who has incorporated himself into the JORDAN brand, whose agent coined the term "superbrand" to describe him. But no discussion of Michael Jordan's brand potential can begin without the brand that branded him: Nike.

Nike has successfully upstaged sports on a scale that makes the breweries' rock-star aspirations look like amateur night. Now of course pro sports, like big-label music, is in essence a profit-driven enterprise, which is why the Nike story has less to teach us about the loss of unmarketed space – space that, arguably, never even existed in this context – than it does about the mechanics of branding and its powers of eclipse. A company that swallows cultural space in giant gulps, Nike is the definitive story of the transcendent nineties superbrand, and more than any other single company, its actions demonstrate how branding seeks to erase all boundaries between the sponsor and the sponsored. This is a shoe company that is determined to unseat pro sports, the Olympics and even star athletes, to become the very definition of sports itself.

Nike CEO Phil Knight started selling running shoes in the sixties, but he didn't strike it rich until high-tech sneakers became the must-have accessory of America's jogging craze. But when jogging subsided in the mid-eighties and Reebok cornered the market on trendy aerobics shoes, Nike was left with a product destined for the great dustbin of yuppie fads. Rather than simply switching to a different kind of sneaker, Knight decided that running shoes should become peripheral in a reincarnated Nike. Leave sneakers to Reebok and Adidas – Nike would transform itself into what Knight calls "the world's best sports and fitness company."[22]

The corporate mythology has it that Nike is a sports and fitness company because it was built by a bunch of jocks who loved sports and were fanatically devoted to the worship of superior athletes. In reality, Nike's project was a little more complicated and can be separated into three guiding principles. First, turn a select group of athletes into Hollywood-style superstars who are associated not with their teams or even, at times, with their sport, but instead with certain pure ideas about athleticism as transcendence and perseverance – embodiments of the Graeco-Roman ideal of the perfect male form. Second, pit Nike's "Pure Sports" and its team of athletic superstars against the rule-obsessed established sporting world. Third, and most important, brand like mad.

Step 1: Create Sport Celebrities

It was Michael Jordan's extraordinary basketball skill that catapulted Nike to branded heaven, but it was Nike's commercials that made Jordan a global superstar. It's true that gifted athletes like Babe Ruth and Muhammad Ali were celebrities before Nike's time, but they never reached Jordan's otherworldly level of fame. That stratum was reserved for movie and pop stars, who had been transformed by the special effects, art direction and careful cinematography of films and music videos. Sport stars pre-Nike, no matter how talented or worshiped, were still stuck on the ground. Football, hockey and baseball may have been ubiquitous on television, but televised sports were just real-time play-by-plays, which were often tedious, sometimes exciting and high tech only in the slow-mo replay. As for athletes endorsing products, their advertisements and commercials couldn't quite be described as cutting-edge star creation – whether it was Wilt Chamberlain goofily grinning from a box of Wheaties or Rocket Richard being sentenced to "two minutes for looking so good" in Grecian Formula commercials.

I wake up every morning, jump in the shower, look down at the symbol, and that pumps me up for the day. It's to remind me every day what I have to do, which is, "Just Do It."

–Twenty-four-year-old Internet entrepreneur Carmine Collettion on his decision to get a Nike swoosh tattooed on his navel, December 1997

Nike's 1985 TV spots for Michael Jordan brought sports into the entertainment world: the freeze frame, the close-up and the quick cuts that allowed Jordan to appear to be suspended in mid-jump, providing the stunning illusion that he could actually take flight. The idea of harnessing sport-shoe technology to create a superior being – of Michael Jordan flying through the air in suspended animation – was Nike mythmaking at work. These commercials were the first rock videos about sports and they created something entirely new. As Michael Jordan says, "What Phil [Knight] and Nike have done is turn me into a dream."[23]

Many of Nike's most famous TV commercials have used Nike superstars to convey the *idea* of sports, as opposed to simply representing the best of the athlete's own team sport. Spots often feature famous athletes playing a game other than the one they play professionally, such as tennis pro Andre Agassi showing off his version of "rock-and-roll golf." And then there was the breakthrough "Bo Knows" campaign, which lifted baseball and football player Bo Jackson out of his two professional sports and presented him instead as the perfect all-around cross-trainer. A series of quick-cut interviews with Nike stars – McEnroe, Jordan, Gretzky – ironically suggested that Jackson knew their sports better than they did. "Bo knows tennis," "Bo knows basketball" and so on.

At the 1998 Winter Olympics in Nagano, Nike took this strategy out of the controlled environment of its TV commercials and applied it to a real sports competition. The experiment started in 1995 when Nike's marketing department dreamed up the idea of turning a couple of Kenyan runners into Africa's first Olympic ski team. As Mark Bossardet, Nike's director of global athletics, explained, "We were sitting around the office one day and we said, 'What if we took Kenyan runners and transferred their skills to cross-country skiing?'"[24] Kenyan runners, who have dominated cross-country track-and-field competitions at the Olympics since 1968, have always represented the "idea of sports" at Nike headquarters. ("Where's the Kenyans running?" Phil Knight has been heard to demand after viewing a Nike ad deemed insufficiently inspiring and heroic. In Nike shorthand it means, "Where's the Spirit of Sports?").[25] So according to Nike marketing logic, if two Kenyan runners – living specimens of sports incarnate – were plucked out of their own sport and out of their country and their native climate, and dumped on a frozen mountaintop, and if they were then able to transfer their agility, strength and endurance to cross-country skiing, their success would represent a moment of pure sporting transcendence. It would be a spiritual transformation of Man over nature, birthright, nation and petty sports bureaucrats – brought to the world by Nike, of course. "Nike always felt sports shouldn't have boundaries," the swooshed press release announced. Finally there would be proof.

And if nothing else, Nike would get its name in lots of quirky human-interest sidebar stories – just like the wacky Jamaican bobsled team that hogged the headlines at the 1988 Winter Olympics in Calgary. What sports reporter could resist the heart-warmer of Africa's first ski team?

Nike found its test-tube subjects in two mid-level runners, Philip Boit and Henry Bitok. Since Kenya has no snow, no ski federation and no training facilities, Nike financed the entire extravagant affair, dishing out $250,000 for training in Finland and custom-designed uniforms, and paying the runners a salary to live away from their families. When Nagano rolled around, Bitok didn't qualify and Boit finished last – a full twenty minutes after the gold-medal winner, Bjorn Daehlie of Norway. It turns out that cross-country running and cross-country skiing – despite the similarity of their names – require entirely different sets of skills and use different muscles.

But that was beside the point. Before the race began, Nike held a press conference at its Olympic headquarters, catered the event with Kenyan food and beer and showed reporters a video of the Kenyans encountering snow for the first time, skiing into bushes and falling on their butts. The journalists also heard accounts of how the climate change was so dramatic that the Kenyans' skin cracked and their fingernails and toenails fell off, but "now," as Boit said, "I love snow. Without snow, I could not do my sport." As the *Tampa Tribune* of February 12, 1998, put it, "They're just two kooky Kenyans trying to make it in the frozen tundra."

It was quintessential Nike branding: by equating the company with athletes and athleticism at such a primal level, Nike ceased merely to clothe the game and started to play it. And once Nike was in the game with its athletes, it could have fanatical sports fans instead of customers.

Step 2: Destroy the Competition

Like any competitive sports player, Nike has its work cut out for it: winning. But winning for Nike is about much more than sneaker wars. Of course Nike can't stand Adidas, Fila and Reebok, but more important, Phil Knight has sparred with sports agents, whose individual greed, he claims, puts them "inherently in conflict with the interests of athletes at every turn";[26] the NBA, which he feels has unfairly piggybacked on Nike's star-creation machinery;[27] and the International Olympic Committee, whose elitism and corruption Knight derided long before the organization's 1999 bribery scandals.[28] In Nike's world, all of the official sports clubs, associations and committees are actually trampling the spirit of sports – a spirit Nike alone truly embodies and appreciates.

So at the same time as Nike's myth machine was fabricating the idea of Team Nike, Nike's corporate team was dreaming up ways to play a more central role in pro sports. First Nike tried to unseat the sports agents by starting an agency of its own, not only to represent athletes in contract negotiation but also to develop integrated marketing strategies for its clients that are sure to complement – not dilute – Nike's own branding strategy, often by pushing its own ad concepts on other companies.

Then there was a failed attempt to create – and own – a college football version of the Super Bowl (the Nike Bowl), and in 1992, Nike did buy the Ben Hogan golf tour and rename it the Nike Tour. "We do these things to be in the sport. We're in sports – that's what we do," Knight told reporters at the time.[29] That is certainly what they did when Nike and rival Adidas made up their own sporting event to settle a grudge match over who could claim the title "fastest man alive" in their ads: Nike's Michael Johnson or Adidas's Donovan Bailey. Because the two compete in different categories (Bailey in the 100-meter, Johnson in the 200), the sneaker brands agreed to split the difference and had the men compete in a made-up 150-meter race. Adidas won.

When Phil Knight faces the inevitable criticism from sports purists that he is having an undue influence on the games he sponsors, his stock response is that "the athlete remains our reason for being."[30] But as the company's encounter with star basketball player Shaquille O'Neal shows, Nike is only devoted to a certain kind of athlete. Company biographer Donald Katz describes the tense meeting between O'Neal's manager, Leonard Armato, and Nike's marketing team:

> Shaq had observed the explosion of the sports-marketing scene ("He took sports-marketing courses," Armato says) and the rise of Michael Jordan, and he'd decided that rather than becoming a part of several varied corporate marketing strategies, an array of companies might be assembled as part of a brand presence that was he. Consumer products companies would become part of Team Shaq, rather than the other way around. "We're looking for consistency of image," Armato would say as he began collecting the team on Shaq's behalf. "Like Mickey Mouse."

The only problem was that at Nike headquarters, there is no Team Shaq, only

Team Nike. Nike took a pass and handed over the player many thought would be the next Michael Jordan to Reebok – not "Nike material," they said. According to Katz, Knight's mission "from the beginning had been to build a pedestal for sports such as the world had never seen."[31] But at Nike Town in Manhattan, the pedestal is not holding up Michael Jordan, or the sport of basketball, but a rotating Nike sneaker. Like a prima donna, it sits in the spotlight, the first celebrity shoe.

Step 3: Sell Pieces of the Brand As If It Was the Berlin Wall

Nothing embodies the era of the brand like Nike Town, the company's chain of flagship retail outlets. Each one is a shrine, a place set apart for the faithful, a mausoleum. The Manhattan Nike Town on East Fifty-seventh Street is more than a fancy store fitted with the requisite brushed chrome and blond wood, it is a temple, where the swoosh is worshiped as both art and heroic symbol. The swoosh is equated with Sports at every turn: in reverent glass display cases depicting "The definition of an athlete"; in the inspirational quotes about "Courage," "Honor," "Victory" and "Teamwork" inlaid in the floorboards; and in the building's dedication "to all athletes and their dreams."

I asked a salesperson if there was anything amid the thousands of T-shirts, bathing suits, sports bras or socks that did not have a Nike logo on the outside of the garment. He racked his brain. T-shirts, no. Shoes, no. Track suits? No.

"Why?" he finally asked, sounding a bit hurt. "Is somebody allergic to the swoosh?"

Nike, king of the superbrands, is like an inflated Pac-Man, so driven to consume it does so not out of malice but out of jaw-clenching reflex. It is ravenous by nature. It seems fitting that Nike's branding strategy involves an icon that looks like a check mark. Nike is checking off the spaces as it swallows them: superstores? Check. Hockey? Baseball? Soccer? Check. Check. Check. T-shirts? Check. Hats? Check. Underwear? Check. Schools? Bathrooms? Shaved into brush cuts? Check. Check. Check. Since Nike has been the leader in branding clothing, it's not surprising that it has also led the way to the brand's final frontier: the branding of flesh. Not only do dozens of Nike employees have a swoosh tattooed on their calves, but tattoo parlors all over North America report that the swoosh has become their most popular item. Human branding? Check.

The Branded Star

There is another reason behind Nike's stunning success at disseminating its brand. The superstar athletes who form the building blocks of its image – those creatures invented by Nike and cloned by Adidas and Fila – have proved uniquely positioned to soar in the era of synergy: they are made to be cross-promoted. The Spice Girls can make movies, and film stars can walk the runways but neither can quite win an Olympic medal. It's more practical for Dennis Rodman to write two books, star in two movies and have his own television show than it is for Martin Amis or Seinfeld to play defense for the Bulls, just as it is easier for Shaquille O'Neal to put out a rap album than it is for Sporty Spice to make the NBA draft. Only animated characters – another synergy favorite – are more versatile than sports stars in the synergy game.

But for Nike, there is a downside to the power of its own celebrity endorsers. Though Phil Knight will never admit it, Nike is no longer just competing with Reebok, Adidas and the NBA; it has also begun to compete with another brand: its name is Michael Jordan.

In the three years before he retired, Jordan was easing away from his persona as Nike incarnate and turning himself into what his agent, David Falk, calls a "superbrand." He refused to go along when Nike entered the sports-agent business, telling the company that it would have to compensate him for millions of dollars in lost revenue. Instead of letting Nike manage his endorsement portfolio, he tried to build synergy deals between his various sponsors, including a bizarre attempt to persuade Nike to switch phone companies when he became a celebrity spokesperson for WorldCom.[32] Other highlights of what Falk terms "Michael Jordan's Corporate Partnership Program" include a WorldCom commercial in which the actors are decked out in Oakley sunglasses and Wilson sports gear, both Jordan-endorsed products. And, of course, the movie *Space Jam* – in which the basketball player starred and which Falk executive-produced – was Jordan's coming-out party as his own brand. The movie incorporated plugs for each of Jordan's sponsors (choice dialogue includes "Michael, it's show time. Get your Hanes on, lace up your Nikes, grab your Wheaties and Gatorade and we'll pick up a Big Mac on the way!"), and McDonald's promoted the event with *Space Jam* toys and Happy Meals.

Nike had been playing up Jordan's business ambitions in its "CEO Jordan" commercials, which show him changing into a suit and racing to his office at halftime. But behind the scenes, the company has always resented Jordan's extra-Nike activities. Donald Katz writes that as early as 1992, "Knight believed that Michael Jordan was no longer, in sports-marketing nomenclature, 'clean.'"[33] Significantly, Nike boycotted the co-branding bonanza that surrounded *Space Jam*. Unlike McDonald's, it didn't use the movie in tie-in commercials, despite the fact that *Space Jam* is based on a series of Nike commercials featuring Jordan and Bugs Bunny. When Falk told *Advertising Age* that "Nike had some reservations about the implementation of the movie,"[34] he was exercising considerable restraint. Jim Riswold, the longtime Nike adman who first conceived of pairing Jordan with Bugs Bunny in the shoe commercials, complained to *The Wall Street Journal* that *Space Jam* "is a merchandising bonanza first and a movie second. The idea is to sell lots of product."[35] It was a historic moment in the branding of culture, completely inverting the traditionally fraught relationship between art and commerce: a shoe company and an ad agency huffing and puffing that a Hollywood movie would sully the purity of their commercials.

For the time being at least, a peace has descended between the warring superbrands. Nike has given Jordan more leeway to develop his own apparel brand, still within the Nike empire but with greater independence. In the same week that he retired from basketball, Jordan announced that he would be extending the JORDAN clothing line from basketball gear into lifestyle wear, competing directly with Polo, Hilfiger and Nautica. Settling into his role as CEO – as opposed to celebrity endorser – he signed up other pro athletes to endorse the JORDAN brand: Derek Jeter, a shortstop for the New York Yankees and boxer Roy Jones Jr. And, as of May 1999, the full JORDAN brand is showcased in its own "retail concept shops" – two in New York and one in Chicago, with plans for up to fifty outlets by the end of the year 2000. Jordan finally had his wish: to be his own free-standing brand, complete with celebrity endorsers.

The Age of the Brandasaurus

On the surface, the power plays between millionaire athletes and billion-dollar companies would seem to have little to do with the loss of unmarketed space that is the subject of this section. Jordan and Nike, however, are only the most broad strokes, manifestations of the way in which the branding imperative changes the way we imagine both sponsor and sponsored to the extent that the idea of unbranded space – music that is distinct from khakis, festivals that are not extensions of beer brands, athletic achievement that is celebrated in and of itself – becomes almost unthinkable. Jordan and Nike are emblematic of a new paradigm that eliminates all barriers between branding and culture, leaving no room whatsoever for unmarketed space.

An understanding is beginning to emerge that fashion designers, running-shoe companies, media outlets, cartoon characters and celebrities of all kinds are all more or less in the same business: the business of marketing their brands. That's why in the early nineties, Creative Artists Agency, the most powerful celebrity agency in Hollywood, began to represent not just celebrity people, but celebrity brands: Coke, Apple and even an alliance with Nike. That's why Benetton, Microsoft and Starbucks have leapfrogged over the "magalog" trend and have gone full force into the magazine publishing business: Benetton with *Colors*, Microsoft with the on-line zine *Slate* and Starbucks with *Joe*, a joint venture with Time Inc. That's why teen sensation Britney Spears and sitcom character Ally McBeal each have their own line of designer clothing; why Tommy Hilfiger has helped launch a record label; and rapper Master P has his own sports agency business. It's also why Ralph Lauren has a line of designer household paints, Brooks Brothers has a line of wines, Nike is set to launch a swooshed cruise ship, and auto-parts giant Magna is opening up an amusement park. It is also why market consultant Faith Popcorn has launched her own brand of leather Cocooning armchairs, named after the trend she coined of the same name, and Fashion Licensing of America Inc. is marketing a line of Ernest Hemingway furniture, designed to capture the "brand personality" of the late writer.[36]

As manufacturers and entertainers swap roles and move together toward the creation of branded lifestyle bubbles, Nike executives predict that their "competition in the future [will] be Disney, not Reebok."[37] And it seems only fitting that just as Nike enters the entertainment business, the entertainment giants have decided to try their hand at the sneaker industry. In October 1997, Warner Brothers launched a low-end basketball shoe, endorsed by Shaquille O'Neal. "It's an extension of what we do at retail," explained Dan Romanelli of Warner Consumer products.

It seems that wherever individual brands began – in shoes, sports, retail, food, music or cartoons – the most successful among them have all landed in the same place: the stratosphere of the superbrand. That is where Mick Jagger struts in Tommy Hilfiger, Steven Spielberg and Coke have the same agent, Shaq wants to be "like Mickey Mouse," and everyone has his or her own branded restaurant – from Jordan to Disney to Demi Moore to Puffy Combs and the supermodels.

It was Michael Ovitz, of course, who came up with the blueprint for the highest temple of branding so far, one that would do for music, sports and fashion what Walt Disney long ago did for kids' cartoons: turn the slick world of television into a real-world branded environment. After leaving Creative Artists Agency in August 1995 and being driven out as president of Disney shortly after, Ovitz took his unprecedented $87 million golden handshake and launched a new venture: entertainment- and sports-themed megamalls, a synthesis of pro sports, Hollywood celebrity and shopping. His vision is of an unholy mixture of Nike Town, Planet Hollywood and the NBA's marketing wing – all leading straight to the cash register. The first venture, a 1.5-million-square-foot theme mall in Columbus, Ohio, is scheduled to open in the year 2000. If Ovitz gets his way, another mall, planned for the Los Angeles area, will include an NFL football stadium.

As these edifices of the future suggest, corporate sponsors and the culture they brand have fused together to create a third culture: a self-enclosed universe of brand-name people, brand-name products and brand-name media. Interestingly, a 1995 study conducted by University of Missouri professor Roy F. Fox shows that many kids grasp the unique ambiguities of this sphere intuitively. The study found that a majority of Missouri high-school students who watched Channel One's mix of news and ads in their classrooms thought that sports stars paid shoe companies to be in their commercials.

"I don't know why athletes do that – pay all that money for all them ignorant commercials for themselves. Guess it makes everyone like 'em more and like their teams more."[38]

So opined Debbie, a ninth-grader and one of the two hundred students who participated in the study. For Fox, the comment demonstrates a disturbing lack of media literacy, proof positive that kids can't critically evaluate the advertising they see on television. But perhaps these findings show that kids understand something most of us still refuse to grasp. Maybe they know that sponsorship is a far more complicated process than the buyer/seller dichotomy that existed in previous decades and that to talk of who sold out or bought in has become impossibly anachronistic. In an era in which people are brands and brands are culture, what Nike and Michael Jordan do is more akin to co-branding than straight-up shilling, and while the Spice Girls may be doing Pepsi today, they could easily launch their own Spice Cola tomorrow.

It makes a good deal of sense that high-school kids would have a more realistic grasp of the absurdities of branded life. They, after all, are the ones who grew up sold.

Top: Virgin's Richard Branson, the rock-and-roll CEO. *Bottom:* Revolution Soda Co.'s consumable Che.

CHAPTER THREE

ALT.EVERYTHING

The Youth Market and the Marketing of Cool

It's terrible to say, very often the most exciting outfits are from the poorest people.

— Designer Christian Lacroix in *Vogue*, April 1994

In our final year of high school, my best friend, Lan Ying, and I passed the time with morbid discussions about the meaninglessness of life when everything had already been done. The world stretched out before us not as a slate of possibility, but as a maze of well-worn grooves like the ridges burrowed by insects in hardwood. Step off the straight and narrow career-and-materialism groove and you just end up on another one — the groove for people who step off the main groove. And that groove was worn indeed (some of the grooving done by our own parents). Want to go traveling? Be a modern-day Kerouac? Hop on the Let's Go Europe groove. How about a rebel? An avant-garde artist? Go buy your alterna-groove at the secondhand bookstore, dusty and moth-eaten and done to death. Everywhere we imagined ourselves standing turned into a cliché beneath our feet — the stuff of Jeep ad copy and sketch comedy. To us it seemed as though the archetypes were all hackneyed by the time our turn came to graduate, including that of the black-clad deflated intellectual, which we were trying on at that very moment. Crowded by the ideas and styles of the past, we felt there was no open space anywhere.

Of course it's a classic symptom of teenage narcissism to believe that the end of history coincides exactly with your arrival on earth. Almost every angst-ridden, Camus-reading seventeen-year-old girl finds her own groove eventually. Still, there is a part of my high-school globo-claustrophobia that has never left me, and in some ways only seems to intensify as time creeps along. What haunts me is not exactly the absence of literal space so much as a deep craving for metaphorical space: release, escape, some kind of open-ended freedom.

All my parents wanted was the open road and a VW camper. That was enough escape for them. The ocean, the night sky, some acoustic guitar... what more could you ask? Well, actually, you could ask to go soaring off the side of a mountain on a snowboard, feeling as if, for one moment, you are riding the clouds instead of the snow. You could scour Southeast Asia, like the world-weary twenty-somethings in Alex Garland's novel *The Beach*, looking for the one corner of the globe uncharted by the Lonely Planet to start your own private utopia. You could, for that matter, join a New Age cult and dream of alien abduction. From the occult to raves to riots to extreme sports, it seems that the eternal urge for escape has never enjoyed such niche marketing.

In the absence of space travel and confined by the laws of gravity, however, most of us take our open space where we can get it, sneaking it like cigarettes, outside hulking enclosures. The streets may be lined with billboards and franchise signs, but kids still make do, throwing up a couple of nets and passing the puck or soccer ball between the cars. There is release, too, at England's free music festivals, and in conversions of untended private property into collective space: abandoned factories turned into squats by street kids or ramped entrances to office towers transformed into skateboarding courses on Sunday afternoons.

But as privatization slithers into every crevice of public life, even these intervals of freedom and back alleys of unsponsored space are slipping away. The indie skateboarders and snowboarders all have Vans sneaker contracts, road hockey is fodder for beer commercials, inner-city redevelopment projects are sponsored by Wells Fargo, and the free festivals have all been banned, replaced with the annual Tribal Gathering, an electronic music festival that bills itself as a "strike back against the establishment and clubland's evil empire of mediocrity, commercialism, and the creeping corporate capitalism of our cosmic counter-culture"[1] and where the organizers regularly confiscate bottled water that has not been purchased on the premises, despite the fact that the number-one cause of death at raves is dehydration.

I remember the moment when it hit me that my frustrated craving for space wasn't simply a result of the inevitable march of history, but of the fact that commercial co-optation was proceeding at a speed that would have been unimaginable to previous generations. I was watching the television coverage of the controversy surrounding Woodstock '94, the twenty-fifth-anniversary festival of the original Woodstock event. The baby-boomer pundits and aging rock stars postured about how the $2 cans of Woodstock Memorial Pepsi, festival key chains and on-site cash machines betrayed the anticommercial spirit of the original event and, incredibly, whined that the $3 commemorative condoms marked the end of "free love" (as if AIDS had been cooked up as a malicious affront to their nostalgia).

What struck me most was that the debate revolved entirely around the sanctity of the past, with no recognition of present-tense cultural challenges. Despite the fact that the anniversary festival was primarily marketed to teenagers and college students and showcased then-up-and-coming bands like Green Day, not a single commentator explored what this youth-culture "commodification" might mean to the young people who would actually be attending the event. Never mind about the offense to hippies decades after the fact; how does it feel to have your culture "sold out" now, as you are living it? The only mention that a new generation of young people even existed came when the organizers, confronted with charges from ex-hippies that they had engineered Greedstock or Woodshlock, explained that if the event wasn't shrink-wrapped and synergized, the kids today would mutiny. Woodstock promoter John Roberts explained that today's youth are "used to sponsorship. If a kid went to a concert and there wasn't merchandise to buy, he'd probably go out of his mind."[2]

Roberts isn't the only one who holds this view. *Advertising Age* reporter Jeff Jensen goes so far as to make the claim that for today's young people, "Selling out is not only accepted, it's considered hip."[3] To object would be, well, unhip. There is no need to further romanticize the original Woodstock. Among (many) other things, it was also a big-label-backed rock festival, designed to turn a profit. Still, the myth of Woodstock as a sovereign youth-culture state was part of a vast project of generational self-definition – a concept that would have been wholly foreign to those in attendance at Woodstock '94, for whom generational identity had largely been a prepackaged good and for whom the search for self had always been shaped by marketing hype, whether or not they believed it or defined themselves against it. This is a side effect of brand expansion that is far more difficult to track and quantify than the branding of culture and city spaces. This loss of space happens inside the individual; it is a colonization not of physical space but of mental space.

In a climate of youth-marketing feeding frenzy, all culture begins to be created with the frenzy in mind. Much of youth culture becomes suspended in what sociologists Robert Goldman and Stephen Papson call "arrested development," noting that "we have, after all, no idea of what punk or grunge or hip hop as social and cultural movements might look like if they were not mined for their gold..."[4] This "mining" has not gone unnoticed or unopposed. Both the anticorporate cultural journal *The Baffler* and the now-defunct *Might* magazine brilliantly lampooned the desperation and striving of the youth-culture industry in the mid-nineties. Dozens, if not hundreds, of zines and Web sites have been launched and have played no small part in setting the mood for the kind of brand-based attacks that I chronicle in Part IV of this book. For the most part, however, branding's insatiable cultural thirst just creates more marketing. Marketing that thinks it is culture.

To understand how youth culture became such a sought-after market in the early nineties, it helps to go back briefly to the recession era "brand crisis" that took root immediately preceding this frenzy – a crisis that, with so many consumers failing to live up to corporate expectations, created a clear and pressing need for a new class of shoppers to step in and take over.

During the two decades before the brand crisis, the major cultural industries were still drinking deeply from the river of baby-boomer buying power, and the youth demographic found itself on the periphery, upstaged by the awesome power of classic rock and reunion tours. Of course actual young consumers remained a concern for the industries that narrowly market to teens, but youth culture itself was regarded as a rather shallow and tepid well of inspiration by the entertainment and advertising industries. Sure, there were plenty of young people who considered their culture "alternative" or "underground" in the seventies and eighties. Every urban center maintained its bohemian pockets, where the faithful wrapped themselves in black, listened to the Grateful Dead or punk (or the more digestible New Wave), and shopped at secondhand clothing stores and in dank record stores. If they lived outside urban centers, tapes and accessories of the cool lifestyle could be ordered from the backs of magazines like *Maximum Rock 'n' Roll*, or swapped through networks of friends or purchased at concerts.

While this is a gross caricature of the youth subcultures that rose and fell during these decades, the relevant distinction is that these scenes were only half-heartedly sought after as markets. In part this was because seventies punk was at its peak at the same time as the infinitely more mass-marketable disco and heavy metal, and the gold mine of high-end preppy style. And while rap music was topping the charts by the mid- to late eighties, arriving complete with a fully articulated style and code, white America was not about to declare the arrival of a new youth culture. That day would have to wait a few years until the styles and sounds of urban black youth were fully co-opted by white suburbia.

Where I'm from there wasn't no scene I got my information reading Highlights *magazine*

– Princess Superstar, "I'm White," Strictly Platinum

So there was no mass-marketing machine behind these subcultures: there was no Internet, no traveling alternative-culture shopping malls like Lollapalooza or Lilith Fair, and there certainly weren't slick catalogs like Delia and Airshop, which now deliver body glitter, plastic pants and big-city attitude like pizzas to kids stuck in the suburbs. The industries that drove Western consumerism were still catering to the citizens of Woodstock Nation, now morphed into consumption-crazed yuppies. Most of their kids, too, could be counted on as yuppies-in-training, so keeping track of the trends and tastes favored by style-setting youth wasn't worth the effort.

The Youth Market Saves the Day

All that changed in the early nineties when the baby boomers dropped their end of the consumer chain and the brands underwent their identity crisis.

At about the time of Marlboro Friday, Wall Street took a closer look at the brands that had flourished through the recession, and noticed something interesting. Among the industries that were holding steady or taking off were beer, soft drinks, fast food and *sneakers* – not to mention chewing gum and Barbie dolls. There was something else: 1992 was the first year since 1975 that the number of teenagers in America increased. Gradually, an idea began to dawn on many in the manufacturing sector and entertainment industries: maybe their sales were slumping not because consumers were "brand-blind," but because these companies had their eyes fixed on the wrong demographic prize. This was not a time for selling Tide and Snuggle to housewives – it was a time for beaming MTV, Nike, Hilfiger, Microsoft, Netscape and *Wired* to global teens and their overgrown imitators. Their parents might have gone bargain basement, but kids, it turned out, were still willing to pay up to fit in. Through this process, peer pressure emerged as a powerful market force, making the keeping-up-with-the-Joneses consumerism of their suburban parents pale by comparison. As clothing retailer Elise Decoteau said of her teen shoppers, "They run in packs. If you sell to one, you sell to everyone in their class and everyone in their school."[5]

There was just one catch. As the success of branding superstars like Nike had shown, it was not going to be sufficient for companies simply to market their same products to a younger demographic; they needed to fashion brand identities that would resonate with this new culture. If they were going to turn their lackluster products into transcendent meaning machines – as the dictates of branding demanded – they would need to remake themselves in the image of nineties cool: its music, styles and politics.

Cool Envy: The Brands Go Back to School

Fueled by the dual promises of branding and the youth market, the corporate sector experienced a burst of creative energy. Cool, alternative, young, hip – whatever you want to call it – was the perfect identity for product-driven companies looking to become transcendent image-based brands. Advertisers, brand managers, music, film and television producers raced back to high school, sucking up to the in-crowd in a frantic effort to isolate and reproduce in TV commercials the precise "attitude" teens and twenty-somethings were driven to consume with their snack foods and pop tunes. And as in high schools everywhere, "Am I cool?" became the deeply dull and all-consuming question of every moment, echoing not only through class and locker rooms, but through the high-powered meetings and conference calls of Corporate High.

The quest for cool is by nature riddled with self-doubt ("Is this cool?" one can hear the legions of teen shoppers nervously quizzing each other. "Do you think this is lame?") Except now the harrowing doubts of adolescence are the billion-dollar questions of our age. The insecurities go round and round the boardroom table, turning ad writers, art directors and CEOs into turbo-powered teenagers, circling in front of their bedroom mirrors trying to look blasé. Do the kids think we're cool? they want to know. Are we trying too hard to be cool, or are we really cool? Do we have attitude? The *right* attitude?

The Wall Street Journal regularly runs serious articles about how the trend toward wide-legged jeans or miniature backpacks is affecting the stock market. IBM, out-cooled in the eighties by Apple, Microsoft and pretty well everybody, has become fixated on trying to impress the cool kids, or, in the company's lingo, the "People in Black." "We used to call them the ponytail brigade, the black turtleneck brigade," says IBM's David Gee, whose job it is to make Big Blue cool. "Now they're the PIBs – People in Black. We have to be relevant to the PIBs."[6] For Pepe Jeans, the goal, articulated by marketing director Phil Spur, is this: "They [the cool kids] have to look at your jeans, look at your brand image and say 'that's cool…' At the moment we're ensuring that Pepe is seen in the right places and on the right people."[7]

The companies that are left out of the crowd of successfully hip brands – their sneakers too small, their pant-legs too tapered, their edgy ads insufficiently ironic – now skulk on the margins of society: the corporate nerds. "Coolness is still elusive for us," says Bill Benford, president of L.A. Gear athletic wear,[8] and one half expects him to slash his wrists like some anxious fifteen-year-old unable to face schoolyard exile for another term. No one is safe from this brutal ostracism, as Levi Strauss learned in 1998. The verdict was merciless: Levi's didn't have superstores like Disney, it didn't have cool ads like the Gap, it didn't have hip-hop credibility like Hilfiger and no one wanted to tattoo its logo on their navel, like Nike. In short, it wasn't cool. It had failed to understand, as its new brand developer Sean Dee diagnosed, that "loose jeans is not a fad, it's a paradigm shift."[9]

Cool, it seems, is the make-or-break quality in 1990s branding. It is the ironic sneer-track of ABC sitcoms and late-night talk shows; it is what sells psychedelic Internet servers, extreme sports gear, ironic watches, mind-blowing fruit juices, kitsch-laden jeans, postmodern sneakers and post-gender colognes. Our "aspirational age," as they say in marketing studies, is about seventeen. This applies equally to the forty-seven-year-old baby boomers scared of losing their cool and the seven-year-olds kick-boxing to the Backstreet Boys.

As the mission of corporate executives becomes to imbue their companies with deep coolness, one can even foresee a time when the mandate of our elected leaders will be "Make the Country Cool." In many ways, that time is already here. Since his election in 1997, England's young prime minister, Tony Blair, has been committed to changing Britain's somewhat dowdy image to "Cool Britannia." After attending a summit with Blair in an art-directed conference room in Canary Wharf, French president Jacques Chirac said, "I'm impressed. It all gives Britain the image of a young, dynamic and modern country." At the G-8 summit in Birmingham, Blair turned the august gathering into a basement rec room get-together, where the leaders watched All Saints music videos and then were led in a round of "All You Need Is Love"; no Nintendo games were reported. Blair is a world leader as nation stylist – but will his attempt to "rebrand Britain" really work, or will he be stuck with the old, outdated Brit brand? If anyone can do it, it's Blair, who took a page from the marketers of Revolution Soda and successfully changed the name of his party from an actual description of its loyalties and policy proclivities (that would be "labour") to the brand-asset descriptor "New Labour." His is not the Labour Party but a labor-scented party.

The Change Agents: Cooling the Water Cooler

The journey to our current state of world cool almost ended, however, before it really began. Even though by 1993 there was scarcely a fashion, food, beverage or entertainment company that didn't pine for what the youth market promised, many were at a loss as to how to get it. At the time that cool-envy hit, many corporations were in the midst of a hiring freeze, recovering from rounds of layoffs, most of which were executed according to the last-hired-first-fired policies of the late-eighties recession. With far fewer young workers on the payroll and no new ones coming up through the ranks, many corporate executives found themselves in the odd position of barely knowing anyone under thirty years old. In this stunted context, youth itself looked oddly exotic – and information about Xers, Generation Y and twenty-somethings was suddenly a most precious commodity.

Fortunately, a backlog of hungry twenty-somethings were already in the job market. Like good capitalists, many of these young workers saw a market niche: being professionally young. In so many words, they assured would-be bosses that if they were hired, hip, young countercultures would be hand-delivered at the rate of one per week; companies would be so cool, they would get respect in the scenes. They promised the youth demographic, the digital revolution, a beeline into convergence.

And as we now know, when they got the job, these conduits of cool saw no need to transform themselves into clone-ish Company Men. Many can be seen now, roaming the hallways of Fortune 500 corporations dressed like club kids, skateboard in tow. They drop references to all-night raves at the office water cooler ("Memo to the boss: why not fill this thing with ginseng-laced herbal iced tea?"). The CEOs of tomorrow aren't employees, they are, to use a term favored at IBM, "change agents." But are they impostors – scheming "suits" hiding underneath hip-hop snowboarding gear? Not at all. Many of these young workers are the real deal; the true and committed product of the scenes they serve up, and utterly devoted to the transformation of their brands. Like Tom Cruise in *Jerry Maguire*, they stay up late into the night penning manifestos, revolutionary tracts about the need to embrace the new, to flout bureaucracy, to get on the Web or be left behind, to redo the ad campaign with a groovier, grittier feel, to change quicker, be hipper.

And what do the change agents' bosses have to say about all this? They say bring it on, of course. Companies looking to fashion brand identities that will mesh seamlessly with the zeitgeist understand, as Marshall McLuhan wrote, "When a thing is current, it creates currency." The change agents stroke their bosses' middle-aged egos simply by showing up – how out of touch could the boss be with a radical like this on the same intranet system? Just look at Netscape, which no longer employs a personnel manager and instead has Margie Mader, Director of Bringing in the Cool People. When asked by *Fast Company*, "How do you interview for cool?" she replied, "... there are the people who just exude cool: one guy skateboarded here for his interview; another held his interview in a roller-hockey rink."[10] At MTV, a couple of twenty-five-year-old production assistants, both named Melissa, co-wrote a document known as the "Melissa Manifesto," calling on the already insufferably bubbly channel to become even more so. ("We want a cleaner, brighter, more fun MTV," was among their fearless demands.) Upon reading the tract, MTV president Judy McGrath told one of her colleagues, "I feel like blowing everybody out and putting these people in charge."[11] Fellow rebel Tom Freston, CEO of MTV, explains that "Judy is inherently an anti-establishment person. Anybody who comes along and says, 'Let's off the pig,' has got her ear."[12]

Cool Hunters: The Legal Stalkers of Youth Culture

While the change agents were getting set to cool the corporate world from the inside out, a new industry of "cool hunters" was promising to cool the companies from the outside in. The major corporate cool consultancies – Sputnik, *The L. Report*, Bureau de Style – were all founded between 1994 and 1996, just in time to present themselves as the brands' personal cool shoppers. The idea was simple: they would search out pockets of cutting-edge lifestyle, capture them on videotape and return to clients like Reebok, Absolut Vodka and Levi's with such bold pronouncements as "Monks are cool."[13] They would advise their clients to use irony in their ad campaigns, to get surreal, to use "viral communications."

In their book *Street Trends*, Sputnik founders Janine Lopiano-Misdom and Joanne De Luca concede that almost anyone can interview a bunch of young people and make generalizations, "but how do you know they are the 'right' ones – have you been in their closets? Trailed their daily routines? Hung out with them socially?... Are they the core consumers, or the mainstream followers?"[14] Unlike the market researchers who use focus groups and one-way glass to watch kids as if they were overgrown lab rats, Sputnik is "one of them" – it is in with the in-crowd.

Of course all this has to be taken with a grain of salt. Cool hunters and their corporate clients are locked in a slightly S/M, symbiotic dance: the clients are desperate to believe in a just-beyond-their-reach well of untapped cool, and the hunters, in order to make their advice more valuable, exaggerate the crisis of credibility the brands face. On the off chance of Brand X becoming the next Nike, however, many corporations have been more than willing to pay up. And so, armed with their change agents and their cool hunters, the superbrands became the perennial teenage followers, trailing the scent of cool wherever it led.

In 1974, Norman Mailer described the paint sprayed by urban graffiti artists as artillery fired in a war between the street and the establishment. "You hit your name and maybe something in the whole scheme of the system gives a death rattle. For now your name is over their name...your presence is on their Presence, your alias hangs over their scene."[15] Twenty-five years later, a complete inversion of this relationship has taken place. Gathering tips from the graffiti artists of old, the superbrands have tagged everyone – including the graffiti writers themselves. No space has been left unbranded.

Hip-Hop Blows Up the Brands

As we have seen, in the eighties you had to be relatively rich to get noticed by marketers. In the nineties, you have only to be cool. As designer Christian Lacroix remarked in *Vogue*, "It's terrible to say, very often the most exciting outfits are from the poorest people."[16]

Over the past decade, young black men in American inner cities have been the market most aggressively mined by the brandmasters as a source of borrowed "meaning" and identity. This was the key to the success of Nike and Tommy Hilfiger, both of which were catapulted to brand superstardom in no small part by poor kids who incorporated Nike and Hilfiger into hip-hop style at the very moment when rap was being thrust into the expanding youth-culture limelight by MTV and *Vibe* (the first mass-market hip-hop magazine, founded in 1992). "The hip-hop nation," write Lopiano-Misdom and De Luca in *Street Trends*, is "the first to embrace a designer or a major label, they make that label 'big concept' fashion. Or, in their words, they 'blow it up.'"[17]

Designers like Stussy, Hilfiger, Polo, DKNY and Nike have refused to crack down on the pirating of their logos for T-shirts and baseball hats in the inner cities and several of them have clearly backed away from serious attempts to curb rampant shoplifting. By now the big brands know that profits from logowear do not just flow from the purchase of the garment but also from people seeing your logo on "the right people," as Pepe Jeans' Phil Spur judiciously puts it. The truth is that the "got to be cool" rhetoric of the global brands is, more often than not, an indirect way of saying "got to be black." Just as the history of cool in America is really (as many have argued) a history of African-American culture – from jazz and blues to rock and roll to rap – for many of the superbrands, cool hunting simply means black-culture hunting. Which is why the cool hunters' first stop was the basketball courts of America's poorest neighborhoods.

The latest chapter in mainstream America's gold rush to poverty began in 1986, when rappers Run-DMC breathed new life into Adidas products with their hit single "My Adidas," a homage to their favorite brand. Already, the wildly popular rap trio had hordes of fans copying their signature style of gold medallions, black-and-white Adidas tracksuits and low-cut Adidas sneakers, worn without laces. "We've been wearing them all our lives," Darryl McDaniels (a k a DMC) said of his Adidas shoes at the time.[18] That was fine for a time, but after a while it occurred to Russell Simmons, the president of Run-DMC's label Def Jam Records, that the boys should be getting paid for the promotion they were giving to Adidas. He approached the German shoe company about kicking in some money for the act's 1987 Together Forever tour. Adidas executives were skeptical about being associated with rap music, which at that time was alternately dismissed as a passing fad or vilified as an incitement to riot. To help change their minds, Simmons took a couple of Adidas bigwigs to a Run-DMC show. Christopher Vaughn describes the event in *Black Enterprise*: "At a crucial moment, while the rap group was performing the song ["My Adidas"], one of the members yelled out, 'Okay, everybody in the house, rock your Adidas!' – and three thousand pairs of sneakers shot in the air. The Adidas executives couldn't reach for their checkbooks fast enough."[19] By the time of the annual Atlanta sports-shoe Super Show that year, Adidas had unveiled its new line of Run-DMC shoes: the Super Star and the Ultra Star – "designed to be worn without laces."[20]

Since "My Adidas," nothing in inner-city branding has been left up to chance. Major record labels like BMG now hire "street crews" of urban black youth to talk up hip-hop albums in their communities and to go out on guerrilla-style postering and sticker missions. The L.A.-based Steven Rifkind Company bills itself as a marketing firm "specializing in building word-of-mouth in urban areas and inner cities."[21] Rifkind is CEO of the rap label Loud Records, and companies like Nike pay him hundreds of thousands of dollars to find out how to make their brands cool with trend-setting black youth.

So focused is Nike on borrowing style, attitude and imagery from black urban youth that the company has its own word for the practice: *bro-ing*. That's when Nike marketers and designers bring their prototypes to inner-city neighborhoods in New York, Philadelphia or Chicago and say, "Hey, bro, check out the shoes," to gauge the reaction to new styles and to build up a buzz. In an interview with journalist Josh Feit, Nike designer Aaron Cooper described his bro-ing conversion in Harlem: "We go to the playground, and we dump the shoes out. It's unbelievable. The kids go nuts. That's when you realize the importance of Nike. Having kids tell you Nike is the number one thing in their life – number two is their girlfriend."[22] Nike has even succeeded in branding the basketball courts where it goes bro-ing through its philanthropic wing, P.L.A.Y (Participate in the Lives of Youth). P.L.A.Y sponsors inner-city sports programs in exchange for high swoosh visibility, including giant swooshes at the center of resurfaced urban basketball courts. In tonier parts of the city, that kind of thing would be called an ad and the space would come at a price, but on this side of the tracks, Nike pays nothing, and files the cost under charity.

Tommy Hilfiger: To the Ghetto and Back Again

Tommy Hilfiger, even more than Nike or Adidas, has turned the harnessing of ghetto cool into a mass-marketing science. Hilfiger forged a formula that has since been imitated by Polo, Nautica, Munsingwear (thanks to Puff Daddy's fondness for the penguin logo) and several other clothing companies looking for a short cut to making it at the suburban mall with inner-city attitude.

Like a depoliticized, hyper-patriotic Benetton, Hilfiger ads are a tangle of Cape Cod multiculturalism: scrubbed black faces lounging with their wind-swept white brothers and sisters in that great country club in the sky, and always against the backdrop of a billowing American flag. "By respecting one another we can reach all cultures and communities," the company says. "We promote ... the concept of living the American dream."[23] But the hard facts of Tommy's interracial financial success have less to do with finding common ground between cultures than with the power and mythology embedded in America's deep racial segregation.

Tommy Hilfiger started off squarely as white-preppy wear in the tradition of Ralph Lauren and Lacoste. But the designer soon realized that his clothes also had a peculiar cachet in the inner cities, where the hip-hop philosophy of "living large" saw poor and working-class kids acquiring status in the ghetto by adopting the gear and accoutrements of prohibitively costly leisure activities, such as skiing, golfing, even boating. Perhaps to better position his brand within this urban fantasy, Hilfiger began to associate his clothes more consciously with these sports, shooting ads at yacht clubs, beaches and other nautical locales. At the same time, the clothes themselves were re-designed to appeal more directly to the hip-hop aesthetic. Cultural theorist Paul Smith describes the shift as "bolder colors, bigger and baggier styles, more hoods and cords, and more prominence for logos and the Hilfiger name."[24] He also plied rap artists like Snoop Dogg with free clothes and, walking the tightrope between the yacht and the ghetto, launched a line of Tommy Hilfiger beepers.

Once Tommy was firmly established as a ghetto thing, the real selling could begin – not just to the comparatively small market of poor inner-city youth but to the much larger market of middle-class white and Asian kids who mimic black style in everything from lingo to sports to music. Company sales reached $847 million in 1998 – up from a paltry $53 million in 1991 when Hilfiger was still, as Smith puts it, "Young Republican clothing." Like so much of cool hunting, Hilfiger's marketing journey feeds off the alienation at the heart of America's race relations: selling white youth on their fetishization of black style, and black youth on their fetishization of white wealth.

Indie Inc.

Offering *Fortune* magazine readers advice on how to market to teenage girls, reporter Nina Munk writes that "you have to pretend that they're running things.... Pretend you still have to be discovered. Pretend the girls are in charge."[25] Being a huge corporation might sell on Wall Street, but as the brands soon learned on their cool hunt, "indie" was the pitch on Cool Street. Many corporations were unfazed by this shift, coming out with faux indie brands like Politix cigarettes from Moonlight Tobacco (courtesy of Philip Morris), Dave's Cigarettes from Dave's Tobacco Company (Philip Morris again), Old Navy's mock army surplus (the Gap) and OK Cola (Coke).

In an attempt to cash in on the indie marketing craze, even Coke itself, the most recognizable brand name on earth, has tried to go underground. Fearing that it was too establishment for brand-conscious teens, the company launched an ad campaign in Wisconsin that declared Coke the "Unofficial State Drink." The campaign included radio spots that were allegedly broadcast from a pirate radio station called EKOC: Coke backward. Not to be outdone, Gap-owned Old Navy actually did launch its own pirate radio station to promote its brand – a microband transmitter that could only be picked up in the immediate vicinity of one of its Chicago billboards.[26] And in 1999, when Levi's decided it was high time to recoup its lost cool, it also went indie, launching Red Line jeans (no mention of Levi's anywhere) and K-1 Khakis (no mention of Levi's or Dockers).

> *They sell 501s and they think it's*
> *funny*
> *Turning rebellion into money*
>
> – Chumbawamba,
> "That's How Grateful We Are"

Ironic Consumption: No Deconstruction Required

But Levi's may have, once again, missed a "paradigm shift." It hasn't taken long for these attempts to seriously pitch the most generic of mass-produced products as punk-rock lifestyle choices to elicit sneers from those ever-elusive, trend-setting cool kids, many of whom had already moved beyond indie by the time the brands caught on. Instead, they were now finding ways to express their disdain for mass culture not by opting out of it but by abandoning themselves to it entirely—but with a sly ironic twist. They were watching *Melrose Place*, eating surf 'n' turf in revolving restaurants, singing Frank Sinatra in karaoke bars and sipping girly drinks in tikki bars, acts that were rendered hip and daring because, well, *they* were the ones doing them. Not only were they making a subversive statement about a culture they could not physically escape, they were rejecting the doctrinaire puritanism of seventies feminism, the earnestness of the sixties quest for authenticity and the "literal" readings of so many cultural critics. Welcome to ironic consumption. The editors of the zine *Hermenaut* articulated the recipe:

> Following the late ethnologist Michel de Certeau, we prefer to concentrate our attention on the independent use of mass culture products, a use which, like the ruses of camouflaged fish and insects, may not "overthrow the system," but which keeps us intact and autonomous within that system, which may be the best for which we can hope.... Going to Disney World to drop acid and goof on Mickey isn't revolutionary; going to Disney World in full knowledge of how ridiculous and evil it all is and still having a great innocent time, in some almost unconscious, even psychotic way, is something else altogether. This is what de Certeau describes as "the art of being in-between," and this is the only path of true freedom in today's culture. Let us, then, be in-between. Let us revel in Baywatch, Joe Camel, *Wired* magazine, and even glossy books about the society of spectacle [touché], but let's never succumb to the glamorous allure of these things.[27]

In this complicated context, for brands to be truly cool, they need to layer this uncool-equals-cool aesthetic of the ironic viewer onto their pitch: they need to self-mock, talk back to themselves while they are talking, be used and new simultaneously. And after the brands and their cool hunters had tagged all the available fringe culture, it seemed only natural to fill up that narrow little strip of unmarketed brain space occupied by irony with preplanned knowing smirks, someone else's couch commentary and even a running simulation of the viewer's thought patterns. "The New Trash brands," remarks writer Nick Compton of kitsch lifestyle companies like Diesel, "offer inverted commas big enough to live, love and laugh within."[28]

Pop Up Videos, the VH1 show that adorns music videos with snarky thought bubbles, may be the endgame of this kind of commercial irony. It grabs the punchline before anyone else can get to it, making social commentary – even idle sneering – if not redundant then barely worth the expense of energy.

Irony's cozy, protected, self-referential niche is a much better fit than attempts to earnestly pass off fruit drinks as underground rock bands or sneakers as gangsta rappers. In fact, for brands in search of cool new identities, irony and camp have become so all-purpose that they even work after the fact. It turns out that the so-bad-it's-good marketing spin can be deployed to resuscitate hopelessly uncool brands and failed cultural products. Six months after the movie *Showgirls* flopped in the theaters, for instance, MGM got wind that the sexploitation flick was doing okay on video, and not just as a quasi-respectable porno. It seemed that groups of trendy twenty-somethings were throwing *Showgirls* irony parties, laughing sardonically at the implausibly poor screenplay and shrieking with horror at the aerobic sexual encounters. Not content to pocket the video returns, MGM decided to relaunch the movie in the theaters as the next *Rocky Horror Picture Show*. This time around, the newspaper ads made no pretense that anyone had seriously admired the film. Instead, they quoted from the abysmal reviews, and declared *Showgirls* an "instant camp classic" and "a rich sleazy kitsch-fest." The studio even hired a troupe of drag queens for the New York screenings to holler at the crowd with bullhorns during particularly egregious cinematic moments.

With the tentacles of branding reaching into every crevice of youth culture, leaching brand-image content not only out of street styles like hip-hop but psychological attitudes like ironic detachment, the cool hunt has had to go further afield to find unpilfered space and that left only one frontier: the past.

What is retro, after all, but history re-consumed with a PepsiCo tie-in, and breath-mint and phone-card brand extensions? As the re-release of *Lost in Space*, the *Star Wars* trilogy, and the launch of *The Phantom Menace* made clear, the mantra of retro entertainment seems to be "Once more with synergy!" as Hollywood travels back in time to cash in on merchandising opportunities beyond the imagination of yesterday's marketers.

Sell or Be Sold

After almost a decade of the branding frenzy, cool hunting has become an internal contradiction: the hunters must rarefy youth "microcultures" by claiming that only full-time hunters have the know-how to unearth them – or else why hire cool hunters at all? Sputnik warns its clients that if the cool trend is "visible in your neighborhood or crowding your nearest mall, the learning is over. It's too late.... You need to get down with the streets, to be in the trenches every day."[29] And yet this is demonstrably false; so-called street fashions – many of them planted by brandmasters like Nike and Hilfiger from day one – reach the ballooning industry of glossy youth-culture magazines and video stations without a heartbeat's delay. And if there is one thing virtually every young person now knows, it's that street style and youth culture are infinitely marketable commodities.

Besides, even if there was a lost indigenous tribe of cool a few years back, rest assured that it no longer exists. It turns out that the prevailing legalized forms of youth stalking are only the tip of the iceberg: the Sputnik vision for the future of hip marketing is for companies to hire armies of Sputnik spawns – young "street promoters," "Net promoters" and "street distributors" who will hype brands one-on-one on the street, in the clubs and on-line. "Use the magic of peer-to-peer distribution – it worked in the freestyle sport cultures, mainly because the promoters were their friends.... Street promoting will survive as the only true means of personally 'spreading the word.'"[30] So all arrows point to more jobs for the ballooning industry of "street snitches," certified representatives of their demographic who will happily become walking infomercials for Nike, Reebok and Levi's.

By fall 1998 it had already started to happen with the Korean car manufacturer Daewoo hiring two thousand college students on two hundred campuses to talk up the cars to their friends. Similarly, Anheuser-Busch keeps troops of U.S. college frat boys and "Bud Girls" on its payroll to promote Budweiser beer at campus parties and bars.[31] The vision is both horrifying and hilarious: a world of glorified diary trespassers and professional eavesdroppers, part of a spy-vs.-spy corporate-fueled youth culture stalking itself, whose members will videotape one another's haircuts and chat about their corporate keepers' cool new products in their grassroots newsgroups.

Rock-and-Roll CEOs

There is an amusing irony in the fact that so many of our captains of industry pay cool hunters good money to lead them on the path to brand-image nirvana. The true barometers of hip are not the hunters, the postmodern admen, the change agents or even those trendy teenagers they're all madly chasing. They are the CEOs themselves, who are, for the most part, so damn rich that they can afford to stay on top of all the coolest culture trends. Guys like Diesel Jeans founder Renzo Rosso, who, according to *Business Week*, "rides to work on a Ducati Monster motorcycle."[32] Or Nike's Phil Knight, who only took off his ever-present wraparound Oakley sunglasses after Oakley CEO Jim Jannard refused to sell him the company. Or famed admen Dan Wieden and David Kennedy who built a basketball court – complete with bleachers – in their corporate headquarters. Or Virgin's Richard Branson, who launched a London bridal store in a wedding dress, rappelled off the roof of his new Vancouver megastore while uncorking a bottle of champagne and then later crash-landed in the Algerian desert in his hot-air balloon – all during the month of December 1996. These CEOs are the new rock stars – and why shouldn't they be? Forever trailing the scent of cool, they are full-time, professional teenagers, but unlike real teenagers, they have nothing to distract them from the hot pursuit of the edge: no homework, puberty, college-entrance exams or curfews for them.

Getting Over It

As we will see later on, the sheer voracity of the corporate cool hunt did much to provoke the rise of brand-based activism: through adbusting, computer hacking and spontaneous illegal street parties, young people all over the world are aggressively reclaiming space from the corporate world, "unbranding" it, guerrilla-style. But the effectiveness of the cool hunt also set the stage for anticorporate activism in another way: inadvertently, it exposed the impotence of almost all other forms of political resistance *except* anticorporate resistance, one cutting-edge marketing trend at a time.

When the youth-culture feeding frenzy began in the early nineties, many of us who were young at the time saw ourselves as victims of a predatory marketing machine that co-opted our identities, our styles and our ideas and turned them into brand food. Nothing was immune: not punk, not hip-hop, not fetish, not techno – not even, as I'll get to in Chapter 5, campus feminism or multiculturalism. Few of us asked, at least not right away, why it was that these scenes and ideas were proving so packageable, so unthreatening – and so profitable. Many of us had been certain we were doing something subversive and rebellious but . . . what was it again?

In retrospect, a central problem was the mostly unquestioned assumption that just because a scene or style is different (that is, new and not yet mainstream), it necessarily exists in opposition to the mainstream, rather than simply sitting unthreateningly on its margins. Many of us assumed that "alternative" – music that was hard to listen to, styles that were hard to look at – was also anticommercial, even socialist. In *Hype!*, a documentary about how the discovery of "the Seattle sound" transformed a do-it-yourself hardcore scene into an international youth-culture-content factory, Pearl Jam's Eddie Vedder makes a rather moving speech about the emptiness of the "alternative" breakthrough of which his band was so emblematic:

> If all of this influence that this part of the country has and this musical scene has – if it doesn't do anything with it, that would be the tragedy. If it doesn't do anything with it like make some kind of change or make some kind of difference, this group of people who feel this certain way, who think these sorts of things that the underdogs we've all met and lived with think – if they finally get to the forefront and nothing comes out of it, that would be the tragedy.[33]

But that tragedy has already happened, and Vedder's inability to spit out what he was actually trying to say had more than a little to do with it. When the world's cameras were turned on Seattle, all we got were a few anti-establishment fuck-yous, a handful of overdoses and Kurt Cobain's suicide. We also got the decade's most spectacular "sellout" – Courtney Love's awe-inspiring sail from junkie punk queen to high-fashion cover girl in a span of two years. It seemed Courtney had been playing dress-up all along. What was revealing was how little it mattered. Did Love betray some karmic debt she owed to smudged eyeliner? To not caring about anything and shooting up? To being surly to the press? Don't you need to buy in to something earnestly before you can sell it out cynically?

Seattle imploded precisely because no one wanted to answer questions like those, and yet in the case of Cobain, and even Vedder, many in its scene possessed a genuine, if malleable, disdain for the trappings of commercialism. What was "sold out" in Seattle, and in every other subculture that has had the misfortune of being spotlighted by the cool hunters, was some pure idea about doing it yourself, about independent labels versus the big corporations, about not buying in to the capitalist machine. But few in that scene bothered to articulate these ideas out loud, and Seattle – long dead and forgotten as anything but a rather derivative fad – now serves as a cautionary tale about why so little opposition to the theft of cultural space took place in the early to mid-nineties. Trapped in the headlights of irony and carrying too much pop-culture baggage, not one of its antiheroes could commit to a single, solid political position.

A similar challenge is now being faced by all those ironic consumers out there – a cultural suit of armor many of us are loath to critique because it lets us feel smug while watching limitless amounts of bad TV. Unfortunately, it's tough to hold on to that subtle state of De Certeau's "in-betweenness" when the eight-hundred-pound culture industry gorilla wants to sit next to us on the couch and tag along on our ironic trips to the mall. That art of being in-between, of being ironic, or camp, which Susan Sontag so brilliantly illuminated in her 1964 essay "Notes on Camp," is based on an essential cliquiness, a club of people who get the aesthetic puns. "To talk about camp is therefore to betray it,"[34] she acknowledges at the beginning of the essay, selecting the format of enumerated notes rather than a narrative so as to tread more lightly on her subject, one that could easily have been trampled with too heavy an approach.

Since the publication of Sontag's piece, camp has been quantified, measured, weighed, focus-grouped and test-marketed. To say it has been betrayed, as Sontag had feared, is an understatement of colossal dimensions. What's left is little more than a vaguely sarcastic way to eat Pizza Pops. Camp cannot exist in an ironic commercial culture in which no one is fully participating and everyone is an outsider inside their clothes, because, as

Sontag writes, "In naive, or pure, Camp, the essential element is seriousness, a seriousness that fails."[35]

Much of the early camp culture that Sontag describes involved using an act of imagination to make the marginal – even the despised – glamorous and fabulous. Drag queens, for instance, took their forced exile and turned it into a ball, with all the trappings of the Hollywood balls to which they would never be invited. The same can even be said of Andy Warhol. The man who took the world on a camping trip was a refugee from bigoted small-town America; the Factory became his sovereign state. Sontag proposed camp as a defense mechanism against the banality, ugliness and overearnestness of mass culture. "Camp is the modern dandyism. Camp is the answer to the problem: how to be a dandy in the age of mass culture."[36] Only now, some thirty-five years later, we are faced with the vastly more difficult question, How to be truly critical in an age of mass camp?

Or perhaps it is not that difficult. Yes, the cool hunters reduce vibrant cultural ideas to the status of archeological artifacts, and drain away whatever meaning they once held for the people who lived with them – but this has always been the case. It's a cinch to co-opt a style; and it has been done many times before, on a much grander scale than the minor takeover of drag and grunge. Bauhaus modernism, for example, had its roots in the imaginings of a socialist utopia free of garish adornment, but it was almost immediately appropriated as the relatively inexpensive architecture of choice for the glass-and-steel skyscrapers of corporate America.

On the other hand, though style-based movements are stripped of their original meanings time and time again, the effect of this culture vulturing on more politically grounded movements is often so ludicrous that the most sensible reaction is just to laugh it off. The spring 1998 Prada collection, for instance, borrowed heavily from the struggle of the labor movement. As "supershopper" Karen von Hahn reported from Milan, "The collection, a sort of Maoist/Soviet-worker chic full of witty period references, was shown in a Prada-blue room in the Prada family palazzo to an exclusive few." She adds, "After the show, the small yet ardent group of devotees tossed back champagne cocktails and canapés while urbane jazz played in the background."[37]

Mao and Lenin also make an appearance on a Spring 1999 handbag from Red or Dead. Yet despite these clear co-optations of the class struggle, one hardly expects the labor movements of the world to toss in the towel in a huff, give up on their demands for decent working conditions and labor standards worldwide because Mao is suddenly the It Boy in Milan. Neither are union members everywhere accepting wage rollbacks because Pizza Hut aired a commercial in which the boss delivers pizzas to a picket line and all anti-management animosity is abandoned in favor of free food.

The Tibetan people in the West seem similarly nonplussed by their continued popularity with the Beastie Boys, Brad Pitt and designer Anna Sui, who was so moved by their struggle that she made an entire line of banana-print bikini tops and surfer shorts inspired by the Chinese occupation (*Women's Wear Daily* dubbed the Tibet line "techno beach blanket bingo"[38]). More indifference has met Apple computers' appropriation of Gandhi for their "Think Different" campaign, and Che Guevara's reincarnation as the logo for Revolution Soda (slogan: "Join the Revolution"; see image on page 62) and as the mascot of the upscale London cigar lounge, Che. Why? Because not one of the movements being "co-opted" expressed itself primarily through style or attitude. And so style co-optation – and indeed any outside-the-box brainstorming on Madison Avenue – does not have the power to undo them either.

It may seem cold comfort, but now that we know advertising is an extreme sport and CEOs are the new rock stars, it's worth remembering that extreme sports are not political movements and rock, despite its historic claims to the contrary, is not revolution. In fact, to determine whether a movement genuinely challenges the structures of economic and political power, one need only measure how affected it is by the goings-on in the fashion and advertising industries. If, even after being singled out as the latest fad, it continues as if nothing had happened, it's a good bet it is a real movement. If it spawns an industry of speculation about whether movement X has lost its "edge," perhaps its adherents should be looking for a sharper utensil. And as we will soon see, that is exactly what many young activists are in the process of doing.

Top: Image from 1984 Apple television campaign; Apple has been a major promoter of technology in classrooms. *Bottom:* Channel One is broadcast in 12,000 U.S. schools.

CHAPTER FOUR

THE BRANDING OF LEARNING

Ads in Schools and Universities

A democratic system of education ... is one of the surest ways of creating and greatly extending markets for goods of all kinds and especially those goods in which fashion may play a part.

– Ex-adman James Rorty, *Our Master's Voice*, 1934

Although the brands seem to be everywhere – at kids' concerts, next to them on the couch, on stage with their heroes, in their on-line chat groups, and on their playing fields and basketball courts – for a long time one major unbranded youth frontier remained: a place where young people gathered, talked, sneaked smokes, made out, formed opinions and, most maddeningly of all, stood around looking cool for hours on end. That place is called school. And clearly, the brands had to get into the schools.

"You'll agree that the youth market is an untapped wellspring of new revenue. You'll also agree that the youth market spends the majority of each day inside the schoolhouse. Now the problem is, how do you reach that market?" asks a typically tantalizing brochure from the Fourth Annual Kid Power Marketing Conference.

As we have just seen, marketers and cool hunters have spent the better part of the decade hustling the brands back to high school and pouring them into the template of the teenage outlaw. Several of the most successful brands had even cast their corporate headquarters as private schools, referring to them as "campuses" and, at the Nike World Campus, nicknaming one edifice "the student union building." Even the cool hunters are going highbrow; by the late nineties, the rage in the industry was to recast oneself less as a trendy club-hopper than as a bookish grad student. In fact, some insist they aren't cool hunters at all but rather "urban anthropologists."

And yet despite their up-to-the-minute outfits and intellectual pretensions, the brands and their keepers still found themselves on the wrong side of the school gate, a truly intolerable state of affairs and one that would not last long. American marketing consultant Jack Myers described the insufferable slight like this: "The choice we have in this country [the U.S.] is for our educational system to join the electronic age and communicate to students in ways they can understand and to which they can relate. Or our schools can continue to use outmoded forms of communications and become the daytime prisons for millions of young people, as they have become in our inner cities."[1] This reasoning, which baldly equates corporate access to the schools with access to modern technology, and by extension to the future itself, is at the core of how the brands have managed, over the course of only one decade, to all but eliminate the barrier between ads and education. It was technology that lent a new urgency to nineties chronic underfunding: at the same time as schools were facing ever-deeper budget cuts, the costs of delivering a modern education were rising steeply, forcing many educators to look to alternative funding sources for help. Swept up by info-tech hype, schools that couldn't afford up-to-date textbooks were suddenly expected to provide students with audiovisual equipment, video cameras, classroom computers, desktop publishing capacity, the latest educational software programs, Internet access – even, at some schools, video-conferencing.

As many education experts have pointed out, the pedagogical benefits technology brings to the classroom are dubious at best, but the fact remains that employers are clamoring for tech-trained graduates and chances are the private school down the street or across town is equipped with all the latest gadgets and toys. In this context, corporate partnerships and sponsorship arrangements have seemed to many public schools, particularly those in poorer areas, to be the only possible way out of the high-tech bind. If the price of staying modern is opening the schools to ads, the thinking goes, then parents and teachers will have to grin and bear it.

The fact that more schools are turning to the private sector to finance technology purchases does not mean that governments are relinquishing any role in supplying public schools with computers. Quite the opposite. A growing number of politicians are making a computer on every desk a key plank in their election platforms, albeit in partnership with local businesses. But in the process school boards are draining money out of programs like music and physical education to finance this high-tech dream – and here too they are opening the door to corporate sponsorships and to direct forms of brand promotion in cash-strapped cafeterias and sports programs.

As fast-food, athletic gear and computer companies step in to fill the gap, they carry with them an educational agenda of their own. As with all branding projects, it is never enough to tag the schools with a few logos. Having gained a foothold, the brand managers are now doing what they have done in music, sports and journalism outside the schools: trying to overwhelm their host, to grab the spotlight. They are fighting for their brands to become not the add-on but the subject of education, not an elective but the core curriculum.

Of course the companies crashing the school gate have nothing against education. Students should by all means learn, they say, but why don't they read about our company, write about our brand, research their own brand preferences or come up with a drawing for our next ad campaign? Teaching students and building brand awareness, these corporations seem to believe, can be two aspects of the same project. Which is where Channel One, owned by K-III Communications, and its Canadian counterpart, the Youth News Network, come in, perhaps the best-known example of in-school branding.

At the beginning of the decade, these self-styled in-school broadcasters approached North American school boards with a proposition. They asked them to open their classrooms to two minutes of television advertising a day, sandwiched between twelve minutes of teenybopper current affairs programming. Many schools consented, and the broadcasts soon aired. Turning off the cheerful ad patter is not an option. Not only is the programming mandatory viewing for students, but teachers are unable to adjust the volume of the broadcast, especially during commercials. In exchange, the schools do not receive direct revenue from the stations but they can use the much-coveted audiovisual equipment for other lessons and, in some cases, receive "free" computers.

Channel One, meanwhile, charges advertisers top dollar for accessing its pipeline to classrooms – twice as much as regular TV stations because, with mandatory attendance and no channel-changing or volume control, it can boast something no other broadcaster can: "No audience erosion." The station now boasts a presence in 12,000 schools, reaching an estimated eight million students (see image on page 86).

When those students aren't watching Channel One or surfing with ZapMe!, an in-school Internet browser first offered free to American schools in 1998, they may turn their attention to their textbooks – and those too may be sending out more messages to "Just Do It" or "CK Be." The Cover Concepts company sells slick ads that wrap around books to 30,000 U.S. schools, where teachers use them instead of plastic or tinfoil as protective jackets. And when lunchtime arrives, more ads are literally on the menu at many schools. In 1997, Twentieth Century–Fox managed to get cafeteria menu items named after characters from its film *Anastasia* in forty U.S. elementary schools. Students could dine on "Rasputin Rib-B-Cue on Bartok Bun" and "Dimitri's Peanut Butter Fudge." Disney and Kellogg's have engaged in similar lunch-menu promotions through School Marketing, a company that describes itself as a "school-lunch ad agency."[2]

Competing with the menu sponsors are the fast-food chains themselves, chains that go head-to-head with cafeterias in 13 percent of U.S. schools. In an arrangement that was unheard of in the eighties, companies like McDonald's and Burger King now set up kiosks in lunchrooms, which they advertise around the school. Subway supplies 767 schools with sandwiches; Pizza Hut corners the market in approximately 4,000 schools; and a staggering 20,000 schools participate in Taco Bell's "frozen burrito product line." A Subway sandwich guide about how to access the in-school market advises franchisees to pitch their brand-name food to school boards as a way to keep students from sneaking out at lunch hour and getting into trouble. "Look for situations where the local school board has a closed campus policy for lunch. If they do, a strong case can be made for branded product to keep the students on campus."[3] The argument works for administrators such as Bob Honson, the director of nutritional services for the Portland, Oregon, school district. "Kids come to us with brand preferences," he explains.[4]

Not all students' brand preferences, however, are accommodated with equal enthusiasm. Since the fast-food outposts don't accept vouchers from kids on the federal lunch program and their food is usually twice as expensive as cafeteria fare, kids from poor families are stuck with mystery meat while their wealthier classmates lunch on Pizza Hut pizza and Big Macs. And they can't even look forward to days when the cafeteria serves pizza or cheeseburgers, since many schools have signed agreements with the chains that prohibit them from serving "generic versions" of fast-food items: no-name burgers, it seems, constitute "unfair competition."

Students may also find that brand wars are being waged over the pop machine outside the gym. In Canada and the U.S., many school boards have given exclusive vending rights to the Pepsi-Cola Company in exchange for generally undisclosed lump sums. What Pepsi negotiates in return varies from district to district. In Toronto, it gets to fill the 560 public schools with its vending machines, to block the sales of Coke and other competitors, and to distribute "Pepsi Achievement Awards" and other goodies emblazoned with its logo. In communities like Cayuga, a rural Ontario tobacco-farming town, Pepsi buys the right to brand entire schools. "Pepsi – Official Soft Drink of Cayuga Secondary School" reads the giant sign beside the road. At South Fork High School in Florida, there is a blunt, hard-sell arrangement: the school has a clause in its Pepsi contract committing the school to "make its best effort to maximize all sales opportunities for Pepsi-Cola products."[5]

Similarly bizarre and haphazard corporate promotions arrangements are thrown together on college and university campuses around the world. At almost every university in North America, advertising billboards appear on campus bicycle racks, on benches, in hallways linking lecture halls, in libraries and even in bathroom stalls. Credit-card companies and long-distance phone carriers solicit students from the moment they receive their orientation-week information kit to the instant after they receive their degree; at some schools, diplomas come with an envelope stuffed with coupons, credit offers and advertising flyers. In the U.S. Barnes & Noble is rapidly replacing campus-owned bookstores, and Chapters has similar plans in Canada. Taco Bells, KFCs, Starbucks and Pizza Huts are already fixtures on university campuses, where they are often clumped together in food courts inside on-campus malls.

Not surprisingly, in the U.S and Canada the fiercest scholastic marketing battles are fought over high-school gym class and university athletics. The top high-school basketball teams have sponsorship deals with Nike and Adidas, which deck out teenagers in swoosh- and stripe-festooned shoes, warm-ups and gym bags. At the university level, Nike has sponsorship deals with more than two hundred campus athletics departments in the U.S. and twelve in Canada. As anyone familiar with college ball well knows, the standard arrangement gives the company the right to stamp the swoosh on uniforms, sports gear, official university merchandise and apparel, on stadium seats and, most important, on ad banners in full view of the cameras that televise high-profile games. Since student players can't get paid in amateur athletics, it is the coaches who receive the corporate money to dress their teams in the right logos, and the amounts at stake are huge. Nike pays individual coaches as much as $1.5 million in sponsorship fees at top sports universities like Duke and North Carolina, sums that make the coaches' salaries look like tokens of appreciation.

As educational institutions surrender to the manic march of branding, a new language is emerging. Nike high schools and universities square off against their Adidas rivals: the teams may well have their own "official drink," either Coke or Pepsi. In its daily broadcasts, Channel One makes frequent references to the goings-on at "Channel One schools." William Hoynes, a sociologist at Vassar College who conducted a study on the broadcaster, says the practice is "part of a broader marketing approach to develop a 'brand name' consciousness of the network, including the promotion of the 'Channel One school' identity."[6]

As several critics have pointed out, Channel One isn't just hawking its advertisers' sneakers and candy to school kids, it is also selling the idea that its own programming is an invaluable educational aid, one that modernizes such arid, outmoded educational resources as books and teachers. In the model advanced by these broadcasters, the process of learning is little more than the transferring of "stuff" to a student's brain. Whether that stuff happens to be about a new blockbuster from Disney or the Pythagorean theorem, the net effect, according to this theory, is the same: more stuff stuffed. So Fox's attempts to flog *Anastasia* in schools didn't stop with lunch-menu ads; it also provided teachers with an "*Anastasia* study guide." Jeffrey Godsick, Fox senior vice president of publicity and promotion, explained that Fox was providing a service to the schools, not the other way around. "Public school teachers are desperate for materials that will excite the kids," he said.[7]

It's impossible to know which teachers use these branded materials in class and which ones toss them away, but a report published by the U.S. Consumers Union in 1995 "found that thousands of corporations were targeting school children or their teachers with marketing activities ranging from teaching videos, to guidebooks, and posters to contests, product giveaways, and coupons."[8]

It will come as no surprise that it is the folks at the Nike World Campus who have devised the most advanced hybrid of in-class advertisement, public relations exercise and faux teaching aid: the "Air-to-Earth" lesson kit. During the 1997–98 academic year, elementary school students in more than eight hundred classrooms across the U.S. sat down at their desks to find that today's lesson was building a Nike sneaker, complete with a swoosh and an endorsement from an NBA star. Called a "despicable use of classroom time" by the National Education Association and "the warping of education" by the Consumers Union, the make-your-own-Nike exercise purports to raise awareness about the company's environmentally sensitive production process. Nike's claim to greenness relies heavily on the fact that the company recycles old sneakers to re-cover community center basketball courts, which, in a postmodern marketing spiral, it then brands with the Nike swoosh.[9]

Hey, Kids! Be a *Self*-Promoter!

In a corporate climate obsessed with finding the secret recipe for cool, there are still more in-school resources to tap. After all, if there is one thing the cool hunters have taught us, it's that groups of kids aren't just lowly consumers: they are also card-carrying representatives of their age demographic. In the eyes of the brand managers, every lunchroom and classroom is a focus group waiting to be focused. So getting access to schools means more than just hawking product – it's a bona fide, bargain-basement cool-hunting opportunity.

For this reason, the in-school computer network ZapMe! doesn't merely sell ad space to its sponsors; it also monitors students' paths as they surf the Net and provides this valuable market research, broken down by the students' sex, age and zip code, to its advertisers. Then, when students log on to ZapMe!, they are treated to ads that have been specially "micro-targeted" for them.[10] This kind of detailed market research is exploding in North American schools: weekly focus groups, taste tests, brand-preference questionnaires, opinion polls, panel discussions on the Internet, all are currently being used inside classrooms. And in a feat of peer-on-peer cool hunting, some market researchers have been experimenting with sending kids home from school with disposable cameras to take pictures of their friends and family – returning with documented evidence, in one assignment conducted for Nike, "of their favorite place to hang out." Exercises like these are "educational" and "empowering" the market researchers argue, and some educators agree. In explaining the merits of a cereal taste test, the principal of Our Lady of Assumption elementary school in Lynnfield, Massachusetts, said: "It's a learning experience. They had to read, they had to look, they had to compare."[11]

Channel One is pushing the market-research model even further, frequently enlisting "partner" teachers to develop class lessons in which students are asked to create a new ad campaign for Snapple or to redesign Pepsi's vending machines. In New York and Los Angeles high-school students have created thirty-second animated spots for Starburst fruit candies, and students in Colorado Springs designed Burger King ads to hang in their school buses.[12] Finished assignments are passed on to the companies and the best entries win prizes and may even be adopted by the companies – all subsidized by the taxpayer-funded school system. At Vancouver's Laurier Annex school, students in Grades 3 and 4 designed two new product lines for the British Columbia restaurant chain White Spot. For several months in 1997, the children worked on developing the concept and packaging for "Zippy" pizza burgers, a product that is now on the kids' menu at White Spot. The following year, they designed an entire concept for birthday parties to be held at the chain. The students' corporate presentation included "sample commercials, menu items, party games invented by the students and cake ideas," taking into account such issues as safety, possible food allergies, low costs "and allowing for flexibility."[13] According to nine-year-old Jeffrey Ye, "It was a lot of work."[14]

Perhaps the most infamous of these experiments occurred in 1998, when Coca-Cola ran a competition asking several schools to come up with a strategy for distributing Coke coupons to students. The school that devised the best promotional strategy would win $500. Greenbriar High School in Evans, Georgia, took the contest extremely seriously, calling an official Coke Day in late March during which all students came to school in Coca-Cola T-shirts, posed for a photograph in a formation spelling Coke, attended lectures given by Coca-Cola executives and learned about all things black and bubbly in their classes. It was a little piece of branding heaven until it came to the principal's attention that in an act of hideous defiance, one Mike Cameron, a nineteen-year-old senior, had come to school wearing a T-shirt with a Pepsi logo. He was promptly suspended for the offense. "I know it sounds bad – 'Child suspended for wearing Pepsi shirt on Coke Day,'" said principal Gloria Hamilton. "It really would have been acceptable... if it had just been in-house, but we had the regional president here and people flew in from Atlanta to do us the honor of being resource speakers. These students knew we had guests."[15]

Though all public institutions are starved for new sources of income, most schools and universities do try to set limits. When York University's Atkinson College sent out a call to donors in 1997 stating that "for a gift of $10,000... you or your corporation can become the official sponsor for the development and design of one of our new multi-media, high-tech courses," the college insisted that only the courses' names were for sale – not their content. Roger Trull, who brokers deals with corporations at Ontario's McMaster University, explains where he draws the line: "They have to be things that don't impact on academics," meaning only extracurricular sponsorship. Besides, many point out that before lunchrooms and letter-man sweaters went brand-name, schools weren't exactly corporate-free turf. Advertising historian Stuart Ewen writes that as early as the 1920s, teaching kids to consume was seen as just another way of promoting patriotism and economic well-being.

Back then, toothbrush companies visited American schools to conduct "toothpaste drills" and cocoa producers made cameos in science class to demonstrate "the various stages in the production of cocoa."[16]

And in more recent history, commercialism had already become a major part of campus life before the brands even arrived. For instance, U.S. college sports is a big business in its own right with sales of merchandise generating $2.75 billion in 1997, a higher figure than the merchandising sales of the National Basketball Association, Major League Baseball and the National Hockey League. And well before the fast-food invasion, many cafeterias had already been contracted out to companies like Marriott and Cara, which also specialize in providing airlines and hospitals with institutional glop.

For these catering giants, however, faceless and generic was their calling card – the very antithesis of branding. When the prima-donna brands arrived on campus, they brought their preening and posturing values with them, introducing to schools new concepts like corporate image control, logo visibility, brand-extension opportunities and the fierce protection of trade secrets. And this collision of the dictates of academia with the dictates of branding often proves uncomfortable. At the University of British Columbia, for instance, students have been unable to find out what is in the text of an agreement between their school and the Coca-Cola Company. Despite the fact that UBC is a publicly funded institution, the soft-drink company demanded that the amount it paid for the vending rights be kept secret for reasons of corporate competitiveness. (Coca-Cola also refused to cooperate with requests for information for this book, claiming that all of its campus activities – including the precise number of campuses with which it has agreements – are confidential "for competitive purposes.")

In May 1996, students and faculty at the University of Wisconsin at Madison did find out what was in the text of a sponsorship deal their administration was about to sign with Reebok – and they didn't like what they discovered. The deal contained a "non-disparagement" clause that prohibited members of the university community from criticizing the athletic gear company. The clause stated: "During and for a reasonable time after the term, the University will not issue any official statement that disparages Reebok. Additionally, the University will promptly take all reasonable steps necessary to address any remark by any University employee, agent or representative, including a Coach, that disparages Reebok, Reebok's products or the advertising agency or others connected with Reebok."[17] Reebok agreed to nix the demand after students and faculty members launched an educational campaign about the company's patchy record on labor rights in Southeast Asia. What was exceptional about the Wisconsin clause is that the university community found out about it before the deal was signed. This has not been the case at other universities where athletic departments have quietly entered into multimillion-dollar deals that contained similar gag orders. The University of Kentucky's deal with Nike, for instance, has a clause that states that the company has the right to terminate the five-year $25 million contract if the "University disparages the Nike brand... or takes any other action inconsistent with the endorsement of Nike products."[18] Nike denies that its motivation is to stifle campus critics.[19]

Regardless of the intentions when the deals are inked, the fact is that campus expression is often stifled when it conflicts with the interests of a corporate sponsor. For example, at Kent State University – one of the U.S. campuses at which Coca-Cola has exclusive vending rights – members of the Amnesty International chapter advocated a boycott of the soft drink because Coca-Cola did business with the since-ousted Nigerian dictatorship. In April 1998, the activists made a routine application to their student council for funding to bring in a human-rights speaker from the Free Nigeria Movement. "Is he going to speak negatively about Coca-Cola?" a council member asked. "Because Coca-Cola does a lot of positive things on our campus like helping organizations and sports." The representatives from Amnesty replied that the speaker would indeed have some negative comments to make about the company's involvement in Nigeria and funding for the event was denied.[20]

On some university campuses, protests critical of a corporate sponsor have been effectively blocked. In August 1996, Tennis Canada hosted the DuMaurier Tennis Open Tournament, sponsored by Imperial Tobacco, at York University. Concerned that neither a university nor a sporting event should be seen to be endorsing tobacco products, an anti-smoking group, the Grim Reaper Society, asked York for permission to pass out pamphlets to students and tournament goers near the university stadium. Susan Mann, the president of York University, refused the request, saying the school did not "normally" allow "interest groups" on campus "unless for University purposes." Activists handed out cards and leaflets to motorists at a traffic light just outside the entrance to York and, on the last day of the tournament, they staged a clever culture-jam: the leaflets they handed out were shaped like fans. Clearly amused, many of the tournament goers brought their fans inside the tennis stadium, cooling themselves off with anti-tobacco slogans. After a few hours, police officers hired by the tournament approached the peaceful, off-site protest and, citing traffic problems, ticketed two of the activists and seized all the remaining fans.

These are extreme examples of how corporate sponsorship deals re-engineer some of the fundamental values of public universities, including financial transparency and the right to open debate and peaceful protest on campus. But the subtle effects are equally disturbing. Many professors speak of the slow encroachment of the mall mentality, arguing that the more campuses act and look like malls, the more students behave like consumers. They tell stories of students filling out their course-evaluation forms with all the smug self-righteousness of a tourist responding to a customer-satisfaction form at a large hotel chain. "Most of all I dislike the attitude of calm consumer expertise that pervades the responses. I'm disturbed by the serene belief that my function – and more important, Freud's, or Shakespeare's, or Blake's – is to divert, entertain, and interest," writes University of Virginia professor Mark Edmundson in *Harper's* magazine.[21] A professor at Toronto's York University, where there is a full-fledged mall on campus, tells me that his students slip into class slurping grande lattes, chat in the back and slip out. They're cruising, shopping, disengaged.

Branding U

While brands slowly transform the experience of campus life for undergraduates, another kind of takeover is under way at the institutional research level. All over the world, university campuses are offering their research facilities, and priceless academic credibility, for the brands to use as they please. And in North America today, corporate research partnerships at universities are used for everything: designing new Nike skates, developing more efficient oil extraction techniques for Shell, assessing the Asian market's stability for Disney, testing the consumer demand for higher bandwidth for Bell or measuring the relative merits of a brand-name drug compared with a generic one, to name just a few examples.

Dr. Betty Dong, a medical researcher at the University of California at San Francisco (UCSF), had the misfortune of taking on that last assignment – testing a brand-name drug with brand-name money. Dong was the director of a study sponsored by the British pharmaceutical company Boots (now called Knoll) and UCSF. The fate of that partnership does much to illuminate precisely how the mandate of universities as sites for public-interest research is often squarely at odds with the interests of branded fact-finding missions.

Dr. Dong's study compared the effectiveness of Boots' thyroid drug, Synthroid, with a generic competitor. The company hoped that the research would prove that its much higher priced drug was better or at least substantially different from the generic one – a claim that, if legitimized by a study from a respected university, would increase Synthroid sales. Instead, Dr. Dong found that the opposite was true. The two drugs were bio-equivalent, a fact that represented a potential saving of $365 million a year for the eight million Americans who were taking the name-brand drug, and a potential loss to Boots of $600 million (the revenue from Synthroid). After the results were reviewed by her peers, Dr. Dong's findings were slated to be published in the *Journal of the American Medical Association* on January 25, 1995. At the last minute, however, Boots successfully halted publication of the article, pointing to a clause in the partnership contract that gave the company veto rights over the publication of findings. The university, fearing a costly lawsuit, sided with the drug company and the article was yanked. After the whole ordeal was exposed in *The Wall Street Journal*, Boots backed off and the paper was finally published in April 1997, two years behind schedule.[22] "The victim is obvious: the university," wrote Dorothy S. Zinberg, a faculty member at Harvard's Center for Science and International Affairs. "Each infringement on its unwritten contract with society to avoid secrecy whenever possible and maintain its independence from government or corporate pressure weakens its integrity."[23]

In 1998, a similar case ripped through the University of Toronto and the affiliated Hospital for Sick Children – only this time, the researcher found that the drug being tested might actually be harmful to patients. Dr. Nancy Olivieri, a world-renowned scientist and expert on the blood disorder thalassemia, entered into a research contract with the drug-company giant Apotex. The company wanted Olivieri to test the effectiveness of the drug deferiprone on her young patients suffering from thalassemia major. When Olivieri found evidence that, in some cases, the drug might have life-threatening side effects, she wanted to warn the patients participating in the trial and to alert other doctors in her field. Apotex pulled the plug on the study and threatened to sue Olivieri if she went public, pointing to an overlooked clause in the research contract that gave it the right to suppress findings for one year after the trials ended. Olivieri went ahead and published in *The New England Journal of Medicine* and, once again, the administration of both her university and her hospital failed to defend the sanctity of academic research conducted in the public interest. Adding further insult, in January 1999, they demoted Olivieri from her top-level research position at the hospital.[24] (After a long and public battle, the doctor eventually got her job back.)

Perhaps the most chilling of these cases involves an associate professor at Brown University in Rhode Island, who worked as an occupational health physician at the university-affiliated Memorial Hospital of Rhode Island in Pawtucket. Dr. David Kern was commissioned by a local textile factory to investigate two cases of lung disease that he had treated at the hospital. He found six more cases of the disease in the 150-person plant, a startling occurrence since its incidence in the general population is one in 40,000. Like Dr. Dong and Dr. Olivieri, Dr. Kern was set to present a paper on his findings when the textile company threatened to sue, citing a clause in the agreement that prevented the publication of "trade secrets." Once again, the university and the hospital administration sided squarely with the company, forbidding Dr. Kern to publish his findings and shutting down the one-person clinic where he conducted his research.[25]

The only element out of the ordinary in these three cases of stifled research is that they involved academics with the personal integrity and the dogged tenacity to publicly challenge their corporate "partners" and their own employers – factors that eventually led to the truth coming out through the press. But relying on crusading individuals to protect the integrity of academic research does not provide a foolproof safeguard in every case. According to a 1994 study conducted on industry research partnerships at U.S. universities, most corporate interference occurs quietly and with no protest. The study found that companies maintained the right to block the publication of findings in 35 percent of cases, while 53 percent of the academics surveyed agreed that "publication can be delayed."[26]

> *Kmart's attitude always has been: What did we get from you this year?... Many people at Kmart thought I was employed by Kmart.*
>
> – J. Patrick Kelly, Kmart Chair of Marketing at Wayne State University, *The Chronicle of Higher Education*, April 1998

There is also a more insidious level of interference that takes place at universities every day, interference that occurs before research even begins, prior, even, to proposals being committed to paper. As John V. Lombardi, president of the University of Florida at Gainesville, says: "We have taken the great leap forward and said: 'Let's pretend we're a corporation.'"[27] What such a leap means back on the ground is that studies are designed to fit the mandate of corporate-endowed research chairs with such grand names as the Taco Bell Distinguished Professor of Hotel and Restaurant Administration at Washington State University, the Yahoo! Chair of Information-Systems Technology at Stanford University and the Lego Professorship of Learning Research at Massachusetts Institute of Technology. J. Patrick Kelly, the professor who holds the Kmart Chair of Marketing at Wayne State, estimates that his research has saved Kmart "many more times" the amount of the $2 million donation that created his position.[28] The professor who holds the Kmart-endowed chair at West Virginia University, meanwhile, has such a hands-on relationship with the retailer that he or she is required by contract to spend a minimum of thirty days a year training assistant managers.

Where Was the Opposition?

Many people, upon learning of the advanced stage of branded education, want to know where the university faculty, teachers, school boards and parents were while this transformation was taking place. At the elementary and high-school level, this is a difficult question to answer – particularly since one is hard-pressed to find anyone but the advertisers who is actively *in favor* of allowing ads into schools. Over the course of the decade, all the large teachers' unions in North America have been quite vocal about the threat to independent instruction posed by commercialization, and many concerned parents have formed groups like Ralph Nader's Commercial Alert to make their opposition heard. Despite this, however, there was never one big issue on which parents and educators could band together to fight – and possibly win – a major policy battle on classroom commercialization.

Unlike the very public standoffs over prayer in schools or over explicit sex education, the move to allow advertisements did not take the form of one sweeping decision but, rather, of thousands of little ones. Usually these were made on an ad hoc, school-by-school basis, frequently with no debate, no notice, no public scrutiny at all, because advertising agencies were careful to fashion school promotions that could slip between the cracks of standard school-board regulations.

However, when Channel One and the Youth News Network wanted to bring ads directly into classrooms, there *was* some debate: genuine, heated discussions took place at the school-board level, and most boards across Canada decided to block YNN. Channel One, though far more successful, particularly in poorer districts, has also had to swallow its share of board refusals.

There is, however, another, more ingrained cultural factor that has helped the brands get inside the schools, and it has to do with the effectiveness of branding itself. Many parents and educators could not see anything to be gained by resistance; kids today are so bombarded by brand names that it seemed as if protecting educational spaces from commercialization was less important than the immediate benefits of finding new funding sources. And the hawkers of in-school advertising have not been at all shy about playing upon this sense of futility among parents and educators. As Frank Vigil, president of ZapMe! computer systems, says: "America's youth is exposed to advertising in many aspects of their lives. We believe students are savvy enough to discern between educational content and marketing materials."[29] Thus it became possible for many parents and teachers to rationalize their failure to protect yet another previously public space by telling themselves that what ads students don't see in class or on campus, they will certainly catch on the subway, on the Net or on TV when they get home. What's one more ad in the life of these marked-up and marked-down kids? And then again... what's another?

But while this may explain the brands' inroads in high schools, it still doesn't explain how this process has been able to take such a firm hold on the university campuses. Why have university professors remained silent, passively allowing their corporate "partners" to trample the principles of freedom of inquiry and discourse that have been the avowed centerpieces of academic life? More to the point, aren't our campuses supposed to be overflowing with troublemaking tenured radicals? Isn't the institution of tenure, with its lifelong promise of job security, designed to make it safe for academics to take controversial positions without fear of repercussion? Aren't these people, to borrow a term more readily understood in the halls of academe, *counter-hegemonic*?

As Janice Newson, a York University sociology professor who has published widely on this issue, has noted: "On the surface, it is easier to account for the increasing realization of the corporate-linked university than it is to account for the lack of resistance to it." Newson, who has been sounding the alarm on the corporate threat to academic freedom for more than a decade, writes that she had (wrongly) assumed that

> members of the academic community would become actively concerned about, if not resistant to, this shift in direction. After all, a significant if not transformative pattern of institutional change has occurred over a relatively short period of time. And in many ways, these changes sharply contrast to both the idea and the practices of the university that preceded them, the university in which most current members of the academy began their careers.[30]

Newson's critique could well be expanded to include student activists, who until the mid-nineties were also mysteriously absent from the corporatization non-debate. Sadly, part of the explanation for the lack of campus mobilization is simple self-interest. Until the mid-nineties, the growing corporate influence in education and research seemed to be taking place almost exclusively in the engineering departments, management schools and science labs. Campus radicals had always been prone to dismiss these faculties as hopelessly compromised right-wing bastions: who cared what was happening on that side of campus, so long as the more traditionally progressive fields (literature, cultural studies, political science, history and fine arts) were left alone? And as long as professors and students in the arts and humanities remained indifferent to this radical shift in campus culture and priorities, they were free to pursue other interests – and there were many on offer. For instance, more than a few of those tenured radicals who were supposed to be corrupting young minds with socialist ideas were preoccupied with their own postmodernist realization that truth itself is a construct. This realization made it intellectually untenable for many academics to even participate in a political argument that would have "privileged" any one model of learning (public) over another (corporate). And since truth is relative, who is to say that Plato's dialogues are any more of an "authority" than Fox's *Anastasia*?

This academic trend only accounts for a few of the missing-in-actions, however. Many other campus radicals were still up for a good old political fight, but during the key years of the corporate campus invasion they were tied up in a different battle: the all-consuming gender and race debates of the so-called political correctness wars. As we will see in the next chapter, if the students allowed themselves to be turned into test markets, it was partly because they had other things on their minds. They were busy taking on their professors on the merits of the canon and the need for more stringent campus sexual-harassment policies. And if their professors failed to prevent the very principles of unfettered academic discourse from being traded in for a quick buck, this may also have been because they were too preoccupied with defending themselves against their own "McCarthyite" students. So there they all were, fighting about women's studies and the latest backlash book while their campuses were being sold out from under their feet. It wasn't until the politics of personal representation were themselves co-opted by branding that students and professors alike began to turn away from their quarrels with each other, realizing they had a more powerful foe.

But by then, much had already been lost. More fundamentally than somewhat antiquated notions of "pure" education and research, what is lost as schools "pretend they are corporations" (to borrow a phrase from the University of Florida) is the very idea of unbranded space. In many ways, schools and universities remain our culture's most tangible embodiment of public space and collective responsibility. University campuses in particular – with their residences, libraries, green spaces and common standards for open and respectful discourse – play a crucial, if now largely symbolic, role: they are the one place left where young people can see a genuine public life being lived. And however imperfectly we may have protected these institutions in the past, at this point in our history the argument against transforming education into a brand-extension exercise is much the same as the one for national parks and nature reserves: these quasi-sacred spaces remind us that unbranded space is still possible.

Top: Scene from a "die-in" at a 1990 Act-Up rally. *Bottom:* Diesel 1995 print campaign showing two sailors kissing.

CHAPTER FIVE

PATRIARCHY GETS FUNKY

The Triumph of Identity Marketing

Let's face it, when you're a story line on Friends, *it's hard to keep thinking you're radical.*

— Jay Blotcher, AIDS activist, *New York* magazine, September 1996

As an undergraduate in the late eighties and early nineties, I was one of those students who took a while to wake up to the slow branding of university life. And I can say from personal experience that it's not that we didn't notice the growing corporate presence on campus — we even complained about it sometimes. It's just that we couldn't get particularly worked up about it. We knew the fast-food chains were setting up their stalls in the library and that profs in the applied sciences were getting awfully cozy with pharmaceutical companies, but finding out exactly what was going on in the boardrooms and labs would have required a lot of legwork, and, frankly, we were busy. We were fighting about whether Jews would be allowed in the racial equality caucus at the campus women's center, and why the meeting to discuss it was scheduled at the same time as the lesbian and gay caucus — were the organizers implying that there were no Jewish lesbians? No black bisexuals?

In the outside world, the politics of race, gender and sexuality remained tied to more concrete, pressing issues, like pay equity, same-sex spousal rights and police violence, and these serious movements were — and continue to be — a genuine threat to the economic and social order. But somehow, they didn't seem terribly glamorous to students on many university campuses, for whom identity politics had evolved by the late eighties into something quite different. Many of the battles we fought were over issues of "representation" – a loosely defined set of grievances mostly lodged against the media, the curriculum and the English language. From campus feminists arguing over "representation" of women on the reading lists to gays wanting better "representation" on television, to rap stars bragging about "representing" the ghettos, to the question that ends in a riot in Spike Lee's 1989 film *Do the Right Thing* – "Why are there no brothers on the wall?" – ours was a politics of mirrors and metaphors.

These issues have always been on the political agendas of both the civil-rights and the women's movements, and later, of the fight against AIDS. It was accepted from the start that part of what held back women and ethnic minorities was the absence of visible role models occupying powerful social positions, and that media-perpetuated stereotypes – embedded in the very fabric of the language – served to not so subtly reinforce the supremacy of white men. For real progress to take place, imaginations on both sides had to be decolonized.

But by the time my generation inherited these ideas, often two or three times removed, representation was no longer one tool among many, it was the key. In the absence of a clear legal or political strategy, we traced back almost all of society's problems to the media and the curriculum, either through their perpetuation of negative stereotypes or simply by omission. Asians and lesbians were made to feel "invisible," gays were stereotyped as deviants, blacks as criminals and women as weak and inferior: a self-fulfilling prophecy responsible for almost all real-world inequalities. And so our battlefields were sitcoms with gay neighbors who never got laid, newspapers filled with pictures of old white men, magazines that advanced what author Naomi Wolf termed "the beauty myth," reading lists that we expected to look like Benetton ads, Benetton ads that trivialized our reading-list demands. So outraged were we media children by the narrow and oppressive portrayals in magazines, in books and on television that we convinced ourselves that if the typecast images and loaded language changed, so too would the reality. We thought we would find salvation in the reformation of MTV, CNN and Calvin Klein. And why not? Since media seemed to be the source of so many of our problems, surely if we could only "subvert" them to better represent us, they could save us instead. With better collective mirrors, self-esteem would rise and prejudices would magically fall away, as society became suddenly inspired to live up to the beautiful and worthy reflection we had retouched in its image.

For a generation that grew up mediated, transforming the world through pop culture was second nature. The problem was that these fixations began to transform us in the process. Over time, campus identity politics became so consumed by personal politics that they all but eclipsed the rest of the world. The slogan "the personal is political" came to replace the economic as political and, in the end, the Political as political as well. The more importance we placed on representation issues, the more central a role they seemed to elbow for themselves in our lives – perhaps because, in the absence of more tangible political goals, any movement that is about fighting for better social mirrors is going to eventually fall victim to its own narcissism.

Soon "outing" wasn't about AIDS, but became a blanket demand for gay and lesbian "visibility" – all gays should be out, not just right-wing politicians but celebrities as well. By 1991, the radical group Queer Nation had broadened its media critique: it didn't just object to portrayals of homicidal madmen with AIDS, but any non-straight killer at all. The group's San Francisco and L.A. chapters held protests against *The Silence of the Lambs*, objecting to its transvestite serial-killer villain, and they disrupted filming on *Basic Instinct* because it featured ice-pick-wielding killer lesbians. GLAAD (Gay and Lesbian Alliance Against Defamation) had moved from lobbying the news media about its use of terms like "gay plague" to describe AIDS, and had begun actively pushing the networks for more gay and lesbian characters in TV shows. In 1993, Torie Osborn, a prominent U.S. lesbian rights activist, said that the single biggest political issue facing her constituency was not same-sex spousal benefits, the right to join the military or even the right of two women to marry and adopt children. It was, she told a reporter, "Invisibility. Period. End of sentence."[1]

Much like a previous generation of anti-porn feminists who held their rallies outside peep shows, many of the political demonstrations of the early nineties had shifted from the steps of government buildings and courthouses to the steps of museums with African art exhibits that were deemed to celebrate the colonial mindset. They massed at the theater entrances showing megamusicals like *Showboat* and *Miss Saigon*, and they even crept right up to the edge of the red carpet at the 1992 Academy Awards.

These struggles may seem slight in retrospect, but you can hardly blame us media narcissists for believing that we were engaged in a crucial battle on behalf of oppressed people everywhere: every step we took sparked a new wave of apocalyptic panic from our conservative foes. If we were not revolutionaries, why, then, were our opponents saying that a revolution was under way, that we were in the midst of a "culture war"? "The transformation of American campuses is so sweeping that it is no exaggeration to call it a revolution," Dinesh D'Souza, author of *Illiberal Education*, informed his readers. "Its distinctive insignia can be witnessed on any major campus in America today, and in all aspects of university life."[2]

Despite their claims of living under Stalinist regimes where dissent was not tolerated, our professors and administrators put up an impressively vociferous counteroffensive: they fought tooth and nail for the right to offend us thin-skinned radicals; they lay down on the tracks in front of every new harassment policy, and generally acted as if they were fighting for the very future of Western civilization. An avalanche of look-alike magazine features bolstered the claim that ID politics constituted an international emergency: "Illiberal Education" (*Atlantic Monthly*), "Visigoths in Tweed" (*Fortune*), "The Silences" (*Maclean's*), "The Academy's New Ayatollahs" (*Outlook*), "Taking Offense" (*Newsweek*). In *New York* magazine, writer John Taylor compared my generation of campus activists with cult members, Hitler Youth and Christian fundamentalists.[3] So great was the threat we allegedly posed that George Bush even took time out to warn the world that political correctness "replaces old prejudices with new ones."

The Marketing of ID

The backlash that identity politics inspired did a pretty good job of masking for us the fact that many of our demands for better representation were quickly accommodated by marketers, media makers and pop-culture producers alike – though perhaps not for the reasons we had hoped. If I had to name a precise moment for this shift in attitude, I would say August of 1992:

the thick of the "brand crisis" that peaked with Marlboro Friday. That's when we found out that our sworn enemies in the "mainstream" – to us a giant monolithic blob outside of our known university-affiliated enclaves – didn't fear and loathe us but actually thought we were sort of interesting. Once we'd embarked on a search for new wells of cutting-edge imagery, our insistence on extreme sexual and racial identities made for great brand-content and niche-marketing strategies. If diversity is what we wanted, the brands seemed to be saying, then diversity was exactly what we would get. And with that, the marketers and media makers swooped down, airbrushes in hand, to touch up the colors and images in our culture.

The five years that followed were an orgy of red ribbons, Malcolm X baseball hats and Silence = Death T-shirts. By 1993, the stories of academic Armageddon were replaced with new ones about the sexy wave of "Do-Me Feminism" in *Esquire* and "Lesbian Chic" in *New York* and *Newsweek*. The shift in attitude was not the result of a mass political conversion but of some hard economic calculations. According to *Rocking the Ages*, a book produced in 1997 by leading U.S. consumer researchers Yankelovich Partners, "Diversity" was the "defining idea" for Gen-Xers, as opposed to "Individuality" for boomers and "Duty" for their parents.

> Xers are starting out today with pluralistic attitudes that are the strongest we have ever measured. As we look towards the next twenty five years, it is clear that acceptance of alternative lifestyles will become even stronger and more widespread as Xers grow up and take over the reins of power, and become the dominant buying group in the consumer marketplace.... *Diversity is the key fact of life for Xers, the core of the perspective they bring to the marketplace.* Diversity in all of its forms – cultural, political, sexual, racial, social – is a hallmark of this generation [italics theirs]...[4]

The Sputnik cool-hunting agency, meanwhile, explained that "youth today are one big sample of diversity" and encouraged its clients to dive into the psychedelic "United Streets of Diversity" and not be afraid to taste the local fare. Dee Dee Gordon, author of *The L. Report*, urged her clients to get into Girl Power with a vengeance: "Teenage girls want to see someone who kicks butt back";[5] and, sounding suspiciously like me and my university friends, brand man Tom Peters took to berating his corporate audiences for being "OWMs – Old White Males."

As we have seen, this information was coming hot on the heels of two other related revelations. The first was that consumer companies would only survive if they built corporate empires around "brand identities." The second was that the ballooning youth demographic held the key to market success. So, of course, if the market researchers and cool hunters all reported that diversity was the key character trait of this lucrative demographic, there was only one thing to be done: every forward-thinking corporation would have to adopt variations on the theme of diversity as their brand identities.

Which is exactly what most brand-driven corporations have attempted to do. In an effort to understand how Starbucks became an overnight household name in 1996 without a single national ad campaign, *Advertising Age* speculated that it had something to do with its tie-dyed, Third World aura. "For devotees, Starbucks' 'experience' is about more than a daily espresso infusion; it is about immersion in a politically correct, cultured refuge...."[6] Starbucks, however, was only a minor player in the P.C. marketing craze. Abercrombie & Fitch ads featured guys in their underwear making goo-goo eyes at each other; Diesel went further, showing two sailors kissing (see image on page 106); and a U.S. television spot for Virgin Cola depicted "the first-ever gay wedding featured in a commercial," as the press release proudly announced. There were also gay-targeted brands like Pride Beer and Wave Water, whose slogan is "We label bottles not people," and the gay community got its very own cool hunters – market researchers who scoured gay bars with hidden cameras.[7]

The Gap, meanwhile, filled its ads with racially mixed rainbows of skinny, childlike models. Diesel harnessed frustration at that unattainable beauty ideal with ironic ads that showed women being served up for dinner to a table of pigs. The Body Shop harnessed the backlash against both of them by refusing to advertise and instead filled its windows with red ribbons and posters condemning violence against women. The rush to diversity fitted in neatly with the embrace of African-American style and heroes that companies like Nike and Tommy Hilfiger had already pinpointed as a powerful marketing source. But Nike also realized that people who saw themselves as belonging to oppressed groups were ready-made market niches: throw a few liberal platitudes their way and, presto, you're not just a product but an ally in the struggle. So the walls of Nike Town were adorned with quotes from Tiger Woods declaring that "there are still courses in the U.S. where I am not allowed to play, because of the color of my skin." Women in Nike ads told us that "I believe 'babe' is a four-letter word" and "I believe high heels are a conspiracy against women."

And everyone, it seemed, was toying with the fluidity of gender, from the old-hat story of MAC makeup using drag queen RuPaul as its spokesmodel to tequila ads that inform viewers that the she in the bikini is really a he; from Calvin Klein's colognes that tell us that gender itself is a construct to Sure Ultra Dry deodorant that in turn urges all the gender benders to chill out: "Man? Woman? Does it matter?"

Oppression Nostalgia

Fierce debates still rage about these campaigns. Are they entirely cynical or do they indicate that advertisers want to evolve and play more positive social roles? Benetton's mid-nineties ads careered wildly between witty and beautiful challenges to racial stereotypes on the one hand, and grotesque commercial exploitation of human suffering on the other. They were, however, indisputably part of a genuine attempt to use the company's vast cultural real estate to send a message that went beyond "Buy more sweaters"; and they played a central role in the fashion world's embrace of the struggle against AIDS. Similarly, there is no denying that the Body Shop broke ground by proving to the corporate sector that a multinational chain can be an outspoken and controversial political player, even while making millions on bubble bath and body lotion. The complicated motivations and stark inconsistencies inside many of these "ethical" businesses will be explored in greater depth in a later chapter. But for many of the activists who had, at one point not so long ago, believed that better media representation would make for a more just world, one thing had become abundantly clear: identity politics weren't fighting the system, or even subverting it. When it came to the vast new industry of corporate branding, they were feeding it.

The crowning of sexual and racial diversity as the new superstars of advertising and pop culture has understandably created a sort of Identity Identity Crisis. Some ex-ID warriors are even getting nostalgic about the good old days, when they were oppressed, yes, but the symbols of their radicalism weren't for sale at Wal-Mart. As music writer Ann Powers observed of the much-vaunted ascendancy of Girl Power, "at this intersection between the conventional feminine and the evolving Girl, what's springing up is not a revolution but a mall… Thus, a genuine movement devolves into a giant shopping spree, where girls are encouraged to purchase whatever identity fits them best off the rack."[8] Similarly, Daniel Mendelsohn has written that gay identity has dwindled into "basically, a set of product choices.… At least culturally speaking, oppression may have been the best thing that could have happened to gay culture. Without it, we're nothing."

The nostalgia, of course, is absurd. Even the most cynical ID warrior will admit, when pressed, that having Ellen Degeneres and other gay characters out on TV has some concrete advantages. Probably it is good for the kids, particularly those who live outside of larger urban settings – in rural or small-town environments, where being gay is more likely to confine them to a life of self-loathing. (The attempted suicide rate in 1998 among gay and bisexual male teens in America was 28.1 percent, compared with 4.2 percent among straight males of the same age group.)[9] Similarly, most feminists would concede that although the Spice Girls' crooning, "If you wanna be my lover, you have to get with my friends" isn't likely to shatter the beauty myth, it's still a step up from Snoop Dogg's 1993 ode to gang rape, "It ain't no fun if my homies can't have none."

And yet, while raising teenagers' self-esteem and making sure they have positive role models is valuable, it's a fairly narrow achievement, and from an activist perspective, one can't help asking, Is this it? Did all our protests and supposedly subversive theory only serve to provide great content for the culture industries, fresh new lifestyle imagery for Levi's new "What's True" ad campaign and girl-power-charged record sales for the music business? Why, in other words, were our ideas about political rebellion so deeply non-threatening to the smooth flow of business as usual?

The question, of course, is not Why, but Why on earth not? Just as they had embraced the "brands, not products" equation, the smart businesses quickly realized that short-term discomfort – whether it came from a requirement to hire more women or to more carefully vet the language in an ad campaign – was a small price to pay for the tremendous market share that diversity promised. So while it may be true that real gains have emerged from this process, it is also true that Dennis Rodman wears dresses and Disney World celebrates Gay Day less because of political progress than financial expediency. The market has seized upon multiculturalism and gender-bending in the same ways that it has seized upon youth culture in general – not just as a market niche but as a source of new carnivalesque imagery. As Robert Goldman and Stephen Papson note, "White-bread culture will simply no longer do."[10] The $200 billion culture industry – now America's biggest export – needs an ever-changing, uninterrupted supply of street styles, edgy music videos and rainbows of colors. And the radical critics of the media clamoring to be "represented" in the early nineties virtually handed over their colorful identities to the brandmasters to be shrink-wrapped.

The need for greater diversity – the rallying cry of my university years – is now not only accepted by the culture industries, it is the mantra of global capital. And identity politics, as they were practiced in the nineties, weren't a threat, they were a gold mine. "This revolution," writes cultural critic Richard Goldstein in *The Village Voice*, "turned out to be the savior of late capitalism."[11] And just in time, too.

Market Masala: Diversity and the Global Sales Pitch

About the same time that my friends and I were battling for better cultural representation, the advertising agencies, broadcasters and global brands were preoccupied with some significant problems of their own. Thanks to freer trade and other forms of accelerated deregulation, the global marketplace was finally becoming a reality, but new, urgent questions were being asked: What is the best way to sell identical products across multiple borders? What voice should advertisers use to address the whole world at once? How can one company accommodate cultural differences while still remaining internally coherent?

For certain corporations, until recently, the answer was simple: force the world to speak *your* language and absorb *your* culture. In 1983, when global reach was still a fantasy for all but a handful of corporations, Harvard business professor Theodore Levitt published the essay "The Globalization of Markets," in which he argued that any corporation that was willing to bow to some local habit or taste was an unmitigated failure. "The world's needs and desires have been irrevocably homogenized," he wrote in what instantly became the manifesto of global marketing. Levitt made a stark distinction between weak *multinational* corporations, which change depending on which country they are operating in, and swaggering *global* corporations, which are, by their very definition, always the same, wherever they roam. "The multinational corporation operates in a number of countries, and adjusts its products and practices to each – at high relative costs. The global corporation operates with resolute constancy – at low relative cost – as if the entire world (or major regions of it) were a single entity; it sells the same things in the same way everywhere.... Ancient differences in national tastes or modes of doing business disappear."[12]

Levitt's "global" corporations were, of course, American corporations and the "homogenized" image they promoted were the images of America: blond, blue-eyed kids eating Kellogg's cereal on Japanese TV; the Marlboro Man bringing U.S. cattle country to African villages; and Coke and McDonald's selling the entire world on the taste of the U.S.A. As globalization ceased to be a somewhat kooky dream and became a reality, these cowboy-marketing antics began to step on a few toes. The twentieth century's familiar bogeyman – "American cultural imperialism" – has, in more recent years, incited cries of "cultural Chernobyl" in France, prompted the creation of a "slow-food movement" in Italy and led to the burning of chickens outside the first KFC outlet in India.

Americans in particular have never been known for their cultural sensitivity and so, not surprisingly, the road to Levitt's global marketing is paved with cultural *faux pas*. The most serious of these took place after the collapse of European communism, when media moguls fell over one another to take the credit for freedom and democracy the world over – a claim they would pay for later on. "We put MTV into East Germany, and the next day the Berlin Wall fell," Viacom International chairman Sumner Redstone said.[13] Ted Turner claimed the credit for CNN and the Goodwill Games. "I said, 'Let's try and undo this. Let's get our young people together, and let's get this cycle together and let's try to get some world peace going and let's end the Cold War.' And, by God, we did it."[14] Rupert Murdoch, meanwhile, told the world that "satellite broadcasting makes it possible for information-hungry residents of many closed societies to bypass state-controlled television."[15]

This post–Cold War bravado didn't go over too well in countries like China, where standing up to so-called Western values remains a sacrosanct political claim. Consequently, several Western media moguls – now hell-bent on penetrating all of Asia with their satellites – have gone to great lengths to distance themselves from their earlier freedom-fighter rhetoric and now actively collaborate with dictatorships to restrict the flow of information, a situation that I'll get to in more detail in Chapter 8.

It was in this minefield that "diversity" marketing appeared, presenting itself as a cure-all for the pitfalls of global expansion. Rather than creating different advertising campaigns for different markets, campaigns could sell diversity itself, to all markets at once. The formula maintained the one-size-fits-all cost benefits of old-style cowboy cultural imperialism, but ran far fewer risks of offending local sensibilities. Instead of urging the world to taste America, it calls out, like the Skittles slogan, to "Taste the Rainbow." This candy-coated multiculturalism has stepped in as a kinder, gentler packaging for the homogenizing effect of what Indian physicist Vandana Shiva calls "the monoculture" – it is, in effect, mono-multiculturalism.

Today the buzzword in global marketing isn't selling America to the world, but bringing a kind of market masala to everyone in the world. In the late nineties, the pitch is less Marlboro Man, more Ricky Martin: a bilingual mix of North and South, some Latin, some R&B, all couched in global party lyrics. This ethnic-food-court approach creates a One World placelessness, a global mall in which corporations are able to sell a single product in numerous countries without triggering the old cries of "Coca-Colonization."

As culture becomes increasingly homogenized globally, the task of marketing is to stave off the nightmare moment when branded products cease to look like lifestyles or grand ideas and suddenly appear as the ubiquitous goods they really are. In its liquid ethnicity, marketing masala has been introduced as the antidote to this horror of cultural homogeneity. By embodying corporate identities that are radically individualistic and perpetually new, the brands attempt to inoculate themselves against accusations that they are in fact selling sameness.

The Global Teen

Of course not everyone is equally amenable to the idea of treating culture and nationality as fashion accessories to be slipped on and off. Those who have fought wars and survived revolutions tend to be more protective of their national traditions. The desolately poor, who constitute one-quarter of the world's population,[16] also have a little trouble getting into the global groove, especially since cable TV and most brand-name products are still just a rumor in those parts of the developing world where a total of 1.3 billion people live on US$1 a day or less.[17] No, it's the young people living in developed and semi-developed countries who are the great global hope. More than anything or anyone else, logo-decorated middle-class teenagers, intent on pouring themselves into a media-fabricated mold, have become globalization's most powerful symbols.

This has happened for several reasons. First of all, just as in the U.S. market, there are a lot of them. The world is crawling with teenagers, especially in southern countries, where the UN estimates that 507 million adults will die before they turn forty.[18] Two-thirds of Asia's population is under thirty and, thanks to years of bloody warfare, about 50 percent of the population in Vietnam was born after 1975. All in all, the so-called global teen demographic is estimated at one billion, and these teenagers consume a disproportionate share of their families' incomes. In China, for instance, conspicuous consumption for all members of the household remains largely unrealistic. But, argue the market researchers, the Chinese make enormous sacrifices for the young – particularly for young boys – a cultural value that spells great news for cell-phone and sneaker companies. Laurie Klein of Just Kid Inc., a U.S. firm that conducted a consumer study on Chinese teens, found that while Mom, Dad and both grandparents may do without electricity, their only son (thanks to the country's one-child policy) frequently enjoys what is widely known as "little emperor syndrome," or what she calls the "4-2-1" phenomenon: four elders and two parents scrimp and save so the one child can be an MTV clone. "When you have the parents and four grandparents spending on one child, it's a no-brainer to know that this is the right market," says one venture capitalist in China.[19] Furthermore, since kids are more culturally absorbent than their parents, they often become their families' dedicated shoppers, even for big household items. Taken together, what this research shows is that while adults may still harbor traditional customs and ways, global teens shed those pesky national hang-ups like last year's fashions. "They prefer Coke to tea, Nikes to sandals, Chicken McNuggets to rice, credit cards to cash," Joseph Quinlan, senior economist at Dean Witter Reynolds Inc. told *The Wall Street Journal*.[20] The message is clear: get the kids and you've got the whole family and the future market.

> *Diversity. Whatever.*
>
> – Slogan for a 1998–99 ad campaign for Eaton's department store, Canada

Inflated by rhetoric like this, the image of the global teen floats over the planet like a euphoric corporate hallucination. These kids, we are repeatedly told, live not in a geographic place but in a global consumer loop: hot-linked from their cellular telephones to Internet newsgroups; bonded together by Sony Playstations, MTV videos and NBA games. The most extensive and widely cited study of the global teen demographic was conducted in 1996 by the New York–based ad agency DMB&B's BrainWaves division. The "New World Teen Study" surveyed 27,600 middle-class fifteen- to eighteen-year-olds in forty-five countries and came up with some resoundingly good news for the agency's clients, a list that includes Coca-Cola, Burger King and Philips. "Despite different cultures, middle-class youth all over the world seem to live their lives as if in a parallel universe. They get up in the morning, put on their Levi's and Nikes, grab their caps, backpacks, and Sony personal CD players, and head for school."[21] Elissa Moses, senior vice president at the advertising agency, called the arrival of the global teen demographic "one of the greatest marketing opportunities of all time."[22]

But before the brands are able to sell the same products in the same way all around the world, the teens themselves must identify with their new demographic. For this reason, what most global ad campaigns are still selling most aggressively is the *idea* of the global teen market – a kaleidoscope of multi-ethnic faces blending into one another: Rasta braids, pink hair, henna hand painting, piercing and tattoos, a few national flags, flashes of foreign street signs, Cantonese and Arabic lettering and a sprinkling of English words, all over the layered samplings of electronic music. Nationality, language, ethnicity, religion and politics are all reduced to their most colorful, exotic accessories, converging to assure us, as Diesel president Renzo Rosso does, there is "never an 'us and them,' but simply one giant 'we.'"[23]

To achieve this state of oneness, global teens must sometimes be pitted against traditional elders who don't appreciate their radical taste in denim. For instance, a TV ad for Diesel jeans shows two Korean teenagers turning into birds after they commit double suicide, finding freedom only in the total surrender to the brand. In these ads, the ultimate product – more than the soft drinks, ice creams, sneakers or jeans – is the global teen, who must exist as a demographic in the minds of young consumers worldwide or the entire exercise of global marketing collapses. For this reason, global youth marketing is a mind-numbingly repetitive affair, drunk on the idea of what it is attempting to engineer: a third notion of nationality – not American, not local, but one that would unite the two, through shopping.

Standing triumphant at the center of the global teen phenomenon is MTV, which, in 1998, was in 273.5 million households worldwide – only 70 million of which were in the U.S. By 1999, MTV's eight global divisions broadcast in 83 countries and territories, fewer than CNN's 212-country reach, but impressive nonetheless. Furthermore, the New World Teen Study found that the single most significant factor contributing to the shared tastes of the middle-class teens it surveyed was TV – in particular, MTV, which 85 percent of them watched every day. Elissa Moses called the station "an all-news bulletin for creating brand-images"[24] and a "public-address system to a generation."[25] This sort of programming reach has been unprecedented since the 1950s when families gathered around the TV set to watch the Ed Sullivan show. Global teens watch so much MTV per day that the only equivalent shared cultural experience among adults occurs during an outbreak of war when all eyes are focused on the same CNN images.

And the more viewers there are to absorb MTV's vision of a tribe of culture swapping, global teen nomads, the more homogeneous a market its advertisers have in which to sell their products. According to Chip Walker, director of the New World Teen Study, "Teens who watch MTV music videos are much more likely than other teens to wear the teen 'uniform' of jeans, running shoes, and denim jacket... They are also much more likely to own electronics and consume 'teen' items such as candy, sodas, cookies and fast food. They are much more likely to use a wide range of personal-care products too."[26] In other words MTV International has become the most compelling global catalog for the modern branded life.

In-Fighting While the Global House Burned Down

The global economy's embrace of Representation Nation suggests that my generation's campus identity politics boiled down, in the end, to a set of modest political goals that were frequently (and deceptively) cloaked in immodest rhetoric and tactics. This isn't a P.C. *mea culpa*—I'm proud of the small victories we won for better lighting on campus, more women faculty members and a less Eurocentric curriculum (to dig up a much-maligned phrase from my P.C. days). What I question is the battles we North American culture warriors never quite got around to. Poverty wasn't an issue that came up much back then; sure, every once in a while in our crusades against the trio of 'isms, somebody would bring up "classism," and, being out-P.C.-ed, we would dutifully add "classism" to the hit list in question. But our criticism was focused on the representation of women and minorities within the structures of power, not on the economics behind those power structures. "Discrimination against poverty" (our understanding of injustice was generally construed as discrimination against something) couldn't be solved by changing perceptions or language or even, strictly speaking, individual behavior. The basic demands of identity politics assumed an atmosphere of plenty. In the seventies and eighties, that plenty had existed and women and non-whites were able to battle over how the collective pie would be divided: would white men learn to share, or would they keep hogging it? In the representational politics of the New Economy nineties, however, women as well as men, and whites as well as people of color, were now fighting their battles over a single, shrinking piece of pie – and consistently failing to ask what was happening to the rest of it. For us, as students, to address the problems at the roots of "classism" we would have had to face up to core issues of wealth distribution – and, unlike sexism, racism or homophobia, that was not what we used to call "an awareness problem."

So class fell off the agenda, along with all serious economic – let alone corporate – analysis. Certainly there were those in the ID ranks with revolutionary goals. Like the sixties counterculture radicals who thought they were shaking the foundations of Western civilization by dropping acid, there were a handful of professors and students of identity politics who believed that "great blows are being struck against capitalism in the realms of theory," as critic Gayatri Spivak put it.[27] And Dinesh D'Souza and his ilk couldn't resist calling the P.C.ers "neo-Marxists" – but in fact, nothing could have been further from the truth. The prospect of having to change a few pronouns and getting a handful of women and minorities on the board and on television posed no real threat to the guiding profit-making principles of Wall Street. "The real guilt of P.C. ...," wrote SUNY professor of literature Tim Brennan in 1991, "is not its supposed intolerance or rigidity, but that it is not political enough – that it is impersonating political struggle."[28]

That failure has turned out to be immeasurably problematic because the economic trends that have so accelerated in the past decade have all been about massive redistribution and stratification of world resources: of jobs, goods and money. Everyone except those in the very highest tier of the corporate elite is getting less.

And what is striking in retrospect is that in the very years when P.C. politics reached their most self-referential peak, the rest of the world was doing something very different: it was looking outward, and expanding. At the moment when the field of vision among most left-wing progressives was shrinking to include only its immediate surroundings, the horizons of global business were growing to encompass the whole globe. While CEOs dreamed of Big Macs in Russia, Benetton in Shanghai and logos projected on the moon, the political lens for far too many activists and theorists was narrowing so dramatically that with the exception of a brief period during the Gulf War, foreign and economic policy were off the radar screen. In North America, even the fight against free trade was all about protecting Canadian or American workers and resources, not about the possible effects of the trade agreement on Mexico, or the effects other rapid liberalization measures were having in the developing world. When the free-trade debate was lost, the left retreated even further into itself, choosing ever more minute disputes over which to go to the wall. This retreat reflected a broader political paralysis in the face of the daunting abstractions of global capitalism – ironically, the very issues that should have been most pressing for anyone concerned with the future of social justice.

In this new globalized context, the victories of identity politics have amounted to a rearranging of the furniture while the house burned down. Yes, there are more multi-ethnic sitcoms and even more black executives – but whatever cultural enlightenment has followed has not prevented the population in the underclass from exploding or homelessness from reaching crisis levels in many North American urban centers. Sure, women and gays have better role models in the media and pop culture – but the ownership in the culture industries has consolidated so rapidly that, according to William Kennard, the chairman of the U.S. Federal Communications Commission, "There are fewer opportunities of entry by minority groups, community groups, small businesses in general."[29] And though girls may indeed rule in North America, they are still sweating in Asia and Latin America, making T-shirts with the "Girls Rule" slogan on them and Nike running shoes that will finally let girls into the game.

This oversight isn't simply a failure of feminism but a betrayal of the feminist movement's own founding principles. Although the gender politics that I grew up with in the eighties were concerned almost exclusively with having women equally represented in the structures of power, the relationship between gender and class have not always been so casually overlooked. Bread and Roses – the rallying cry of the women's movement – has its origin in a slogan on a banner in the 1912 walkout of textile workers in Lawrence, Massachusetts. "What the woman who labors wants," explained historic organizer Rose Schneiderman in a 1912 speech, "is the right to live, not simply exist."[30] And March 8, the date of International Women's Day, was selected to mark the anniversary of a 1908 demonstration in which "women garment workers marched through the streets of New York, protesting dreadful working conditions, child labor, 12-hour working days, minuscule pay."[31] The young women who grew up reading *The Beauty Myth*, and who saw eating disorders and low self-esteem as the most harmful by-products of the fashion industry, tended to forget those women when we marched on March 8, if we ever knew about them to begin with.

As we look back, it seems like willful blindness. The abandonment of the radical economic foundations of the women's and civil-rights movements by the conflation of causes that came to be called political correctness successfully trained a generation of activists in the politics of image, not action. And if the space invaders marched into our schools and our communities unchallenged, it was at least partly because the political models in vogue at the time of the invasion left many of us ill-equipped to deal with issues that were more about ownership than representation. We were too busy analyzing the pictures being projected on the wall to notice that the wall itself had been sold.

If that remained true until recently, however, it is no longer so. As we will see in Part IV, a radical new political culture is emerging in high schools and on college campuses. Rather than calling attention to the house of mirrors that passes for empirical truth (as the postmodern academics did), and rather than fighting for better mirrors (as the ID warriors did), today's media activists are concentrating on shattering the impenetrable shiny surfaces of branded culture, picking up the pieces and using them as sharp weapons in a war of actions, not images.

NO CHOICE

Top: Wal-Mart greeter with the human touch. *Bottom:* The citizens of Warrenton, Virginia, aren't buying it.

CHAPTER SIX

BRAND BOMBING

Franchises in the Age of the Superbrand

MTV is associated with the forces of freedom and democracy around the world.

– Viacom CEO Sumner Redstone, owner of MTV, October 1994

There isn't a lot of angst, it's just unbridled consumerism.

– MTV CEO Tom Freston describes the content on MTV India, June 1997

The branded multinationals may talk diversity, but the visible result of their actions is an army of teen clones marching – in "uniform," as the marketers say – into the global mall. Despite the embrace of polyethnic imagery, market-driven globalization doesn't want diversity; quite the opposite. Its enemies are national habits, local brands and distinctive regional tastes. Fewer interests control ever more of the landscape.

Dazzled by the array of consumer choices, we may at first fail to notice the tremendous consolidation taking place in the boardrooms of the entertainment, media and retail industries. Advertising floods us with the kaleidoscopic soothing images of United Streets of Diversity and Microsoft's wide-open "Where do you want to go today?" enticements. But in the pages of the business section, the world goes monochromatic and doors slam shut from all sides: every other story – whether the announcements of a new buyout, an untimely bankruptcy, a colossal merger – points directly to a loss of meaningful choices. The real question is not "Where do you want to go today?" but "How best can I steer you into the synergized maze of where *I* want you to go today?"

This assault on choice is taking place on several different fronts at once. It is happening structurally, with mergers, buyouts and corporate synergies. It is happening locally, with a handful of superbrands using their huge cash reserves to force out small and independent businesses. And it is happening on the legal front, with entertainment and consumer-goods companies using libel and trademark suits to hound anyone who puts an unwanted spin on a pop-cultural product. And so we live in a double world: carnival on the surface, consolidation underneath, where it counts.

Everyone has, in one form or another, witnessed the odd double vision of vast consumer choice coupled with Orwellian new restrictions on cultural production and public space. We see it when a small community watches its lively downtown hollow out, as big-box discount stores with 70,000 items on their shelves set up on their periphery, exerting their gravitational pull to what James Howard Kunstler describes as "the geography of nowhere."[1] It is there on the trendy downtown main street as yet another favorite café, hardware store, independent bookstore or art video house is cleared away and replaced by one of the Pac-Man chains: Starbucks, Home Depot, the Gap, Chapters, Borders, Blockbuster. It is there inside the big-box retail outlets each time a magazine is taken off a shelf by a manager mindful of his bosses' corporate definition of "family values." You can see it in the messy bedroom of a fourteen-year-old Web master who has just had her fan page shut down by Viacom or EMI, unimpressed by her attempts to create her own little pocket of culture with borrowed snippets of trademarked song lyrics and images. It is there again when protesters are thrown out of shopping malls for handing out political leaflets, told by the security guards that although the edifice may have replaced the public square in their town, it is, in fact, private property.

A decade ago, any attempt to connect the dots among this mess of trends would have seemed strange indeed: what does synergy have to do with the chain-store craze? What does copyright and trademark law have to do with personal fan culture? Or corporate consolidation with freedom of speech? But today, a clear pattern is emerging: as more and more companies seek to be the one overarching brand under which we consume, make art, even build our homes, the entire concept of public space is being redefined. And within these real and virtual branded edifices, options for unbranded alternatives, for open debate, criticism and uncensored art – for real choice – are facing new and ominous restrictions. If the erosion of noncorporate space explored in the last section is feeding a kind of globo-claustrophobia that longs for release, then it is these restrictions on choice – restricted by the same companies that promised a new age of freedom and diversity – that are slowly focusing that potentially explosive longing on the multinational brands, creating the conditions for the anticorporate activism that will be explored later on in the book.

Constant Cloning

There is a distinctive quality to many of the chains that have proliferated during the eighties and nineties – Ikea, Blockbuster, the Gap, Kinko's, the Body Shop, Starbucks – which sets them apart from the fast-food restaurants, strip malls and muffler joints responsible for the sixties and seventies franchise sprawl. They don't flash with the garish, cartoonlike plastic yellow shells and golden arches; they are more apt to glow with a healthy New Age sheen. These crisp royal blue and kelly green boxes snap together like pieces of Lego (the new kind that can make only one thing: the model fire station or spaceship helpfully pictured on the box). The Kinko's, Starbucks and Blockbuster clerks buy their uniform of khakis and white or blue shirts at the Gap; the "Hi! Welcome to the Gap!" greeting cheer is fueled by Starbucks double espressos; the résumés that got them the jobs were designed at Kinko's on friendly Macs, in 12-point Helvetica on Microsoft Word. The troops show up for work smelling of CK One (except at Starbucks, where colognes and perfumes are thought to compete with the "romance of coffee" aroma), their faces freshly scrubbed with Body Shop Blue Corn Mask, before leaving apartments furnished with Ikea self-assembled bookcases and coffee tables.

The cultural transformation these institutions have effected is familiar to everyone, but there are few helpful statistics available on the proliferation of franchises and chains, largely because most research on retailing lumps franchises in with independent businesses. A franchise is technically owned by the franchisee, even if every detail of the outlet – from the sign that hangs out front to the precise temperature of the coffee – is controlled by a head office hundreds or even thousands of miles away. Even without industry-wide figures, it's undeniable that something very dramatic has happened to the face of retail this decade. Take Starbucks, for instance. As recently as 1986, the coffee company was a strictly local phenomenon, with a handful of cafés around Seattle. By 1992, Starbucks had 165 stores with outlets in several U.S. and Canadian cities. By 1993, that number had already gone up to 275, and in 1996, it reached 1,000. In early 1999, Starbucks hit 1,900 stores with outlets in twelve countries, from the U.K. to Kuwait.

Blockbuster, another of the distinctly nineties chains, has enjoyed an even more dramatic expansion rate over precisely the same time period. In 1985, Blockbuster was a lone video store in Dallas, Texas. It was bought by waste-management czar Wayne Huizenga in 1987 and by 1989 there were 1,079 stores. In 1994, the year Huizenga sold Blockbuster to Viacom, there were 3,977. By early 1999, the number had reached 6,000, distributed over twenty-six countries, including 700 outlets in the U.K. alone.

Similar patterns can be tracked for the Gap (and its holdings Banana Republic and Old Navy) and the Body Shop, which averaged between 120 and 150 store openings a year through the mid-eighties to the present. Even Wal-Mart didn't truly find its feet as a retail powerhouse until the late eighties. Although the first Wal-Mart outlet opened in 1962, the superstore model didn't take off until 1988 and it wasn't until 1991 that Wal-Mart – by then opening 150 discount stores a year – surpassed Kmart and Sears to become the most powerful force in American retailing.

This growth spurt was brought about by three industry trends, all of them dramatically favoring big chains with deep cash reserves. The first is price wars, in which the biggest megachains systematically undersell all their competitors; the second is the practice of blitzing out the competition by setting up chain-store "clusters." The third trend, to be explored in the next chapter, is the arrival of the palatial flagship superstore, which appears on prime real estate and acts as a three-dimensional ad for the brand.

Price Wars: The Wal-Mart Model

In mid-1999, Wal-Mart had 2,435 big-box discount stores in nine countries, selling everything from Barbie Dream Homes to Kathie Lee Gifford skirts and handbags to Black & Decker drills to Prodigy CDs. Of those stores, 565 were "Supercenters," a concept that combines Wal-Mart's original discount model with full-service grocery stores, hair salons and banks, as well as 443 Sam's Clubs, which offer even deeper discounts for bulk purchases and big-ticket items like office furniture. (See Tables 6.1 and 6.2, Appendix, page 472.)

The recipe that has made Wal-Mart the largest retailer in the world, hauling in $137 billion in sales in 1998, is straightforward enough. First, build stores two and three times the size of your closest competitors. Next, pile your shelves with products purchased in such great volume that the suppliers are forced to give you a substantially lower price than they would otherwise. Then cut your in-store prices so low that no small retailer can begin to compete with your "everyday low prices."

Because everything about the Arkansas-based retailer is premised on achieving an economy of scale, an average Wal-Mart store measures 92,000 square feet, not including the requisite substantial parking lot. Since discounting is its calling card, Wal-Mart must keep its overhead down, which is why the lots for its windowless stores are purchased on the edges of towns, where land is cheap and taxes are lower. Every year of Wal-Mart's expansion, its new stores have grown bigger in size, and many of its original, comparatively modest discount outlets have been converted and expanded into superstores, some as large as 200,000 square feet.

Another key element in keeping costs down is that Wal-Mart only opens outlets close to its distribution centers. For this reason, Wal-Mart has spread like molasses: slow and thick. It won't move into a new region until it has blanketed the last area with stores—as many as forty in a hundred-mile radius. That way, the company saves money on transportation and shipping costs, and develops such a concentrated presence in an area that advertising its brand is barely necessary.[2] "We would go as far as we could from a warehouse and put in a store. Then we would fill in the map of that territory, state by state, county seat by county seat, until we had saturated the market area," Wal-Mart founder Sam Walton explained.[3] Then the company would open up a new distribution center in a new region and repeat the process.

After Wal-Mart began in the U.S. South, plodding slowly through Arkansas, Oklahoma, Missouri and Louisiana, it took a while before Wall Street and the Eastern-based media grasped the magnitude of Sam Walton's project. For this reason, it wasn't until the early nineties, three decades after the opening of the first Wal-Mart, that opposition to the big boxes began to mount. The argument against Wal-Mart's retail style—by now almost as familiar as Wal-Mart itself—holds that bargain prices lure shoppers to the suburbs, sucking community life and small businesses out of the town centers. Smaller businesses can't compete—in fact, many of Wal-Mart's competitors claim they pay more for their goods wholesale than Wal-Mart charges retail.

By now, there have been several books written about the effect of the big boxes, most notably *In Sam We Trust*, by *Wall Street Journal* reporter Bob Ortega. As Ortega notes, Wal-Mart is not alone in its "size matters" approach to retailing—it is simply the leader in an exploding category of big-box retailers who use their clout to wrangle special treatment. Home Depot, Office Depot and Bed, Bath & Beyond, which are often grouped together in pumped-up strip malls called "power centers," are all known in the retail industry as "category killers" because they enter a category with so much buying power that they almost instantly kill the smaller competitors.[4]

This retail style has always been controversial and was responsible for the first anti-chain movement, which arose in the 1920s. As discounters like A&P and Woolworths proliferated, small merchants tried to make it illegal for chains to use their relative size to extract lower wholesale prices and drive down retail prices. The rhetoric of the time, as Ortega points out, bears a striking resemblance to the language of the grassroots opposition groups that have sprung up in dozens of North American towns when the pending arrival of a new Wal-Mart outlet has been announced.[5]

On the legal front, charges of monopolistic practices have been cropping up with growing regularity, and not just against Wal-Mart. In September 1997, for instance, the U.S. Federal Trade Commission found that Toys 'R' Us was guilty of illegally pressuring manufacturers not to supply popular toys to other chains. Because Toys 'R' Us is the largest toy retailer in the world, the manufacturers agreed; and consumers' options were reduced dramatically, along with their chances to comparison shop. "Many toy manufacturers had no choice but to go along," said William Baer, director of the Federal Trade Commission's Bureau of Competition when the case was decided.[6] This was precisely the type of situation the FTC was hoping to avoid when, in 1997, it blocked a planned merger between two huge office-supply chains – Staples and Office Depot – stating that the consolidation would hurt competition.

Beyond spawning the category killer, Sam Walton's legacy has had other, further-reaching effects. In many ways, it was the inhuman scale of the big boxes and their accompanying sprawl – the streets without sidewalks, the shopping centers only accessible by car, the stores the size of small hamlets with all the design flair of toolsheds – that set the stage for the other significant retail trends of the decade. Discount stores were great for saving money but not for much else. And so, as the big boxes expanded into seas of concrete on the edge of town, they generated a renewed hunger for human-scale development; for the old-fashioned town square, for public gathering places that allowed both large meetings and intimate conversation; for a kind of retail with more interaction and more sensory stimulation. In other words, they laid the groundwork for Starbucks, Virgin Megastores and Nike Town.

Where big boxes used their size to move previously unimaginable amounts of product, the new retailers would use their size to fetishize brand-name goods, placing them on a pedestal as high as Wal-Mart's discounts were low. Where the big boxes had swapped a sense of community values for a discount, the branded chains would re-create it and sell it back – at a price.

Clustering: The Starbucks Model

"A Comforting Third Place" is the phrase Starbucks uses to promote itself in its newsletters and evangelical annual reports. This is not just another non-space like Wal-Mart or McDonald's, it's an intimate nook where sophisticated people can share "coffee . . . community . . . camaraderie . . . connection."[7] Everything about New Age chains like Starbucks is designed to assure us that they are a different breed from the strip-mall franchises of yesterday. This isn't dreck for the masses, it's intelligent furniture, it's cosmetics as political activism, it's the bookstore as an "old-world library," it's the coffee shop that wants to stare deep into your eyes and "connect."

But there's a catch. The need for more intimate spaces designed to tempt people to linger may indeed provide a powerful counterpoint to the cavernous big boxes, but these two retail trends are not as far apart as they appear at first. For instance, the mechanics of Starbucks' dizzying expansion during the past thirteen years has more in common with Wal-Mart's plan for global domination than the brand managers at the folksy coffee chain like to admit. Rather than dropping an enormous big box on the edge of town, Starbucks' policy is to drop "clusters" of outlets in urban areas already dotted with cafés and espresso bars. This strategy relies just as heavily on an economy of scale as Wal-Mart's does and the effect on competitors is much the same. Since Starbucks is explicit about its desire to enter markets only where it can "become the leading retailer and brand of coffee,"[8] the company has concentrated its store-a-day growth in relatively few areas. Instead of opening a few stores in every city in the world, or even in North America, Starbucks waits until it can blitz an entire area and spread, to quote *Globe and Mail* columnist John Barber, "like head lice through a kindergarten."[9] It's a highly aggressive strategy, and it involves something the company calls "cannibalization."

The idea is to saturate an area with stores until the coffee competition is so fierce that sales drop even in individual Starbucks outlets. In 1993, for instance, when Starbucks had just 275 outlets concentrated in a few U.S. states, per-store sales increased by 19 percent from the previous year. By 1994, store sales growth was only 9 percent, in 1996 it dipped to 7 percent, and in 1997 Starbucks saw only a 5 percent sales growth; in new stores, it was as low as 3 percent. (See Table 6.3, Appendix, page 473.) Understandably, the closer the outlets get to each other, the more they begin to poach or "cannibalize" each other's clientele – even in hyper-caffeinated cities like Seattle and Vancouver people can only suck back so many lattes before they float into the Pacific. Starbucks' 1995 annual report explains: "As part of its expansion strategy of clustering stores in existing markets, Starbucks has experienced a certain level of cannibalization of existing stores by new stores as the store concentration has increased, but management believes such cannibalization has been justified by the incremental sales and return on new store investment." What that means is that while sales were slowing at individual stores, the total sales of all the chain's stores combined continued to rise – doubling, in fact, between 1995 and 1997. Put another way, Starbucks the company was expanding its market while its individual outlets were losing market share, largely to other Starbucks outlets (see Table 6.4, Appendix, page 473).

It also helped Starbucks, no doubt, that its cannibalization strategy preys not only on other Starbucks outlets but equally on its real competitors, independently run coffee shops and restaurants. And, unlike Starbucks, these lone businesses can only profit from one store at a time. The bottom line is that clustering, like big-boxing, is a competitive retail strategy that is only an option for a large chain that can afford to take a beating on individual stores in order to reap a larger, long-term branding goal. It also explains why critics usually claim that companies like Starbucks are preying on small businesses, while the chains themselves deny it, admitting only that they are expanding and creating new markets for their products. Both are true, but the chains' aggressive strategy of market expansion has the added bonus of simultaneously taking out competitors.

There have been other, more brazen ways in which Starbucks has used its size and deep pockets to its competitive advantage. Until the practice began creating controversy a few years back, Starbucks' real-estate strategy was to stake out a popular independent café in a well-trafficked, funky location and simply poach the lease from under it. Several independent café owners in prime locations are on record claiming that Starbucks went directly to their landlords and offered to pay them higher rental payments for the same or adjacent spaces. For instance, Chicago's Scenes Coffee House and Drama received an eviction notice after Starbucks rented a space in the shopping complex where it was located. The coffee chain attempted a similar maneuver with Dooney's café in Toronto, though Starbucks claims it was the landlord who made the initial approach. Starbucks did gain control of Dooney's lease but the community protest was so strong that the company ended up having to sublet the space back to Dooney's.

These cutthroat real-estate practices hardly make Starbucks unique as a developer: McDonald's has perfected the scorched-earth approach to franchising, opening neighboring franchises and mini-outlets at gas stations until an area is blanketed. The Gap has also adopted the cluster approach to retailing, brand bombing key neighborhoods with multiple outlets of the Gap, Baby Gap, Gap Kids, Old Navy, Banana Republic and in 1999 Gap Body stores. The idea is to make Gap's family of brands synonymous with clothing in the same way that McDonald's is synonymous with hamburgers and Coke is synonymous with soft drinks. "If you go to a supermarket, you would expect to find some fundamental items. You would expect to find milk: nonfat, 1 percent, 2 percent, whole milk. You would expect dates to be fresh.... I don't know why apparel stores should be any different," says Mickey Drexler, Gap CEO.[10] It's fitting that Drexler's model for the Gap's ubiquity is the supermarket, since it was the first supermarket chains that pioneered the clustering expansion model. After A&P launched its "economy stores" in 1913 (the prototype of the modern supermarket), it quickly opened 7,500 outlets, then closed half of them after saturation had been achieved and many competitors were forced out of business.

The Gap welcomes these comparisons with Coke, McDonald's and A&P, but Starbucks, because of the nature of its brand image, strenuously rejects them.[11] After all, the Gap's project is to take a distinctive product – clothing – and brand it so completely that purchasing it from the Gap is as easy as buying a quart of milk or a can of Coke. Starbucks, on the other hand, is in the business of taking a much more generic product – a cup of coffee – and branding it so completely that it becomes a spiritual/designer object. So Starbucks doesn't want to be known as a blockbuster, it wants, as its marketing director Scott Bedbury says, to "align ourselves with one of the greatest movements towards finding a connection with your soul."[12]

Yet no matter how urbane the original concept may have been, the business of chains has a logic and a momentum of its own, having very little to do with what it sells. It breaks down each of a brand's elements – no matter how progressive and homespun – into a kit of easy-to-assemble bits and parts. Just as the chains snap together like Lego, each chain outlet is made up of hundreds of its own snappable parts. Within the logic of chains, it matters little whether those snappable parts are a McDonald's deep frier and a Hamburglar mannequin or the "four elemental icons" that form the building blocks for each Starbucks store design: "Earth to grow. Fire to roast. Water to brew. Air for aroma." A clone is a clone, whether it is molded in the shape of an arch or a peace symbol, and its purpose is still replication.

This process is even more apparent when the chains expand on the global stage. When retailers move outside their countries of origin, Starbucks-style clustering melds with Wal-Mart–style price wars to create a kind of "bulk clustering strategy." To keep prices low in a new market, chains like Wal-Mart, Home Depot and McDonald's must carry with them their trump card of being volume buyers; and in order to have the market clout to get lower prices than their competitors, they can't dribble into countries one store at a time. Instead, it has become a favored expansion tactic to buy out an existing chain and simply move into its stores in one dramatic entrance, as Wal-Mart did when it bought out 120 Woolco stores in Canada in 1994 and when it purchased the Wertkauf GmbH hypermarket chain in Germany in 1997. Similarly, when Starbucks moved into the U.K. in 1998, it acquired the already existing Seattle Coffee Company and retrofitted its 82 stores as Starbucks outlets.

For national companies looking to avoid becoming the prey of the global giants, it has become an increasingly popular strategy to initiate preemptive mergers of their own between two or more large national brands. In the name of nationalism and global competitiveness, they consolidate, lay off staff and mimic American retail formulas. Not surprisingly, they generally end up transforming themselves into copies of the global brands they were attempting to block. That's what happened in Canada when fear of Wal-Mart prompted the country's oldest department store chain, the Hudson's Bay Company, to buy Kmart Canada, fold it in with Zellers, lay off six thousand workers and open several lines of big-box discount outlets: one for furniture, one for home and bath and one for discount clothing. "Wal-Mart executed better than either Kmart or Zellers. By merging the two operations, we're going to learn how to execute better," said George Heller, president of Kmart.[13]

Selection versus Choice

The combination of the big-box and clustering approaches to retailing is having a transformative effect on the retail landscape. Though they represent very different retail trends, the combined effect of the Wal-Mart and Starbucks models has been to gradually erode the market share of small business in what was one of the few fields remaining where independent operators stood a solid chance of competing head-to-head with multinationals. With the chains able to outbid smaller competitors for space and supplies with barely a second thought, retail has become a battle of the big spenders. Whether they are using their clout to drive prices down to impossibly low levels, to keep them artificially high or simply to seize near monopolistic market shares, the net effect is the same: a retail arena in which size is a prerequisite and small companies can barely maintain a toehold. Like sumo wrestlers, the competitors in this game must push the limits of their weight category; bigness begets bigness.

Of course independent stores and restaurants continue to open and thrive, but more and more, these are high-end, specialty retailers in gentrified neighborhoods, while the suburbs, small towns and working-class neighborhoods get blanketed in – and blasted by – the self-replicating clones. This shift affects not only who can afford to stay in business but also (as I'll get into in Chapter 8) what makes it onto the store's shelves.

There is another retail trend that is in many ways exerting an even more significant influence than the two just discussed: the branded superstore, a marketplace marriage of the size power of big boxes with the branding clout of the store clusters. As I'll show in the next chapter, the superstore is the logical result of the corporate preoccupation with synergy: part marketing, part brand-extension supermarket, part theme park.

All of these three retail phenomena, and the impact they are having on consumer choice, are about much more than changes to the way we shop. They are key pieces of the branding puzzle that is transforming everything, from the way we congregate to the way we work. In fact, the divide between the bland big boxes at the edge of town and the branded castles and clusters in the center of town can be traced back to Marlboro Friday and its aftermath. These parallel developments are the physical embodiment of the split that opened up between the lowly price-slashers and the spiritual brand-builders. For its part, Wal-Mart stands as the single most powerful symbol of the decline in brand value that sent Wall Street into a tailspin on that Friday in April 1993. The year before the so-called brand crash was a record one for Wal-Mart, during which it opened 161 new discount stores – unheard-of growth for the end of a recession. Wal-Mart's shoppers were the new "value generation" in motion, flocking to the suburbs to avoid paying premium prices for heavily marketed brands. If Wal-Mart was selling Tide at deep discounts, so much the better, but these formerly brand-conscious shoppers were just as happy to take home detergent from Wal-Mart's own private label, Great Value.

At the same time, the proliferation of Nike Towns, Disney Stores and Starbucks clusters is powerful evidence of a renewed reverence for a handful of élite lifestyle brands. For many of their loyal consumers, no price is too high to pay for these branded goods and, in fact, merely buying the products provides an insufficient relationship. Brand-obsessed shoppers have adopted an almost fetishistic approach to consumption in which the brand name acquires a talismanic power.

Not surprisingly, capitalizing on the urge for this sort of brand cocooning has become the central preoccupation of the fashion, athletics and entertainment corporations selling these fetish brands. Theme-park-inspired superstores are one part of this process, but as the successive waves of mergers and attendant synergies continue, they are only the beginning.

Top: Michael Eisner (Walt Disney Co. CEO) seals merger with Thomas Murphy (Capital Cities/ABC Chairman). *Bottom:* Ted Turner (Turner Broadcasting Chairman and President) does the same with Gerald Levin (Time Warner Chairman and CEO).

CHAPTER SEVEN

MERGERS AND SYNERGY

The Creation of Commercial Utopias

I would prefer ABC not cover Disney.

– Disney CEO Michael Eisner, September 29, 1998, National Public Radio

Commenting on the future of poetry and art in a democratic society, Alexis de Tocqueville wrote that he was not worried about a lapse into safe realism so much as a flight into unanchored fantasy. "I fear that the productions of democratic poets may often be surcharged with immense and incoherent imagery, with exaggerated descriptions and strange creations; and that the fantastic beings of their brain may sometimes make us regret the world of reality."[1]

We are surrounded now by the realization of Tocqueville's predictions: gleaming, bulbous golden arches; impossibly smooth backlit billboards; squishy cartoon characters roaming fantastically fake theme parks. When I was growing up, these strange creations awakened something in me that I've since come to think of as deep longing for the seductions of fake; I wanted to disappear into shiny, perfect, unreal objects.

Maybe this condition was brought on by television, maybe it was a too-early trip to Disneyland, maybe it was malls, but just as Tocqueville predicted in 1835, the world of reality looked pretty dingy by comparison. The humiliating spectacle of my all-too-real family, so sixties authentic, set against the cascade of inviting plasticity that was the seventies and eighties, was simply too much to bear. "Stop it, guys, you're embarrassing me!" was the near-hysterical *cri de coeur* of my youth. Even when there was no one but family around, I could feel the plastic world's reproachful gaze.

My parents, part of a wave of American hippies who moved to Canada to dodge the Vietnam War draft, were terribly disturbed by these tendencies of mine. In their newly adopted country, they had imagined themselves to be breeding a new kind of postrevolutionary child, blessed with the benefits of Canada's humane social services, public health-care system and solid subsidies to the arts. Hadn't they diligently mushed their own baby food? Read *Parent Effectiveness Training*? Banned war toys and other "gendered" play?

In an effort to save me from corruption, my parents were forever dragging me out of the city to appreciate the Canadian wilderness and experience the joys of real-time family interaction. I was distinctly unimpressed. The only thing that saved me on these reality excursions was my dreams of fakeness, unfolding in the back seat of our station wagon as it sped past verdant farmland and majestic mountains. At five or six, I would eagerly await the molded plastic of franchise signs on the side of the road, craning my neck as we passed McDonald's, Texaco, Burger King. My favorite was the Shell sign, so bright and cartoon-like I was convinced that if I could climb up and touch it, it would be like touching something from another dimension—from the world of TV. During these family trips, my brother and I would beg to stop for fast food packed in shiny laminated boxes, and sometimes my parents would relent, if they were feeling particularly defeated that day. But more often than not, lunch would be another ponchoed picnic at a national park, with dry cheddar cheese, autumnal fruit and other distressingly unpackaged foodstuffs.

By the time I was eight or nine, my back-seat daydreams grew more intricate. I spent an entire journey through the Rockies conducting covert makeovers on everyone in the car. My father would lose the sandals and get a sharp, dignified suit, my mother a helmet hairdo and a wardrobe of smart pastel blazers, skirts and matching pumps. As for me, the possibilities were endless: kitchen cupboards filled with fake foods, closets overflowing with designer labels, unlimited access to eyeliner and perms. I wasn't allowed to have a Barbie ("a racket," my parents ruled, "first it's a doll, then a camper van, then the whole mansion") but I had Barbie in my brain.

It seemed as if the vanguard feminist-socialist child-rearing experiment was doomed to failure. Not only was I crazy for Shell signs, but by the age of six, my older brother had developed an uncanny knack for remembering the jingles from television commercials and would tear around the house in his Incredible Hulk T-shirt declaring himself "cuckoo for Cocoa Puffs." At the time, I couldn't understand why my parents were so upset about these stupid rhymes, but now I've come to feel their pain: despite their very best efforts, they had somehow given birth to an advertisement for General Mills – in other words, to regular kids.

Cartoons and fast-food franchises speak to children in a voice too seductive for mere mortal parents to compete with. Every kid wants to hold a piece of the cartoon world between his or her fingers – that's why the licensing of television and movie characters for toys, cereals and lunchboxes has spawned a $16.1 billion annual industry.[2] It is also why so-called family entertainment companies have been going to greater and greater lengths to extend their television and movie fantasies into real-world experiential extravaganzas: branded museum exhibits, high-tech superstores, and, the old standard, theme parks. Back in the 1930s, Walt Disney, the grandfather of modern synergy, understood the desire to crawl inside the screen when he fantasized about building a self-enclosed Disney city and remarked that every Mickey Mouse product or toy doubled as an advertisement for his cartoons. Mattel has long grasped this as well, but if Disney's project has been extending the fantasy of its films into toys, then Mattel's was extending its toys into ever more elaborate fantasy worlds. This vision is perhaps best understood as the "Zen of Barbie": Barbie is One. Barbie is all things.

Which is to say that the corporate synergy mania consuming so much of pop culture today is not all new. Barbie and Mickey Mouse are miniature branding trailblazers – those two have always wanted more extensions for their brands, more lateral monopolies to control. What has changed in the past decade is that almost everyone in the corporate world now recognizes that the urge to disappear into the cross-promotional tie-ins of cherished consumer products (be they toys, TV shows or sneakers) does not magically disappear when children outgrow sugar cereal. Plenty of Saturday-morning-cartoon kids have grown up into Saturday-night-club kids, fulfilling their longing for plastic fantasy with earnestly ironic Hello Kitty backpacks and Japanimation-inspired helmets of blue hair. You can see some of them at the Sega Playdiums, which are filled with grown-up gamers on weekend nights – no one under eighteen is even allowed to enter these roaring carnivals of virtual reality, especially on *South Park* theme nights.

It is this insistent desire to become one with your favorite pop-culture products that every one of the superbrands – from Nike to Viacom to the Gap to Martha Stewart – is trying to harness and expand upon, exporting Walt Disney's synergy principles from kid culture and transplanting them into every aspect of both teen and adult mass culture. Michael J. Wolf, a management consultant to such major players as Viacom, Time Warner, MTV and Citigroup, can attest to that fact. "I can't begin to count the number of times that people who run consumer businesses have confided to me that their goal is to create the broad-based success that Disney seems to bring to every project and every business it touches," he writes.[3]

This goal didn't materialize out of thin air. Rather, it can be traced back once again to the corporate "brands, not products" epiphany sparked by Marlboro Friday: if brands are about "meaning," not product attributes, then the highest feat of branding comes when companies provide their consumers with opportunities not merely to shop but to fully experience the meaning of their brand. Sponsorship, as seen in Chapter 2, is a good start, but synergy and lifestyle branding are the logical conclusion. Just as companies like Molson and Nike have sought to build celebrity brands by upstaging the concerts and sports matches they sponsored, so are many of these same companies also attempting to overthrow local retailers by creating branded superstores, then, further down the road, branded hotels and miniature villages. As two sides of the same project, synergy and branding are both about creating cross-promotional brand-based experiences that combine buying with elements of media, entertainment and professional sports to create an integrated branded loop. Disney and Mattel have always known this – now everyone else is learning it too.

A true branded loop cannot be created overnight, which is why the process usually begins with the simplest form of brand extension, a giant merger: Bell Atlantic and Nynex; Digital Equipment and Compaq; WordCom Inc. and MCI; Time Warner and Turner; Disney and ABC; Cineplex and Loews; Citicorp and Travelers; Bertelsmann and Random House; Seagram and

PolyGram; America Online and Netscape; Viacom and CBS...the list grows each day. Usually, the companies cite the Wal-Mart principle: everyone else in the industry is merging and only the biggest and strongest will survive. But size for its own sake is only the beginning of the story. Once the perimeter of the brand has expanded, corporate attention inevitably shifts to ways of making it more self-sufficient, through various internally coordinated cross-promotions. In a word, through synergy.

Sometime in the early nineties, writes Michael J. Wolf, the attitude of his media industry clients underwent a philosophical change. "Companies were no longer interested in merely being the biggest studio or the most successful TV network. They had to be more. Theme parks, cable networks, radio, consumer products, books, and music all became prospects for their potential empires. Media land was gripped by merger mania. If you weren't everywhere...you were nowhere."[4]

This sort of reasoning lies behind virtually all the major mergers of the mid- to late nineties. Disney buys ABC, which then broadcasts its movies and cartoons. Time Warner purchases Turner Broadcasting, which then cross-promotes its magazines and films on CNN. George Lucas buys block stocks in Hasbro and Galoob before he sells the toy companies the licensing rights for the new *Star Wars* films, at which point Hasbro promptly buys Galoob to consolidate its hold on the toy market. Time Warner opens a division devoted to turning its films and cartoons into Broadway musicals. Nelvana, a Canadian-based producer of kids' cartoons, purchases Kids Can Press, a publisher of children's books upon which such lucrative Nelvana cartoons as *Franklin the Turtle* are based. The merger transforms Nelvana into an "integrated company," in which future books can get their genesis in the company's marketable TV cartoons and lucrative lines of toys.[5]

In the broader book world, after purchasing Random House (this book's primary publisher), Bertelsmann AG buys 50 percent of Barnesandnoble.com, giving the largest English-language publishing company in the world a significant stake in the exploding on-line book retail market. Barnes & Noble, meanwhile, bids to buy Ingram, a major American book distributor, which also services the chain's competitors. If the Ingram deal had gone through (it was abandoned amid public outcry), the potential synergies among these three companies would have stretched to include the entire book publishing process, from contracting and editing to distributing, publicizing and, finally, retailing.

Perhaps the purest expression of synergy's market goals was Viacom's 1994 purchase of Blockbuster Video and Paramount Pictures. The deal gave Viacom the opportunity not only to profit from Paramount films when they played in its Paramount theaters but when they came out on video as well. "The combination of Viacom and Paramount, in my view, is the whole essence of the multimedia revolution," says Sumner Redstone, the billionaire mogul behind Viacom.[6] And this ability to keep cash flows inside a corporate family carries for these moguls its own kind of reward. Virgin's Richard Branson, for instance, laughs in the face of the accusation that his far-flung branding forays are stretching the Virgin name in too many directions. "It may be right that Mars sticks to the chocolate bar and Nike keeps its feet on the ground. But if their executives cross the Atlantic on a Virgin plane, listen to Virgin records and keep their money with a Virgin bank, then at least Britain will have one new global brand for the next century."

What the Virgin case clearly shows is that in the aftermath of the synergy revolution, brand extensions are no longer adjuncts to the core product or main attraction; rather, these extensions form the foundation upon which entire corporate structures are being built. Synergy, as Branson suggests, is about much more than old-style cross-promotion; it is about using ever-expanding networks of brand extensions to spin a self-sustaining lifestyle web. Branson and others are stretching the fabric of their brands in so many directions that they are transformed into tent-like enclosures large enough to house any number of core activities, from shopping to entertainment to holidays. Starbucks, upon announcing that it would begin selling furniture over the Internet, calls this a "brand canopy." This is the true meaning of a lifestyle brand: you can live your whole life inside it.

The concept is key to understanding not only synergy but also the related blurring of boundaries between sectors and industries. Retail is blurring with entertainment, entertainment with retail. Content companies (like film studios and book publishers) are leaping into distribution; distribution networks (like phone and Internet companies) are leaping into content production.

And all the while, the people previously pigeonholed as pure content – the stars themselves – are charging into production, distribution and, of course, retail. So the "if you aren't everywhere, you're nowhere" sentiment described by Wolf reaches well beyond the media conglomerates. Everyone, it seems, wants to be everywhere – whether they started as home decorators, sneaker manufacturers, record companies or basketball stars, they are all ending up, as Shaquille O'Neal and his people so aptly put it, "like Mickey Mouse."

In this fluid context, the branded tent of tents might be Disney or Viacom, but it could just as easily be Tommy Hilfiger, America Online, Martha Stewart or Microsoft. Quite simply, every company with a powerful brand is attempting to develop a relationship with consumers that resonates so completely with their sense of self that they will aspire, or at least consent, to be serfs under these feudal brandlords. This explains why marketing talk of pitch and product has been usurped so completely by the more intimate discourse of "meaning" and "relationship building" – brand-based companies are no longer interested in a consumer fling. They want to move in together.

And so the fiercest marketplace battles are taking place not between warring products but between warring branded camps that are constantly redrawing the borders around their enclaves, pushing the boundaries to include ever more complete lifestyle packages: if music, why not food, asks Puff Daddy. If clothes, why not retail, asks Tommy Hilfiger. If retail, why not music, asks the Gap. If coffee houses, why not publishing, asks Starbucks. If theme parks, why not towns, asks Disney.

Superstores: Stepping Inside the Brand

Not surprisingly, it was the Walt Disney Company, the inventor of modern branding, that created the model for the branded superstore, opening the first Disney Store in 1984. There are now close to 730 outlets worldwide. Coke followed shortly after with a store sporting all manner of branded paraphernalia, from key chains to cutting boards. But if Disney and Coke paved the way, it was Barnes & Noble that created the model that would forever change the face of retailing, introducing the first superstore to its chain of bookstores in 1990. The prototype for the new construct, according to company documents, was "old-world library ambiance and a wood and green palette" complemented by "comfortable seating, restrooms and extended hours" – and, of course, by a little co-branding in the form of in-store Starbucks coffee shops. The formula affected not only the chain's ability to sell books but also the role it occupied in pop culture; it became a celebrity, a source of endless media controversy, and eventually the thinly veiled inspiration for a Hollywood movie, *You've Got Mail*. In less than a decade, Barnes & Noble became the first bookstore that was also a superbrand in its own right.

Little wonder, then, that virtually all the consumer and entertainment companies that have been building up their brand images through marketing, synergy and sponsorship are now intent on having their own retail temples. Nike, Diesel, Warner Brothers, Tommy Hilfiger, Sony, Virgin, Microsoft, *Hustler* and the Discovery Channel have all leaped into branded retail. For these companies, stores that sell mul-tiple brands have become antithetical to the very principles of sound brand management. They want nothing to do with venues in which their products are sold side by side with their competitors'. "The multi-brand store is disappearing, and companies like us need stores that reflect our personality," explains Maurizio Marchiori, advertising director at Diesel, which has opened twenty branded stores since 1996.[7]

I'm really very, very disappointed that I didn't move into the retail business years ago, because I never realized the marketing power of the Hustler *name and logo.*

– *Hustler* owner Larry Flynt, *The New York Times*, March 21, 1999

The superstores constructed to reflect these corporate personalities are exploring the boundaries of what Nike refers to as "inspirational retail." As Nike president Thomas Clarke explains, large-scale "event" outlets "give retailers the opportunity to romance products better."[8] How this seduction takes place varies from brand to brand, but the general idea is to create a venue that is part shopping center, part amusement park, part multimedia extravaganza – an advertisement more potent and evocative than a hundred billboards. Popular superstore attractions include deejays spinning live from their own in-house broadcast booths, giant screens and star-studded launch parties. A cut above are the listening booths at the Virgin Megastores, the indoor waterfalls and rock-climbing walls at Seattle's Recreational Equipment, Inc., the interactive digital foot-measuring stations at Nike Town, the complimentary foot massages and reflexology at Rockport stores and the arcade-style computer games at the San Francisco Microsoft Store. And then, of course, there is that fixture of branded retail: the in-store coffee bar – even the Hustler superstore has one of those. Describing his vision for the 9,000-square-foot branded sex emporium in West Hollywood, *Hustler* owner Larry Flynt explained that he wanted to create a retail space "more comfortable for women, more like Barnes & Noble."[9]

"Creating a destination" is the key buzz-phrase for the superstore builder: these are places not only to shop but also visit, places to which tourists make ritualistic pilgrimages. For this reason, the locations chosen for the stores are far more upmarket than those to which the hawkers of Disney key chains, Nike sneakers and Tommy jeans are accustomed. In fact, so many mass-market brand meccas have made their home on New York's Fifth Avenue and L.A.'s Rodeo Drive that the neighbors – the exclusive Gucci, Cartier and Armani brands – have begun to complain about the popularizing presence of Daffy Duck and Air Jordan.

Selling mass-market consumer goods and doodads on the most expensive pieces of real estate in the world, in the most costly, high-tech, art-directed retail environments ever imagined, doesn't always add up on paper. But to look at the superstore as a break-even business enterprise is to miss the point entirely. No expense is spared in the building of the stores because, while the Time Square Disney Store or the Fifth Avenue Warner Brothers outlet may be money losers in and of themselves, they serve a much higher purpose in the overall branding picture. As Dan Romanelli, president of Warner Brothers consumer products division, says of the company's flagship, "Fifth and 57th is probably the best retail location in the world. It has helped immensely in building our international business and in making a statement about our brand."[10] Discovery Communication takes a similar attitude. Spinning off from its four television channels, the media company has launched thirty-five Discovery shops since 1996, hybrids of department stores, amusement parks and museums. The jewel in the crown is a $20 million flagship store in Washington, D.C., that features a full-scale model of a T. rex dinosaur skeleton and a World War II fighter plane. According to Michela English, president of Discovery Enterprises Worldwide, these outlets are not expected to make money until at least 2001. That, however, isn't stopping the company from adding dozens more stores. "There is a billboard impact to having the Discovery name on stores," she explains.[11]

Generally, this "billboard impact" is favored by companies whose primary source of sales is still multibrand venues: department stores, Cineplex Theaters, HMV record stores, Foot Locker and so on. Even without being able to control their entire distribution networks, branded superstores provide these companies with a kind of spiritual homeland for their brands, one so recognizable and grand that no matter where the individual products roam they will carry that grandness with them like a halo. It is as if a homing device had been implanted in the brand, so that, for instance, stalls selling Virgin merchandise at Virgin movie theaters aren't stalls selling merchandise at movie theaters – they are "Virgin mini-megastores," a satellite of something much deeper and more important than what meets the eye. And when consumers go to the local Foot Locker and are confronted with pairs of Nikes unceremoniously lined up next to the Reeboks, Filas and Adidas, they will, with any luck, remember the sensory overload they experienced on their pilgrimage to Nike Town. As Michael Wolf writes, branded retail is about "imprinting an experience on you as surely as the farmer's wife imprints good feelings in a clutch of baby geese when she feeds them a handful of grain every day."[12]

Branded Villages: Moving into the Brand

The stores are only the beginning – the first phase in an evolution from experiential shopping to living the fully branded experience. In a superstore, writes Wolf, "the lights, the music, the furniture, the cast of clerks create a feeling not unlike a play in which you, the shopper, are given a leading role."[13] But in the scheme of things that play is rather short: an hour or two at the most. Which is why the next phase after retail-as-tourist-destination has been the creation of branded holidays: never mind Disney World, Disney has launched the *Disney Magic* cruise ship and among its destinations is Disney's privately owned island in the Bahamas, Castaway Cay. Nike has its own sports-themed cruise ship in the works and Roots Canada, shortly after introducing a homewear line and opening a flagship store in Manhattan, launched the Roots Lodge, a branded hotel in British Columbia.

I visited the Roots development at the construction phase in Ucluelet, a small town on the west coast of Vancouver Island. The site is called the Reef Point Resort and it is here that branding is being taken to the next level. In April 1999, the Roots Lodge wasn't yet open, but construction was far enough along to make the concept perfectly clear: a high-end, fully branded summer camp for adults. Instead of canoes, an "adventure station" rents out ocean kayaks and surfboards; instead of outhouses, each cabin has its own hot tub; instead of the communal campfire, individual gas fireplaces. The lodge restaurant is set up mess-hall style, but the food is pure Pacific Coast gourmet. Most important, the rough-hewn wooden cabins are equipped with the entire Roots home furniture line.

"Like living in a billboard," one visitor observes as we receive our official tour, and that is no exaggeration. A cross between a catalog showroom and an actual living room, the resort has a Roots logo on display in the cabins on pillows, towels, cutlery, plates and glasses. The chairs, sofas, rugs, blinds and shower curtains are all Roots. On the wooden Roots coffee table is a brown leather Roots blotter, gently cradling a flattering book about the Roots story – and you can buy it all to take with you at the Roots store across the way. At the lodge, the "play" Wolf refers to lasts not a few hours but a weekend, maybe even a week or two. And the set at the company's disposal includes not only the architecture and design of the buildings (as is the case with superstores), but the entire Canadian wilderness around the lodge: the eagle in the cedar outside the window, the old-growth forest that guests walk through to reach the cabins, the crashing waves of the Pacific.

There is a strong symmetry at work in this branding exercise. The Roots clothing line got its genesis in a place not unlike this one. Company founders Don Green and Michael Budman both went to summer camp in Algonquin Park, Ontario, and were so moved by their experience of active living in the Canadian outdoors that they designed a line of clothing to capture the very best of that feeling: comfortable walking shoes, cozy sweatshirts, Canadian Workman socks, and, of course, the beaver logo. "Algonquin's majestic hills, sparkling lakes and forest primeval inspired

Roots," states an early print advertisement. "Its golden summer days, cold starry nights, autumn blaze and still winter white are now recreated in the colours and spirit of Roots Algonquin."[14]

The pitch was anything but subtle, as journalist Michael Posner observed in 1993 when he wrote, "Here's the truth: Roots is less a company than a summer camp."[15] The clothing manufacturer has been expanding on that carefully crafted image since the beginning. First it built retail outlets that, with the help of wall-mounted canoe paddles and exposed beams, conjure not a chain store but, as journalist Geoff Pevere writes, "summer-camp mess halls and cottages built by caring and callused hands."[16] Then came the homewear line, featuring blankets and pillowcases designed to look like oversized workmen's socks. And now, full circle, comes the Roots Lodge, where the original "inspiration" for a line of clothing becomes a fully realized extension of the Roots brand: from summer camp to branded camp; from lifestyle marketing to the lifestyle itself.

Mark Consiglio, the fast-talking, fleece-wearing developer of the resort, has bigger plans still for Reef Point, of which the Roots Lodge represents only a fraction of the available property. He shows me a model for a 250-cabin complex and explains his vision: a retail town center with brand-name stores and services. The Roots store, of course, but perhaps an Aveda Spa as well, and maybe stores like Club Monaco and the Body Shop too. Each retail outlet will be attached by boardwalk to its very own branded lodge, which, like the Roots Lodge, will be kitted with all the logo-festooned accessories the company can supply. Consiglio can't name names yet – "still in negotiations" – but he does tell me pointedly that "Roots isn't the only clothing company getting into homewear, you know. Everyone is doing it."

The problem with branded vacation destinations, however, is that they only provide temporary opportunities for brand convergence, an oasis from which families, at the end of the trip, are abruptly yanked and dumped back into their old lives, no doubt a poorly managed mishmash of competing logos and brand identities. Which is where Celebration, Florida, comes in – that very first Disney town. The meticulously planned development arrives complete with picket fences, a Disney-appointed homeowners' association and a phony water tower. For the families who live there year-round, Disney has achieved the ultimate goal of lifestyle branding: for the brand to become life itself.

Except the life on offer is perhaps not the one we might have expected from the Mouse. When Walt Disney first conceived of a branded city, it was meant to be an artificiality bonanza, a temple to the mid-fifties futuristic gods of technology and automation. The city never was built in Walt's lifetime, though some of the ideas went into the Epcot Center sixteen years after his death. When Disney CEO Michael Eisner decided to pick up on Walt's old dream and build a branded town, he opted against the *Jetsons*-inspired fantasy world his predecessor had imagined. Though wired with every modern technology and convenience, Celebration is less futurism than homage, an idealized re-creation of the livable America that existed before malls, big-box sprawl, freeways, amusement parks and mass commercialization. Oddly enough, Celebration is not even a sales vehicle for Mickey Mouse licensed products; it is, in contemporary terms, an almost Disney-free town – no doubt the only one left in America. In other words, when Disney finally reached its fully enclosed, synergized, self-sufficient space, it chose to create a pre-Disneyfied world – its calm, understated aesthetics are the antithesis of the cartoon world for sale down the freeway at Disney World.

Like the gated communities that have sprung up across the U.S., on Celebration's tranquil, tree-lined, billboard-free streets inhabitants are not subject to any of the stimulations or ravages of contemporary life. No Levi Strauss has bought up all the storefronts on Main Street to sell a new style of wide-legged pants, and no graffiti artists have defaced the ads; no Wal-Mart has left the downtown boarded up and twisted, and no community group has formed to fight the big boxes; no factory closures have eroded the tax base and pumped up the welfare rolls and no quarrelsome critics are around to point fingers. What is most striking about Celebration, however, particularly when compared with most North American suburban communities, is the amount of public space it offers – parks, communal buildings and village squares. In a way, Disney's branding breakthrough is a celebration of brandlessness, of the very public spaces the company has always been so adept at getting its brands on in the rest of its endeavors.

Of course this is an illusion. The families who have chosen to make Celebration their home are leading the first branded lives. As social historian Dieter Hassenpflug has remarked, "Even the streets are under Disney's control – private space that pretend[s] to be public."[17] So Celebration is an intricate inversion of Tocqueville's prediction: an "authenticity" bunker, specially retrofitted by the founder of fake.

The whole idea reminds me of a place on Vancouver Island called Cathedral Grove, about an hour and half's drive from the Roots Lodge and the mouth of Clayoquot Sound, Canada's most cherished old-growth forest. The drive through this part of the world has converted thousands of unsuspecting tourists into environmental activists, and it's easy to see why. After driving uphill for miles you reach a vista of mountains covered with lush cedars, sparkling lakes and drifting eagles – the wilderness that soothes and reassures the soul. The planet is as strong and rich as it ever was, it tells us – we just have to drive farther north to see it. But the serenity doesn't last long. The next dip and climb brings a radically different view: two huge bald gray mountains so burned and scarred they look more like the moon's surface than the earth. Nothing but death and asphalt for miles.

Nestled in the folds of this psychic roller coaster is the entrance to Cathedral Grove. Every day, hundreds of cars pull over to the side of the road, and their passengers embark on foot, glossy brochures in hand, to see the only old-growth trees left in the area. The largest tree has a rope around it and a plaque mounted on a stick. The irony, not lost on most residents of the area, is that this miniature park is owned and operated by MacMillan Bloedel, the logging company responsible for clear-cutting Vancouver Island and much of Clayoquot Sound. Cathedral Grove isn't a forest but a tree museum – just as Celebration is a town museum.

It's tempting to dismiss Celebration and the idea of the branded town as the particular neurotic obsession of the Disney corporation: this isn't a harbinger of the future privatization of public space, it's just Walt playing God again from beyond the grave. But with virtually every superbrand openly modeling itself after Disney, Celebration should not be too readily dismissed. Of course Disney is ahead of the game – Disney invented the game – but as is always the case with the Mouse, there are many would-be imitators trailing behind, taking notes. From his perch as adviser to the top media conglomerates, Michael J. Wolf observes that theme-park-style shopping locations like Minneapolis's Mall of America may be precursors to the live-in malls of the future. "Maybe the next step in this evolution is to put housing next to the stores and megaplexes and call it a small town. People living, working, shopping, and consuming entertainment in one place. What a concept," he enthuses.[18]

Setting aside, for a moment, the Brave New World/Stepford Wives associations such a vision inevitably evokes, there is something undeniably seductive about these branded worlds. It has to do, I think, with the genuine thrill of utopianism, or the illusion of it at any rate. It's worth remembering that the branding process begins with a group of people sitting around a table trying to conjure up an ideal image; they toss around words like "free," "independent," "rugged," "comfortable," "intelligent," "hip." Then they set out to find real-world ways to embody those ideas and attributes, first through marketing, then through retail environments like superstores and coffee chains, then – if they are really cutting edge – through total lifestyle experiences like theme parks, lodges, cruise ships and towns.

Why wouldn't these creations be seductive? We live in a time when expectations for building real-world commons and monuments with pooled public resources – schools, say, or libraries or parks – are consistently having to be scaled back or excised completely. In this context, these private branded worlds are aesthetically and creatively thrilling in a way that is totally foreign to anyone who missed the postwar boom. For the first time in decades, groups of people are constructing their own ideal communities and building actual monuments, whether it's the marriage of work and play at the Nike World Campus, the luxurious intellectualism of the Barnes & Noble superstores or the wilderness fantasy of the Roots Lodge. The emotional power of these enclaves rests in their ability to capture a nostalgic longing, then pump up the intensity: a school gym equipped with NBA-quality equipment; summer camp with hot tubs and gourmet food; an old-world library with designer furniture and latte; a town with no architectural blunders and no crime; a museum with the deep pockets of Hollywood. Yes, these creations can be vaguely spooky and sci-fi, but they should not be dismissed as just more crass commercialism for the unthinking masses: for better or for worse, these are privatized public utopias.

Shrinking Options in the Privatized Town Square

The terrible irony of these surrogates, of course, is how destructive they are proving to be to the real thing: to actual town centers, to independent business, to the non-Disney version of public spaces, to art as opposed to synergized cultural products and to a free and messy expression of ideas. Commercial climates are being dramatically altered by the expanding size and ambitions of these large players, and nowhere more so than in retail, where, as we have seen, companies like Discovery and Warner Brothers are in it for the "billboard effect" as much as for the sales. Independent shopkeepers, on the other hand, generally lack the resources to turn shopping into performance art, let alone into a destination vacation spot.

As superstores adopt the production values and special effects of Hollywood, small business is getting caught between, on the one hand, the deep discounting of the Wal-Marts and on-line retailers like Amazon.com, and on the other the powerful draw of the theme-park-infused retail environments. These market trends are combining to drastically undermine the traditional concepts of value and individual service that small business is known for offering. The staff at the indies may be more experienced and knowledgeable than the assistants at the superstores (the high turnover doesn't allow clerks to gain experience: more on that in the next section, "No Jobs"), but even that relative advantage can often get drowned out by the pure entertainment value of the superstores.

As many have commented, this phenomenon has been particularly pronounced in the book industry, where membership in the American Booksellers Association has fallen startlingly from 5,132 in 1991 to 3,400 in 1999.[19] Part of the problem is the Wal-Mart effect: the superstore chains have negotiated discounts on wholesale books with many publishers, making it nearly impossible for the independents to compete on price. The other difficulty is the retail standard set by the superstores. Bookstores are now expected to play the role of the university library, theme park, playground, pickup joint, community center, literary salon and coffee house all in one – a pricey undertaking even for the big players, which often involves taking a loss in the interest of future brand equity and market share. That has been the experience here in Canada, where the Canadian equivalent of Barnes & Noble, the bookstore chain Chapters, was able to open ten superstores in prime locations in 1997, while running at a loss of $2.1 million.[20]

It is here, once again, that the economy of scale comes powerfully into play. Of course some independent bookstores have held their own against the chains by adding cafés, cozy reading chairs and cooking demonstrations, but there is only so far most independents can travel down the road of experiential shopping before they experience financial stress. If, on the other hand, they do nothing to compete, single, independent stores can all too soon begin to look like poor cousins next to the brandstravaganza unfolding across the street. The end result is a retail playing field where more books are being sold, but it is becoming as difficult for small retailers to compete as it is for independent film producers to go up against the major studios on the multiplex circuit. Retail has become a vastly unequal playing field; yet another industry – like film, television or software – where you have to be huge to stay in the game. Here once again is the strange combination of a sea of product coupled with losses in real choice: the signature of our branded age.

A great deal of critical attention has been lavished on the effects of superstores on the book industry – partly because bookstore consolidation has clear implications for freedom of speech, and partly because media types tend to care more passionately about where they buy their books than where they buy their socks. In many ways, however, the bookstores are an anomaly in the superstore universe: they are multibrand stores, carrying books from thousands of book publishers, and they are primary business ventures, as opposed to being extensions, synergy schemes or 3-D billboards for brands primarily invested elsewhere. To see the animosity toward marketplace diversity most directly, one has to look not to the bookstores but to the pure branded superstores like those built by Virgin, Sony and Nike. It is there that the quest for total brand reach is revealed most starkly as the antithesis of marketplace diversity: like synergy itself, these stores seek name-brand cohesion, a safe logo cocoon apart from the warring messages of other brands.

The Virgin megastores provide perhaps the clearest displays of this kind of brand cohesion, employing various intra-brand synergies to leapfrog over entire stages of consumer choice. In the past, record labels, no matter how much money they sank into promoting new artists, were still at the mercy of record-store owners and radio- and music-video station programmers (which is why the labels got themselves into so much legal trouble in the fifties for bribing deejays). No more. Virgin's 122 megastores are wired up to be synergy machines, equipped with building-sized mural ads, listening stations for customers to sample new CDs, huge video screens, deejay booths, and satellite dishes to beam live concerts into the stores. This is par for the course in the age of the superstore, but since Virgin is also a record label, all of this technology can be harnessed to create a sense of breaking excitement about a new Virgin artist. "We'll be featuring certain artists every month. That means we play them in the store, we can do live shows via satellite from another location and we can give them store presence," says Christos Garkinos, vice president of marketing for Virgin Entertainment Group. "Think of what we can do for a developing artist."[21] More to the point, why wait around for something as temperamental as audience demand or radio play when by controlling all the variables you can create the illusion of a blockbuster success before it even happens?

That is synergy, in a nutshell. Microsoft uses the term "bundling" to describe the expanding package of core goods and services included in its Windows operating system, but bundling is simply the software industry's word for what Virgin calls synergy and Nike calls brand extensions. By bundling the Internet Explorer software within Windows, one company, because of its near monopoly in system software, has attempted to buy its way in as the exclusive portal to the Internet. What the Microsoft case so clearly demonstrates is that the moment when all the synergy wheels are turning in unison and all's right in the corporate universe is the very moment when consumer choice is at its most rigidly controlled and consumer power at its feeblest. Similarly, in the entertainment and media industries, synergy nirvana has been attained when all of a conglomerate's arms have been successfully coordinated to churn out related versions of the same product, like molded Play-Doh, into different shapes: toys, books, theme parks, magazines, television specials, movies, candies, CDs, CD-ROMs, superstores, comics and megamusicals.

Because synergy's efficiency is not measured by the success of any one "product," whether a film or a book, but rather on how well any one of those products travels through the conglomerate's multimedia channels, synergy projects tend to grow out of freewheeling meetings in which agents, clients, brand managers and producers riff on how next to leverage their flagship brands. And so the market is flooded with the mutant progeny of these brainstorming sessions: Planet Hollywood restaurants, Disney-published books written by ABC sitcom stars, Starbucks coffee-flavored beer, *Lost in Space* breath mints, a chain of airport bars modeled after the deceased set of the sitcom *Cheers*, Taco Bell–flavored Doritos...

It seems fitting, then, that Sumner Redstone calls his Viacom entertainment products "software" since there is so little that is firm at the center of these synergy schemes. By software, Redstone means branded entertainment products that he pats and molds to fit his various media holdings. "We have created a software-driven media global powerhouse," he says. "Our mission is to drive that software in every application here in the U.S. and to every region on the earth. We're going to do it." Redstone prides himself on the "absolute open communication" between his holdings. "We are coordinating various aspects of the business so each takes advantage of the opportunities provided by the other."[22]

The New Trusts: The Assault on Choice

In less enthusiastic eras than our own, other words besides "synergy" were commonly used to describe attempts to radically distort consumer offerings to benefit colluding owners; in the U.S., illegal trusts were combinations of companies that secretly agreed to fix prices while pretending to be competitive. And what else is a monopoly, after all, but synergy taken to the extreme? Markets that respond to the tyranny of size have always had a tendency toward monopoly. Which is why much of what has taken place in the entertainment industry during the last decade of merger mania would have been outlawed as recently as 1982, before President Ronald Reagan's all-out assault on U.S. anti-trust laws.

Although many media empires have long had the capacity to coordinate their holdings to promote their various offerings, most were held in check from aggressively doing so by laws designed to put up barriers between media production and media distribution. For example, U.S. regulations passed between 1948 and 1952 limited the ability of film studios to own first-run movie theaters because lawmakers feared a vertical monopoly in the industry. Though the regulations were loosened in 1974, the U.S. government was at that point in the midst of implementing a similar series of anti-trust actions designed to keep the three major U.S. television networks (CBS, ABC and NBC) from producing entertainment shows and movies for their own stations. The Justice Department charged that the three networks had an illegal monopoly that was blocking the work of outside producers. According to the Justice Department, the networks should act as programming "conduits," not programmers themselves. During this government anti-trust campaign, CBS was forced to sell off its programming arm – which, ironically, is now the synergy-obsessed Viacom. Another irony is that the interest that pushed most aggressively for the Federal Trade Commission investigation was Westinghouse Broadcasting, the same company that merged with CBS in 1995 and now enjoys all the attendant synergies between production and distribution. Full circle arrived in September 1999 when Viacom and CBS announced their merger, worth an estimated $80 billion. The companies, reunited after all these years apart, converged into an entity far more powerful than before the divorce took place.

In the seventies and early eighties, however, the majors were under so much scrutiny that according to Jack Myers, then a sales executive at CBS-TV, his network was reluctant to coordinate the sales departments of its television, radio, music and publishing divisions for cross-promotional purposes. "The idea," writes Myers, "is one that several major media companies are today attempting to follow, but in 1981 concerns about anti-trust regulations prevented direct divisional interaction."[23]

Those concerns were alleviated when, in 1983, Reagan began the not-so-gradual dismantling of U.S. anti-trust laws, first opening the door to joint research between competitors, then removing the roadblocks to giant mergers. He yanked the teeth out of the Federal Trade Commission, dramatically limiting its ability to impose fines for anticompetitive actions, cutting the staff from 345 to 134 and appointing an FTC chairman who prided himself on reducing the agency's "excessively adversarial role."[24] A former FTC regional director, Carlton Eastlake, commented in 1983 that "if the policies of the current chairman are permitted to govern for a sufficient period of time, some of our most basic liberties will be jeopardized."[25] Not only were the policies continued, but in 1986 even more dismantling legislation was passed with the explanation that American companies needed greater flexibility to compete with the Japanese. Reagan's term saw the ten biggest mergers in American history up until that point—and not one was challenged by the FTC. The number of FTC anti-trust cases against corporations dropped by half during the eighties, and the cases that were prosecuted tended to target such ultra-powerful forces as the Oklahoma Optometric Association, at the same time as Reagan stepped in personally to protect the world's ten largest airlines from a pending anti-trust investigation by his own government.[26] For the culture industries, the final piece of the new-world jigsaw fell into place in 1993 when Federal Judge Manuel Real lifted the anti-trust restrictions that had been imposed on the three major television networks in the seventies. The decision opened the door for the majors to once again produce their own prime-time entertainment shows and movies and neatly paved the way for the Disney-ABC merger.[27]

However, even in today's climate of weak anti-trust laws, some of the more audacious synergy dreams have begun to wake up the long-dormant FTC. In addition to the high-profile case against Microsoft, Barnes & Noble's bid to buy the book distributor Ingram created such rage in the book industry that the FTC was forced to set up a dedicated phone line to deal with the complaints and Barnes & Noble abandoned the bid. That these controversies are fiercest in the book and software industries is no coincidence: what is at stake is not the availability of cheap staplers, toys or non-branded towels but the free publication of, and access to, a healthy diversity of ideas. It doesn't help that the concentration of ownership among Internet, publishing and book retail companies has come hot on the heels of what must now seem an incautious level of hype about the openness and personal empowerment of the so-called Information Revolution.

In an open E-mail to Bill Gates, Andrew Shapiro, a Fellow at Harvard Law School's Center for Internet and Society, voices an opinion that has surely occurred to most thoughtful observers of modern mergers and synergy schemes. "If the whole idea of this revolution is to empower people, Bill, why are you locking up the market and restricting choices? Synergizing your way from one biz to another every month?"[28]

This contradiction represents a much larger betrayal than the usual double-speak of advertising that we are all accustomed to. What is being betrayed is no less than the central promises of the information age: the promises of choice, interactivity and increased freedom.

CHAPTER EIGHT

CORPORATE CENSORSHIP

Barricading the Branded Village

Every other week I pull something off the shelf that I don't think is of Wal-Mart quality.

– Teresa Stanton, manager of Wal-Mart's store in Cheraw, South Carolina, on the chain's practice of censoring magazines with provocative covers, in *The Wall Street Journal*, October 22, 1997

In some instances, the assault on choice has moved beyond predatory retail and monopolistic synergy schemes and become what can only be described as straightforward censorship: the active elimination and suppression of material. Most of us would define censorship as a restriction of content imposed by governments or other state institutions, or instigated – particularly in North American societies – by pressure groups for political or religious reasons. It is rapidly becoming evident, however, that this definition is drastically outdated. Although there will always be a Jesse Helms and a Church Lady to ban a Marilyn Manson concert, these little dramas are fast becoming sideshows in the context of larger threats to free expression.

Corporate censorship has everything to do with the themes of the last two chapters: media and retail companies have inflated to such bloated proportions that simple decisions about what items to stock in a store or what kind of cultural product to commission – decisions quite properly left to the discretion of business owners and culture makers – now have enormous consequences: those who make these choices have the power to reengineer the cultural landscape. When magazines are pulled from Wal-Mart's shelves by store managers, when cover art is changed on CDs to make them Kmart-friendly, or when movies are refused by Blockbuster Video because they don't conform to the chain's "family entertainment" image, these private decisions send waves through the culture industries, affecting not just what is readily available at the local big box but what gets produced in the first place.

Both Wal-Mart and Blockbuster Video have their roots in the southern U.S. Christian heartland – Blockbuster in Texas, Wal-Mart in Arkansas. Both retailers believe that being "family" stores is at the core of their financial success, the very key to their mass appeal. The model (also adopted by Kmart), is to create a one-size-fits-all family-entertainment center, where Mom and Dad can rent the latest box-office hit and the new Garth Brooks release a few steps away from where Johnny can get *Tomb Raider 2* and Melissa can co-angst with Alanis.

To protect this formula, Blockbuster, Wal-Mart, Kmart and all the large supermarket chains have a policy of refusing to carry any material that could threaten their image as a retail destination for the whole family. The one-stop-shopping recipe is simply too lucrative to risk. So magazines are rejected by Wal-Marts and supermarket chains – which together account for 55 percent of U.S. newsstand sales – for offenses ranging from too much skin on the cover girls, to articles on "His & Her Orgasms" or "Coming Out: Why I Had to Leave My Husband for Another Woman."[1] Wal-Mart's and Kmart's policy is not to stock CDs with cover art or lyrics deemed overly sexual or touching too explicitly on topics that reliably scandalize the heartland: abortion, homosexuality and satanism. Meanwhile, Blockbuster Video, which controls 25 percent of the home-video market in the U.S., carries plenty of violent and sexually explicit movies but it draws the line at films that receive an NC-17 rating, a U.S. designation meaning that nobody under seventeen can see the film, even accompanied by an adult.

To hear the chains tell it, censoring art is simply one of several services they provide to their family-oriented customers, like smiling faces and low prices. "Our customers understand our music and video merchandising decisions are a common-sense attempt to provide the type of material they might want to purchase," says Dale Ingram, Wal-Mart director of corporate relations. Blockbuster's line is: "We respect the needs of families as well as individuals."[2]

Wal-Mart can afford to be particularly zealous since entertainment products represent only a fraction of its business anyway. No one hit record or movie has the power to make a dent in Wal-Mart's bottom line, a fact that makes the retailer unafraid to stand up to the entertainment industry's best-selling artists and defend its vision of a shopping environment where power tools and hip-hop albums are sold in adjoining aisles. The most well known of these cases involved the chain's refusal to carry Nirvana's second hit album, *In Utero*, even though the band's previous album had gone quadruple platinum, because it objected to the back-cover artwork portraying fetuses. "Country artists like Vince Gill and Garth Brooks are going to sell much better for Wal-Mart than Nirvana," Wal-Mart spokesperson Trey Baker blithely said at the time.[3] Facing a projected loss of 10 percent (Wal-Mart's then share of U.S. music sales), Warner and Nirvana backed down and changed the artwork. They also changed the title of the song "Rape Me" to "Waif Me." Kmart Canada took a similar attitude to the Prodigy's 1997 release *Fat of the Land*, on the basis that the cover art and the lyrics in the songs "Smack My Bitch Up" and "Funky Shit" just wouldn't fit in at the Mart. "Our typical customer is a married working mother and we felt it was inappropriate for a family store," said manager Allen Letch.[4] Like Nirvana, the British bad boys complied with their label's subsequent request and issued a cleaned-up version.

Such censorship, in fact, has become so embedded in the production process that it is often treated as simply another stage of editing. Because of Blockbuster's policy, some major film studios have altogether stopped making films that will be rated NC-17. If a rare exception is made, the studios will cut two versions—one for the theaters, one sliced and diced for Blockbuster. What producer, after all, would be willing to forgo 25 percent of video earnings before their project is even out of the gate? As film director David Cronenberg told *The New Yorker*, "The assumption now seems to be that every movie should be watchable by a kid.... So the pressure on anyone who wants to make a grownup movie is enormous."[5]

Many magazines, including *Cosmopolitan* and *Vibe*, have taken to showing advance copies of new issues to big boxes and supermarkets before they ship them out. Why risk having to deal with the returns if the issue is deemed too risqué? "If you don't let them know in advance, they will delist the title and never carry it again," explains Dana Sacher, circulation director of *Vibe*.

"This way, they don't carry one issue, but they might carry the next one."[6]

Since bands put out a record every couple of years – not one a month – they don't have the luxury of warning Wal-Mart about a potentially contentious cover and hoping for better luck on the next release. Like film producers, record labels are instead acting preemptively, issuing two versions of the same album – one for the big boxes, bleeped, airbrushed, even missing entire songs. But while that has been the strategy for multi-platinum-selling artists like the Prodigy and Nirvana, bands with less clout often lose the opportunity to record their songs the way they intended, preempting the objections of family-values retailers by issuing only pre-sanitized versions of their work.

In large part, the complacency surrounding the Wal-Mart and Blockbuster strain of censorship occurs because most people are apt to think of corporate decisions as non-ideological. Businesses make business decisions, we tell ourselves – even when the effects of those decisions are clearly political. And when retailers dominate the market to the extent that these chains do today, their actions can't help raising questions about the effect on civil liberties and public life. As Bob Merlis, a spokesperson for Warner Brothers Records explains, these private decisions can indeed have very public effects. "If you can't buy the record then we can't sell it," he says. "And there are some places where these mass merchandisers are the only game in town."[7] So in much the same way that Wal-Mart has used its size to get cheaper prices out of suppliers, the chain is also using its heft to change the kind of art that its "suppliers" (i.e., record companies, publishers, magazine editors) provide.

Censorship in Synergy

While the instances of corporate censorship discussed so far have been a direct by-product of retail concentration, they represent only the most ham-fisted form of corporate censorship. More subtly – and perhaps more interestingly – the culture industry's wave of mergers is breeding its own blockages to free expression, a kind of censorship in synergy.

One of the reasons that producers are not standing up to puritanical retailers is that those retailers, distributors and producers are often owned, in whole or in part, by the same companies. Nowhere is this conflict of interest more in play than in the relationship between Paramount Films and Blockbuster Video. Paramount is hardly positioned to lead the charge against Blockbuster's conservative stocking policy, because if indeed such a policy is the most cost-effective way to draw the whole family into the video store, then who is Paramount to take money directly out of mutual owner Viacom's pockets? Similar conflicts arise in the aftermath of Disney's 1993 purchase of Miramax, the formerly independent film company. On the one hand, Miramax now has deep resources to throw behind commercially risky foreign films like Roberto Benigni's *Life Is Beautiful*; on the other, when the company decides whether or not to carry a politically controversial and sexually explicit work like Larry Clarke's *Kids*, it cannot avoid weighing how that decision will reflect on Disney and ABC's reputations as family programmers, with all the bowing to pressure groups that that entails.

Such potential conflicts become even more disturbing when the media holdings involved are not only producing entertainment but also news or current affairs. When newspapers, magazines, books and television stations are but one arm of a conglomerate bent on "absolute open communication" (as Sumner Redstone puts it), there is obvious potential for the conglomerate's myriad financial interests to influence the kind of journalism that is produced. Of course, newspaper publishers meddling in editorial content to further their own financial interests is as old a story as the small-town paper owner who uses the local *Herald* or *Gazette* to get his buddy elected mayor. But when the publisher is a conglomerate, its fingers are in many more pots at once. As multinational conglomerates build up their self-enclosed, self-promoting worlds, they create new and varied possibilities for conflict of interest and censorship. Such pressures range from pushing the magazine arm of the conglomerate to give a favorable review to a movie or sitcom produced by another arm of the conglomerate, to pushing an editor not to run a critical story that could hurt a merger in the works, to newspapers being asked to tiptoe around judicial or regulatory bodies that award television licenses and review anti-trust complaints. And what is emerging is that even tough-minded editors and producers who unquestioningly stand up to external calls for censorship – whether from vocal political lobbies, Wal-Mart managers or their own advertisers—are finding these intracorporate pressures much more difficult to resist.

The most publicized of the synergy-censorship cases occurred in September 1998 when ABC News killed a Disney-related story prepared by its award-winning investigative team of correspondent Brian Ross and producer Rhonda Schwartz. The story began as a broad investigation of allegations of lax security at theme parks and resorts, leading to the inadvertent hiring of sex offenders, including pedophiles, as park employees.

Because Disney was to be only one of several park owners under the microscope, Ross and Schwartz got the go-ahead on the story. After all, it wasn't the first time the team had faced the prospect of reporting on their parent company. In March 1998, ABC newsmagazine *20/20* had aired their story about widespread sweatshop labor in the U.S. territory of Saipan. Though it focused its criticism on Ralph Lauren and the Gap, the story did mention in passing that Disney was among the other American companies contracting to the offending factories.

But reporting has a life of its own and as Ross and Schwartz progressed on the theme-park investigation, they found that Disney wasn't on the periphery, but was at the center of this story. When they handed in two drafts of what had turned into a sex-and-scandal exposé of Disney World, David Westin, president of ABC News, rejected the drafts. "They didn't work," said network spokeswoman Eileen Murphy.[8] Even though Disney denies the allegations of lax security, first made in the book *Disney: The Mouse Betrayed*, and even though CEO Michael Eisner is on record saying "I would prefer ABC not cover Disney,"[9] ABC denies the story was killed because of pressure from its parent company. Murphy did say, however, that "we would generally not embark on an investigation that focused solely on Disney, for a whole variety of reasons, one of which is that whatever you come up with, positive or negative, will seem suspect."[10]

The most vocal criticism of the affair came from *Brill's Content*, the media-watch magazine founded in 1998 by Steven Brill. The publication lambasted ABC executives and journalists for their silence in the face of censorship, accusing them of caving in to their own internalized "Mouse-Ke-Fear." In his previous incarnation as founder of the Court TV cable network and *American Lawyer* magazine, Steven Brill had some firsthand experience with censorship in synergy. After selling his miniature media empire to Time Warner in 1997, Brill claims that he faced pressure on several different stories that brushed up against the octopus-like tentacles of the Time Warner/Turner media empire. In a memo excerpted in *Vanity Fair*, Brill writes that company lawyers tried to suppress a report in *American Lawyer* about a Church of Scientology lawsuit against *Time* magazine (owned by Time Warner) and asked Court TV to refrain from covering a trial involving Warner Music. He also claims to have received a request from Time Warner's chief financial officer, Richard Bressler, to "kill a story" about William Baer, the director of the Federal Trade Commission's Bureau of Competition – ironically, the very body charged with reviewing the Time Warner-Turner merger for any violation of anti-trust law.[11]

Despite the alleged meddling, all the stories in question made it to print or to air, but Brill's experience still casts a shadow over the future of press freedom inside the merged giants. Individual crusading editors and producers have always carried the flag for journalists' right to do their job, but in the present climate, for every crusader there will be many more walking on eggs for fear of losing their job. And it's not surprising that some have begun to see trouble everywhere, second-guessing the wishes of top executives in ways more creative and paranoid than the executives may even dare to imagine themselves. This is the truly insidious nature of self-censorship: it does the gag work more efficiently than an army of bullying and meddling media moguls could ever hope to accomplish.

China Chill

As we have seen in recent years, journalists, producers and editors are not only finding reason to walk carefully when dealing with judicial and regulatory bodies (not to mention theme parks), but – in the case of China – we have watched an entire country become a tiptoe zone. A wave of China-chill incidents has swept through the Western media and entertainment industries since Deng Xiaoping tentatively lifted the Communist Party monopoly on news and began slowly to open his country's borders to some censor-approved foreign media and entertainment.

Now the global culture industry faces the possibility that it is the West that may have to play by China's rules – outside as well as inside its borders. Those rules were neatly summed up in a 1992 article in *The South China Morning Post*: "Provided they do not break the law or go against party line, journalists and cultural personnel are guaranteed freedom from interference by commissars and censors."[12] And with 100 million cable subscribers expected in China by the year 2000, several cultural empire builders have already begun exercising their freedom to agree with the Chinese government.

An early incident involved Rupert Murdoch's notorious decision to drop the BBC's World Service news from the Asian version of Star TV. Chinese authorities had objected to a BBC broadcast on Mao Tse-tung, sending a clear warning about the types of reporting that will be welcome and profitable in China's wired world. More recently, HarperCollins Publishers (this book's publisher in the United Kingdom), also owned by Murdoch's News Corp, decided to drop *East and West: China, Power, and the Future of Asia*, written by Hong Kong's last British governor, Chris Patten. At issue was the possibility that the views expressed by Patten – who had called for more democracy in Hong Kong and criticized human-rights abuses in China – would enrage the Chinese government upon which Murdoch's satellite ventures are dependent. In the storm of controversy that followed, more allegations of censorship for the sake of global synergy came out of the woodwork, including one by Jonathan Mirsky, former East Asia editor for the Murdoch-owned London *Times*. He claimed that the paper "has simply decided, because of Murdoch's interests, not to cover China in a serious way."[13]

Fears of retaliation from the Chinese are not without basis. Famous for punishing media organizations that don't toe the government line and rewarding those that do, the Chinese government banned the sale and ownership of private satellite dishes in October 1993: the dishes were picking up more than ten foreign stations, including CNN, BBC and MTV. Liu Xilian, vice minister for radio, film and television, would only say, "Some of the satellite programs are suitable and some are not suitable for the normal public."[14] The Chinese government fired another salvo in December 1996 after learning of Disney's plans to release *Kundun*, a Martin Scorsese film about Tibet's Dalai Lama. "We are resolutely opposed to the making of this movie. It is intended to glorify the Dalai Lama, so it is an interference in China's internal affairs," stated Kong Min, an official at the Ministry of Radio, Film and Television.[15] When the studio went ahead with the film anyway, Beijing instituted a ban on the release of all Disney films in China, a ban that stayed in place for two years.

Since China only lets in ten foreign films a year and puts controls on their distribution, the *Kundun* incident sent a chill through the film industry, which had several other China-related projects in the works, including MGM's *Red Corner* and Sony's *Seven Years in Tibet*. To their credit, none of the studios pulled the plug on these films in progress, and in fact many in the film community rallied around Scorsese and *Kundun*. However, both MGM and Sony made official statements that attempted to depoliticize their China films, even if it meant contradicting their lead actors and directors. MGM went ahead with *Red Corner*, a movie about China's corrupt criminal justice system, starring Richard Gere, but while Gere maintained that the film is "a different angle of dealing with Tibet,"[16] MGM's worldwide marketing president, Gerry Rich, told a different story: "We're not pursuing a political agenda. We're in the business of selling entertainment." *Seven Years in Tibet* got a similar sell from Sony: "You don't want to convey that it's a movie about a political cause," a studio executive said.[17] Disney, meanwhile, finally managed to get the Chinese government to lift the ban on its films with the release of *Mulan*, a feel-good animated tale based on a 1,300-year-old legend from the Sui Dynasty. *The South China Morning Post* described the depiction of Chinese heroism and patriotism as an "olive branch" and "the most China-friendly movie Hollywood has made in years." It also served its purpose: *Mulan* flopped at the box office but it opened the door to discussions between Disney and Beijing for a planned $2 billion Disney theme park in Hong Kong.

The medium will change from a mass-produced and mass-consumed commodity to an endless feast of niches and specialties.... A new age of individualism is coming and it will bring an eruption of culture unprecedented in human history.

– George Gilder, *Life After Television*, 1990

If anything, the Western lust for access to the Chinese entertainment market has only become more intense in recent years, despite worsening relationships between the U.S. and Chinese governments over such issues as access to China's securities and telecommunications industries, more revelations of espionage and, most disastrous of all, the accidental bombing of the Chinese embassy in Belgrade during the Kosovo war. The reason, in part, is that in the past, the desire to enter China was based on projected earnings, but in 1998, those projections became a reality. James Cameron's *Titanic* broke all the records for foreign releases and earned $40 million at the box office in China, even in the midst of an economic downturn.

China chill is significant above all in what it tells us about the priorities and power wielded today by the multinationals. Financial self-interest in business is nothing new, nor is it in itself destructive. What *is* new is the reach and scope of these megacorporations' financial self-interest, and the potential global consequences, in both international and local terms. These consequences will be felt not in boisterous celebrity standoffs between such players as Rupert Murdoch, Michael Eisner, Martin Scorsese and Chris Patten, all of whom have the resources and clout to advance their positions regardless of minor setbacks. Disney and News Corp are moving swiftly ahead in China, yet Tibet remains a cause célèbre among movie stars and musicians, while Patten's book, after quickly finding another publisher, certainly sold more copies as a result of the controversy. Rather, the lasting effects, once again, will be in the self-censorship that the media conglomerates are now in a position to seed down through the ranks of their organizations. If news reporters, editors and producers have to take into account their moguls' expansionist agendas when reporting on foreign affairs, why stop at China? Wouldn't coverage of the Indonesian government's genocide in East Timor raise concerns for any multinational doing, or hoping to do, business in populous Indonesia? What if a conglomerate has deals in the works in Nigeria, Colombia or Sudan? This is a long way from the rhetoric following the fall of the Berlin Wall, when the media moguls claimed that their cultural products would carry the torch of freedom to authoritarian regimes. Not only does that mission appear to have been swiftly abandoned in favor of economic self-interest, but it seems that it may be the torch of authoritarianism that is being carried by those most determined to go global.

Copyright Bullies

After NATO's 1999 air strikes provoked Serbian "rock rallies" where teens in Chicago Bulls caps defiantly burned the American flag, few would be naive enough to reassert the tired old refrain that MTV and McDonald's are bringing peace and democracy to the world. What was crystallized in those moments when pop culture bridged the wartime divide, however, was that even if there exists no other cultural, political or linguistic common ground, Western media have made good on the promise of introducing the first truly global lexicon of imagery, music and icons. If we agree on nothing else, virtually everyone knows that Michael Jordan is the best basketball player that ever lived.

That may seem a minor achievement compared with the grand "global village" pronouncements made after the collapse of Communism, but it is an accomplishment sufficiently vast to have revolutionized both the making of art and the practicing of politics. Verbal or visual references to sitcoms, movie characters, advertising slogans and corporate logos have become the most effective tool we have to communicate across cultures – an easy and instant "click." The depth of this form of social branding came into sharp focus in March 1999 when a scandal erupted over a popular textbook used in American public schools. The Grade 6 math text was riddled with mentions and photographs of well-known brand-name products: Nike shoes, McDonald's, Gatorade. In one instance, a word problem taught students to calculate diameters by measuring an Oreo cookie. Predictably, parents' groups were furious over this milestone in the commercialization of education; here was a textbook, it seemed, with paid advertorial. But McGraw-Hill, the book's publisher, insisted that the critics had it all wrong. "You're trying to get into what people are familiar with, so they can see, hey, mathematics is in the world out there," Patricia S. Wilson, one of the book's authors, explained. The brand-name references weren't paid advertisements, she said, but an attempt to speak to students with their own references and in their own language – to speak to them, in other words, in brands.[18]

Nobody is more acutely aware of how enmeshed language and brands have become than the brand managers themselves. Cutting-edge trends in marketing theory encourage companies not to think of their brands as a series of attributes but to look at the psychosocial role they play in pop culture and in consumers' lives. Cultural anthropologist Grant McCracken teaches corporations that to understand their own brands they have to set them free. Products like Kraft Dinner, McCracken argues, take on a life of their own when they leave the store – they become pop-culture icons, vehicles for family bonding, and creatively consumed expressions of individuality.[19] The most recent chapter in this school of brand theory comes from Harvard professor Susan Fournier, whose paper, "The Consumer and the Brand: An Understanding within the Framework of Personal Relationships," encourages marketers to use a human-relationship model in conceptualizing the brand's place in society: is it a wife through an arranged marriage? A best friend or a mistress? Do customers "cheat" on their brand or are they loyal? Is the relationship a "casual friendship" or a "master/slave engagement"? As Fournier writes, "this connection is driven not by the image the brand 'contains' in the culture, but by the deep and significant psychological and socio-cultural meanings the consumer bestows on the brand in the process of meaning creation."[20]

So here we are, for better or for worse, having meaningful committed relationships with our toothpaste and co-dependencies on our conditioner. We have almost two centuries' worth of brand-name history under our collective belt, coalescing to create a sort of global pop-cultural Morse code. But there is just one catch: while we may all have the code implanted in our brains, we're not really allowed to use it. In the name of protecting the brand from dilution, artists and activists who try to engage with the brand as equal partners in their "relationships" are routinely dragged into court for violating trademark, copyright, libel or "brand disparagement" laws – easily abused statutes that form an airtight protective seal around the brand, allowing it to brand us, but prohibiting us from so much as scuffing it.

Much of this comes back to synergy. The definition of trademark in U.S. law is "any word, name, symbol, or device, or combination thereof, used ... to identify and distinguish goods from those manufactured or sold by others." Many alleged violators of copyright are not trying to sell a comparable good or pass themselves off as the real thing. As branding becomes more expansionist, however, a competitor is anyone doing anything remotely related, because anything remotely related has the potential to be a spin-off at some point in the synergistic future.

And so, when we try to communicate with each other by using the language of brands and logos, we run the very real risk of getting sued. In the U.S., copyright and trademark laws – strengthened by Ronald Reagan in the same piece of 1983 legislation that loosened anti-trust law – are being invoked in ways that have far more to do with brand control than market competition. Of course there are many uses of these laws that are absolutely crucial if artists are to have a hope of making a living, particularly with the growing ease of digital and electronic distribution. Artists need to be protected from outright thievery of their work by competitors and from its use for commercial profit without permission. I do know a few anti-copyright radicals who walk around in "All Copyright Is Theft" and "Information Wants to Be Free" T-shirts, though it seems to me that those positions are more provocative than practical. But what they do serve to highlight, if only rhetorically, is the climate of cultural and linguistic privatization being advanced through outright copyright and trademark harassment.

Copyright and trademark harassment is a massive and growing industry, and though its effects are too sweeping to fully document, here are a few random examples. Dairy Queen bakers won't squirt Bart Simpson onto frozen birthday cakes for fear of a lawsuit from Fox; in 1991, Disney forced a group of New Zealand parents in a remote country town to remove their amateur renditions of Pluto and Donald Duck from a playground mural; and Barney has been breaking up children's birthday parties across the U.S., claiming that any parent caught dressed in a purple dinosaur suit is violating its trademark. The Lyons Group, which owns the Barney character, "has sent 1,000 letters to shop owners" renting or selling the offending costumes. "They can have a dinosaur costume. It's when it's a purple dinosaur that it's illegal, and it doesn't matter what shade of purple, either," says Susan Elsner Furman, Lyons' spokesperson.[21]

McDonald's, meanwhile, continues busily to harass small shopkeepers and restaurateurs of Scottish descent for that nationality's uncompetitive predisposition toward the Mc prefix on its surnames. The company sued the McAllan's sausage stand in Denmark; the Scottish-themed sandwich shop McMunchies in Buckinghamshire; went after Elizabeth McCaughey's McCoffee shop in the San Francisco Bay Area; and waged a twenty-six-year battle against a man named Ronald McDonald whose McDonald's Family Restaurant in a tiny town in Illinois had been around since 1956.

These types of cases may seem trivial, but the same aggressive ownership rules apply to artists and cultural producers who are attempting to comment on our shared branded world. Increasingly, musicians are sued not only for sampling, but for attempting to sing about a patented common dream. That's what happened to the San Francisco "audio-collage" band Negativland when it called one of its albums *U2*, and sampled out-takes from Casey Kasem's *American Top 40* radio show. It happened, also, to Toronto avant-garde musician John Oswald when he used his "plunderphonics" method to remix Michael Jackson's song "Bad" on a 1989 album that he distributed free. Negativland was sued successfully by U2's label, Island Records, and Jackson's label, CBS Records, sued Oswald for copyright violation. As part of the settlement Oswald had to hand over all the CDs to be destroyed.

Artists will always make art by reconfiguring our shared cultural languages and references, but as those shared experiences shift from firsthand to mediated, and the most powerful political forces in our society are as likely to be multinational corporations as politicians, a new set of issues emerges that once again raises serious questions about out-of-date definitions of freedom of expression in a branded culture. In this context, telling video artists that they can't use old car commercials, or musicians that they can't sample or distort lyrics, is like banning the guitar or telling a painter he can't use red. The underlying message is that culture is something that happens to you. You buy it at the Virgin Megastore or Toys 'R' Us and rent it at Blockbuster Video. It is not something in which you participate, or to which you have the right to respond.

The rules of this one-way dialogue went unchallenged for a long time, mostly because until the eighties, copyright and trademark cases were largely between corporate competitors suing each other for infringing on their market share. Artists like REM, the Clash, Dire Straits and k.d. lang were free to sing about such trademarked products as Orange Crush, Cadillacs, MTV and *Chatelaine* magazine, respectively. Moreover, the average consumer didn't have the means to cut and click into mass-produced culture and incorporate it into something new of their own – a zine, a High-8 video or an electronic recording. It wasn't until scanners, cheap photocopiers, digital editing machines and computer programs like Photoshop appeared on the market as fairly inexpensive consumer goods that copyright and trademark law became a concern for independent culture-makers assembling their own basement publications, Web sites and recordings. "I think that culture has always cyclically reiterated itself.... Technology makes it possible to have access to and easily manipulate and store information from distant places and times," says audio pirate Steev Hise. "People will do what they can do."[22]

Doing what he could do is what produced John Oswald's plunderphonics method. As Oswald explains, it grew out of the fact that he had access to technology that enabled him to listen to records at different speeds. "I was doing a kind of manipulative listening in fairly complex ways, and as my interactive listening habits grew more complex, I began to think of ways to preserve them for other people to hear."[23]

What most bothers Oswald and other artists like him is not that their work is illegal – it's that it is illegal only for some artists. When Beck, a major-label artist, makes an album parked with hundreds of samples, Warner Music clears the rights to each and every piece of the audio collage and the work is lauded for capturing the media-saturated, multi-referenced sounds of our age. But when independent artists do the same thing, trying to cut and paste together art from their branded lives and make good on some of the info-age hype about DIY culture, it's criminalized – defined as theft, not art. This was the point made by the musicians on the 1998 *Deconstructing Beck* underground CD, produced entirely by electronically recontextualizing Beck's already recontextualized sounds. Their point was simple: if Beck could do it, why shouldn't they? Right on cue, Beck's label sent out threatening lawyers' letters that quieted down abruptly when the musicians made it clear that they were gunning for a media fight. Their point, however, had been made: the prevailing formula for copyright and trademark enforcement is a turf war over who is going to get to make art with the new technologies. And it seems that if you're not on the team of a company large enough to control a significant part of the playing field, and can't afford your very own team of lawyers, you don't get to play.

This is the lesson, it would seem, of Mattel's copyright suit against the Danish pop band Aqua and its label MCA. Mattel charged that the band's hit song "Barbie Girl" – which contains lyrics like "Kiss me here, touch there, hanky panky" – wrongfully sexualizes its wholesome blonde. Mattel went to court in September 1997 charging Aqua with trademark infringement and unfair competition. The toy manufacturer asked for damages and for the album to be removed from stores and destroyed. Aqua won the dispute but not because its case was any stronger than Negativland's or John Oswald's (it might have been weaker) but rather because, unlike these independent musicians, Aqua had behind it MCA's team of lawyers, willing to fight tooth and nail to make sure the hit single was allowed to stay on the charts and the shelves. It was, like Jordan versus Nike, a battle of the brands.

Although the music itself is pure cotton candy, the Aqua case is worth considering because it pushed the envelope on copyright bullying, introducing the idea that musicians must now be wary not only of direct sampling but of so much as mentioning any trademarked products. It also highlighted the uncomfortable tension between the expansive logic of branding – the corporate desire for full cultural integration – and the petty logic of these legal crusades. Who if not Barbie is as much cultural symbol as product? Barbie, after all, is the archetypal space invader, a cultural imperialist in pink. She is the one who paints entire towns fuchsia to celebrate "Barbie Month." She is the Zen mistress who for the past four decades has insisted on being everything to young girls – doctor, bimbo, teenager, career girl, Unicef ambassador....

The people at Mattel weren't interested in talking about Barbie the cultural icon when they launched the Aqua suit, however. "This is a business issue, not a freedom of speech issue," a Mattel spokesperson told *Billboard*. "This is a $2 billion company, and we don't want it messed around with, and situations like this gradually lead to brand erosion."[24] Barbie is a for-profit enterprise, it's true. And brands such as Barbie, Aspirin, Kleenex, Coca-Cola and Hoover have always walked a fine line between wanting to be ubiquitous but not wanting to become so closely associated with a product category that the brand name itself becomes generic – as easily invoked to sell a competing brand as their own.

But while this fight against erosion seems reasonable in the context of brands competing with each other, it's a different matter when looked at through the lens of aggressive lifestyle branding – and from that perspective, a re-examination of the public's right to respond to these "private" images seems urgently required. Mattel, for instance, has reaped huge profits by encouraging young girls to build elaborate dream lives around their doll, but it still wants that relationship to be a monologue. The toy company, which boasts of having "as many as 100 different [trademark] investigations going on at any time throughout the world,"[25] is almost comically aggressive in protecting this formula. Among other feats, its lawyers have shut down a riot girl zine called *Hey There, Barbie Girl!* and successfully blocked the distribution of Todd Haynes's documentary *Superstar: The Karen Carpenter Story*, a reenactment of the life of the anorexic pop star using Barbies as puppets (legal pressure also came from Carpenter's family).

It seems fitting that Aqua member Sren Rasted says he got the idea for the song "Barbie Girl" after visiting "an art-museum exhibition for kids on Barbie."[26] In an effort to have its star doll inaugurated as a cultural artifact, Mattel has in recent years been mounting traveling exhibits of old Barbies, which claim to tell the history of America through "America's favorite doll." Some of these shows are put on directly by Mattel, others by private collectors working closely with the company, a relationship that ensures that unpleasant chapters in Barbie's history – the feminist backlash against the doll, say, or Barbie the cigarette model – are mysteriously absent. There is no question that Barbie, like a handful of other classic brands, is an icon and artifact in addition to being a children's toy. But Mattel – and Coca-Cola, Disney, Levi's and the other brands that have launched similar self-curatorial projects – wants to be treated as an important pop-culture artifact at the same time as it seeks to maintain complete proprietary control over its historical and cultural legacy. It's a process that ultimately gags cultural criticism, using copyright and trademark laws as effective tools to silence all unwanted attention. The editors of *Miller's*, a magazine for Barbie collectors, are convinced that Mattel targeted them with a copyright suit because, unlike the uncritical collectors mounting Barbie art shows, the publication criticized Mattel's high prices and ran old photographs of Barbie posing with packs of Virginia Slims cigarettes. Mattel is by no means unique in its employment of this strategy. Kmart, for instance, shut down the Kmart Sucks Web site mounted by a disgruntled employee, not by using libel or defamation law, which would have required that the chain prove the allegations were false, but by suing for unauthorized use of its trademark K.

When copyright or trademark law can't be invoked to prevent an unwanted brand portrayal, many corporations do rely on libel and defamation law to keep their practices from being debated in the public realm. The high-profile "McLibel" case in Britain, in which the fast-food chain sued two environmentalists for libel, was one such attempt. (The issue will be discussed in detail in Chapter 16.) Regardless of which legal tactic they choose, the impossibly contradictory message sent out by the producers of these iconic products is the same: we want our brands to be the air you breathe in – but don't dare exhale.

The more corporations like Mattel and McDonald's succeed in their goal of building self-enclosed branded worlds, the more culturally asphyxiating that demand may become. Copyright and trademark laws are perfectly justifiable if the brand in question is just a brand, but increasingly that's like saying that Wal-Mart is just a store. The brand in question may well represent a corporation with a budget larger than that of many countries, and a logo that is among the world's most transcendent symbols, one that has aggressively sought to replace the role played by art and media. When we lack the ability to talk back to entities that are culturally and politically powerful, the very foundations of free speech and democratic society are called into question.

Privatizing the Town Square

There is an unavoidable parallel between the privatization of language and cultural discourse occurring through copyright and trademark bullying, and the privatization of public space taking place through the proliferation of superstores, theme-park malls and branded villages like Celebration, Florida. Just as privately owned words and images are being adopted as a de facto international shorthand, so too are private branded enclaves becoming de facto town squares – once again, with troubling implications for civil liberties.

The conflation of shopping and entertainment found at the superstores and theme-park malls has created a vast gray area of pseudo-public private space. Politicians, police, social workers and even religious leaders all recognize that malls have become the modern town square. But unlike the old town squares, which were and still are sites for community discussion, protests and political rallies, the only type of speech that is welcome here is marketing and other consumer patter. Peaceful protestors are routinely thrown out by mall security guards for interfering with shopping, and even picket lines are illegal inside these enclosures. The town-square concept has recently been picked up by the superstores, many of which now claim that they too are providing public space. "Essentially, we want people to use the store as a meeting place. A place where people can get their fix of pop culture and hang out for a while. It's not just a place to shop, it's a place to be," said Christos Garkinos, vice president of marketing for the Virgin Entertainment Group, on the occasion of the opening of Vancouver's 40,000-square-foot Virgin Megastore.[27]

The building in which Virgin set up shop previously housed the public library, an apt metaphor for the way brand expansion is altering the way we congregate, not just as shoppers but as citizens. Barnes & Noble describes its superstores as "a center for cultural events and gatherings," and some of these stores, particularly in the United States, do play the part well, housing everything from pop concerts to poetry readings.[28] Book superstores, with their plush chairs, faux fireplaces, book clubs and coffee bars, have slowly come to replace libraries and university lecture halls as locales of choice for author readings on the book-tour circuit. But, as with the ban on protests in malls, a different set of rules applies in these quasi-public spaces. For example, when promoting his book, *Downsize This!*, filmmaker Michael Moore was confronted with a picket line outside a Philadelphia outlet of Borders bookstore, where he was scheduled to read. He told the store he wouldn't go in unless the striking employees were allowed inside and given some time at the microphone. The manager complied, but Moore's future Borders readings were canceled. "I couldn't believe I was being censored in a bookstore,"

Moore wrote of the incident.[29]

As good as the superstores are at dressing up like town halls, no one mimics public space like America Online, the virtual community of chat rooms, message boards and discussion groups where there are no customers – only netizens. But AOL subscribers have, in the past two years, learned some harsh lessons about their virtual community and the limits on the rights of its citizens. AOL, though part of the publicly owned Internet, is a sort of privatized mini-Net inside the larger Web. The company collects the toll on the way in and, like mall security guards, it can set the rules while customers are inside its domain. That was the message that echoed through the virtual commons when AOL's so-called Community Action Team began deleting messages from discussion groups deemed harassing, profane, embarrassing or just "unwanted." In addition to screening messages, the team also has the right to forbid virtual sparring partners from ever trading messages again and to suspend or expel repeat offenders from the service and from access to their own E-mail accounts. Some lists – like a particularly heated one on Irish politics – have been shut down for extended "cooling-off" periods.

The company's rationale is strikingly similar to Wal-Mart's shelving policy (and Blockbuster's video rental policy). Katherine Boursecnik, AOL's vice president for network programming, told *The New York Times*, "We are a service that prides ourselves on having a wide-ranging appeal to a wide range of individuals. But at the same time we're also a family service."[30] While few contest that on-line discussion is a breeding ground for all sorts of antisocial behavior (from chronic overposting to sexual harassment), the sheer power that the company has to regulate the tone and content of on-line discourse has raised the specter of the "AOL Thought Police."[31] The issue, as with Wal-Mart, is AOL's commanding market share: in mid-1999 it had 15 million subscribers – 43 percent of the U.S. Internet service market. Its closest competitor, Microsoft, had only 6.4 percent.[32]

Complicating matters further, Internet discussion is a hybrid medium, falling somewhere between making a personal telephone call and watching cable television. So while its subscribers may view AOL as a phone company, with no more right to intercept their communications than AT&T has to disconnect unsavory phone discussions, the company has another view entirely.

"Virtual community" babble aside, AOL is, above all, a branded media empire over which it exercises as much control as Disney does over the fence colors in Celebration, Florida.

It seems that no matter how successfully the private sphere emulates or even enhances the look and feel of public space, the restrictive tendencies of privatization have a way of peeking through. And the same applies not only to corporate-owned space, like AOL or Virgin Megastores, but even to publicly owned space that is sponsored or branded. That point was graphically made in Toronto in 1997 when antitobacco activists were forcibly removed from the open-air du Maurier Downtown Jazz Festival, just as student protestors had been removed from the du Maurier Tennis Open on their campus. The irony was that the festival happened to be taking place in the city's actual town square – Nathan Phillips Square, just in front of Toronto City Hall. The protestors learned that while the square may be as public a space as one can find, it becomes, during jazz festival week, the property of the tobacco sponsor. No critical material was permitted on the premises.

When any space is bought, even if only temporarily, it changes to fit its sponsors. And the more previously public spaces are sold to corporations or branded by them, the more we as citizens are forced to play by corporate rules to access our own culture. Does this mean that free speech is dead? Of course not, but it does call to mind Noam Chomsky's view that "freedom without opportunity is a devil's gift."[33] In a context of media and marketing overload, meaningful opportunities to express our freedom – at levels loud enough to break through the barrage of commercial sound effects and disturb the corporate landlords – are disappearing fast around us. Yes, dissenting voices have their Web pages, zines, posters, picket signs and independent newspapers, as well as plenty of cracks in the corporate armor to exploit – and as we will see in Part IV, they are exploiting them as never before. But when corporate speech is increasingly expressed in multiplatform synergy and in ever more extraordinary displays of branded "meaning," popular speech comes to look like the tiny independent retailer next to the superstore. As consumer advocate Ralph Nader puts it: "There is a decibel-level quality to the exercise of our first amendment rights."[34]

Perhaps the most disturbing manifestation of corporate censorship takes place when the space that is sold is not a place but a person. As we have seen, the high-stakes sponsorship agreements in the sports world first exerted their influence by deciding what logo athletes wore and what teams they played on. Now that control has expanded to what political views they may hold publicly. Daring political stands like Muhammad Ali's opposition to the Vietnam War have long since been replaced by the soft-drink radicalism of NBA cross-dresser Dennis Rodman, as sponsors push their athletes to be little more than billboards with attitude. As Michael Jordan once commented, "Republicans buy sneakers too."

Canadian sprinter Donovan Bailey learned that lesson the hard way. Days before he won the Olympic race that would make him the fastest man alive, Bailey came under attack for telling *Sports Illustrated* that Canadian society "is as blatantly racist as the United States." Adidas, horrified that its branded property would risk alienating so many white sneaker buyers with such an unpopular opinion, rushed in to shut Bailey up. Adidas vice president Doug Hayes told *The Globe and Mail* that the comments "have nothing to do with Donovan the athlete or the Donovan we know"[35] – seemingly attributing the views to a fictional alter-athlete who had possessed Bailey temporarily.

A similar case of branding censorship involved British soccer star Robbie Fowler. After the twenty-one-year-old scored the second goal against the Norwegian team Brann Bergen in March 1997, Fowler turned to the crowd, pulled up his official jersey, and revealed a red political T-shirt: "500 Liverpool dockers sacked since 1995," the shirt said. The dockers have been on strike for years, fighting hundreds of layoffs and the shift to contract work. Fowler, a Liverpool boy himself, decided to publicize the cause when the world was watching. Ingenuously he commented: "I thought it would be just a simple statement."[36]

He was, of course, mistaken. The Liverpool Football Club, which collects the toll on the branded messages that appear on the players' official jerseys, raced in to stem any copycat actions. "We will be pointing out to all our players that comments on matters outside football are not acceptable on the field of play," the club said in a hastily issued statement.[37] And just to make extra sure that the only message on the athletes' shirts would be from Umbro or Adidas, the European football governing body UEFA followed up by slapping Fowler with a fine of 2,000 Swiss francs.

There was yet another twist in this branded tale. The shirt Fowler revealed didn't bear just any political slogan, it was also an ad bust: in a not-so-subtle subversion of a ubiquitous brand, the letters "c" and "k" in the word "dockers" had been enlarged and designed to look like Calvin Klein's logo: docKers. When photographs of the T-shirt were splashed all over British newspapers, the designer threatened to sue for trademark violation.

When piled on together, such examples give a picture of corporate space as a fascist state where we all salute the logo and have little opportunity for criticism because our newspapers, television stations, Internet servers, streets and retail spaces are all controlled by multinational corporate interests. And considering the speed with which these trends are developing, we clearly have good reason for alarm. But a word of caution: we may be able to see a not-so-brave new world on the horizon, but that doesn't mean we are already living in Huxley's nightmare.

In drawing up octopus-like charts of corporate ownership structures and quoting CEOs on their dreams of world domination, we may easily lose sight of the fact that censorship is not nearly as absolute as many a newly converted Noam Chomsky acolyte might like to believe. Instead of an airtight formula, it is a steady trend, clearly intensified by synergy and the mounting stakes of brand-name protection, but riddled with exceptions. It's true, for example, that Viacom is coating the world in bubble gum through its Blockbuster and MTV holdings, but Viacom-owned Simon & Schuster has published some of the best critiques of unregulated economic globalization: Richard J. Barnet and John Cavanagh's *Global Dreams* and William Greider's *One World, Ready or Not*, among others. NBC and Fox did, however briefly, run Michael Moore's series *TV Nation*, which gleefully went after advertisers and even targeted NBC's parent company, General Electric. And while Disney's purchase of Miramax inspired dark foreboding about the future of independent film, it was Miramax that distributed Moore's anticorporate documentary *The Big One* – a film based on his similarly critical book, published by Random House, now owned by Bertelsmann. As, I hope, the book you are holding helps to prove, there is clearly still room for corporate critiques within the media giants.

In a sense, the shift that is taking place is at once less totalitarian and more dangerous. We haven't lost the possibility for non-synergistic art, and serious critical work has a greater potential to reach wide audiences at this time than ever before in the history of art and culture. But we are losing the spaces in which the noncorporate-minded can flourish – those spaces are there, but they are shrinking as the captains of the culture industry become more enraptured by the dream of global cross-promotions. Much of this is a matter of simple economics: there are limited numbers of movies, books, magazine articles and programming hours that can be economically produced, published, broadcast, etc., and the window for the ones that don't fit into the reigning corporate strategy narrows with every merger and consolidation.

There is a chance, however, that the current mania for synergy will collapse under the weight of its unfulfilled promises. Already, Blockbuster has become a dead weight around Viacom's debt-ridden neck. The stock-market analysts blame "the quality of products coming through its stores"[38] – and it probably doesn't help that the chain has had to devote entire wings of its stores to showcasing some thirty-four copies of Kevin Kline's unwatchable *In & Out* (or some other Paramount flop) because the folks at Viacom were determined to make back some of the millions they lost in theaters. And after its "eatertainment" outlets hemorrhaged money for two years, Planet Hollywood announced in August 1999 that it would file for bankruptcy protection. Another synergy scheme that looked foolproof on paper was the 1998 release of *Godzilla*. Sony thought it had its blockbuster status sewn up: it had a Madison Square Garden premiere, a made-for-Toys 'R' Us star, a $60 million marketing budget orchestrating a year-long "teaser" campaign, and a heavy-handed legal team cracking down on all unwanted publicity on the Internet. Most important, thanks to Sony's newly consolidated movie theater holdings, the movie played on more screens than any film ever before: on launch day, 20 percent of all U.S. movie theater screens were playing *Godzilla*. Yet none of this could compensate for the simple fact that nearly everyone who saw *Godzilla* warned their friends to stay away, and they did, in droves.[39]

Even branding evangelist Tom Peters acknowledges that there is such a thing as too much brand, and impossible though it is to predict when we will reach that point, when we pass it, it will be unmistakable. "How much is enough?" asks Peters. "Nobody knows for sure. It's pure art. Leverage is good. Too much leverage is bad."[40] MTV founder Tom Freston, the man who made marketing history by turning a television station into a brand, admitted in June 1998 that "you can beat a brand to death."[41]

Indeed, by early 1998, Wall Street was declaring the unthinkable: Nike had outswooshed itself; its ubiquity had ceased to be a branding success story and had become a liability. "Nike's biggest challenge is itself. They need to come up with another identity that they can still say, 'This is Nike,' but it's something beyond the swoosh," Josie Esquivel, a stock analyst with Morgan Stanley told *The New York Times*.[42]

Nike has attempted to respond to this challenge, as we shall see. But if such a backlash is possible against a single brand, then perhaps it's conceivable that a similar phenomenon can apply equally to the act of branding as a whole: that after a certain amount of branding mania is stamped on a culture, those of us who have been branded – by Nike, Wal-Mart, Hilfiger, Microsoft, Disney, Starbucks, *et al.* – will begin to turn not just against these specific logos, but also against the control that corporate power as a whole exerts over our spaces and choices. Maybe there is a moment when the idea of branding reaches a saturation point and the backlash is directed not at a product that suddenly finds itself on the wrong side of a fad but at the multinationals behind the brands.

There is some evidence that this process is already under way. As we will see in Part IV, "No Logo," communities around the world, and at various generational levels, are no longer being blinded by the brands' shiny promises of newness and of endless selection. Instead of swinging open their doors, they are organizing at community levels to block the arrival of big-box retailers; they are participating in street-level campaigns against Nike's Third World labor practices and Shell Oil's human-rights record. They are launching movements, like Britain's Reclaim the Streets, to regain some fleeting public control over public space; and they are supporting anti-trust actions against companies such as Microsoft. Given the relative suddenness of the backlash, this wave of anticorporate hostility is understandably taking its targets by surprise. "A few months ago, everyone I met seemed to think that working for Microsoft was a pretty cool thing to do. Now, strangers treat us like we work for Philip Morris," wrote *Slate* columnist Jacob Weisberg. The bewildered sentiment is shared by multinational employees across many sectors. "I don't know how we are offending people," said Starbucks regional marketing director Donna Peterson in May 1999. "But sometimes it seems we are."[43] And Royal Dutch/Shell head Mark Moody-Stuart told *Fortune* magazine, "Previously, if you went to your golf club or church and said, 'I work for Shell,' you'd get a warm glow. In some parts of the world that changed a bit." And (as we will see in the examination of the Shell boycott in Chapter 16), that in itself is a bit of an understatement.

Mounting disillusionment in the face of the forces described here in "No Space" and "No Choice" is not, however, sufficiently widespread or deep to spark a genuine backlash against the power of the brands. In all likelihood, resentment at invasive advertising, the corporate takeover of public space, and monopolistic business practices would have festered as little more than run-of-the-mill cynicism had many of the same companies gobbling up both space and choice not decided simultaneously to bankroll their innovative branding forays by slashing jobs. It is this essential economic, human concern that has been a major force in contributing to the rise in anticorporate activism: No Good Jobs.

NO JOBS

EMPLOYEES
TO DRY

CHAPTER NINE

THE DISCARDED FACTORY

Degraded Production in the Age of the Superbrand

Our strategic plan in North America is to focus intensely on brand management, marketing and product design as a means to meet the casual clothing wants and needs of consumers. Shifting a significant portion of our manufacturing from the U.S. and Canadian markets to contractors throughout the world will give the company greater flexibility to allocate resources and capital to its brands. These steps are crucial if we are to remain competitive.

– John Ermatinger, president of Levi Strauss Americas division, explains the company's decision to shut down twenty-two plants and lay off 13,000 North American workers between November 1997 and February 1999

Many brand-name multinationals, as we have seen, are in the process of transcending the need to identify with their earthbound products. They dream instead about their brands' deep inner meanings – the way they capture the spirit of individuality, athleticism, wilderness or community. In this context of strut over stuff, marketing departments charged with the managing of brand identities have begun to see their work as something that occurs not in conjunction with factory production but in direct competition with it. "Products are made in the factory," says Walter Landor, president of the Landor branding agency, "but brands are made in the mind."[1] Peter Schweitzer, president of the advertising giant J. Walter Thompson, reiterates the same thought: "The difference between products and brands is fundamental. A product is something that is made in a factory; a brand is something that is bought by a customer."[2] Savvy ad agencies have all moved away from the idea that they are flogging a product made by someone else, and have come to think of themselves instead as brand factories, hammering out what is of true value: the idea, the lifestyle, the attitude. Brand builders are the new primary producers in our so-called knowledge economy.

This novel idea has done more than bring us cutting-edge ad campaigns, ecclesiastic superstores and utopian corporate campuses. It is changing the very face of global employment. After establishing the "soul" of their corporations, the superbrand companies have gone on to rid themselves of their cumbersome bodies, and there is nothing that seems more cumbersome, more loathsomely corporeal, than the factories that produce their products. The reason for this shift is simple: building a superbrand is an extraordinarily costly project, needing constant managing, tending and replenishing. Most of all, superbrands need lots of space on which to stamp their logos. For a business to be cost-effective, however, there is a finite amount of money it can spend on all of its expenses – materials, manufacturing, overhead *and* branding – before retail prices on its products shoot up too high. After the multimillion-dollar sponsorships have been signed, and the cool hunters and marketing mavens have received their checks, there may not be all that much money left over. So it becomes, as always, a matter of priorities; but those priorities are changing. As Hector Liang, former chairman of United Biscuits, has explained: "Machines wear out. Cars rust. People die. But what lives on are the brands."[3]

According to this logic, corporations should not expend their finite resources on factories that will demand physical upkeep, on machines that will corrode or on employees who will certainly age and die. Instead, they should concentrate those resources in the virtual brick and mortar used to build their brands; that is, on sponsorships, packaging, expansion and advertising. They should also spend them on synergies: on buying up distribution and retail channels to get their brands to the people.

This slow but decisive shift in corporate priorities has left yesterday's non-virtual producers – the factory workers and craftspeople – in a precarious position. The lavish spending in the 1990s on marketing, mergers and brand extensions has been matched by a never-before-seen resistance to investing in production facilities and labor. Companies that were traditionally satisfied with a 100 percent markup between the cost of factory production and the retail price have been scouring the globe for factories that can make their products so inexpensively that the markup is closer to 400 percent.[4] And as a 1997 UN report notes, even in countries where wages were already low, labor costs are getting a shrinking slice of corporate budgets. "In four developing countries out of five, the share of wages in manufacturing value-added today is considerably below what it was in the 1970s and early 1980s."[5] The timing of these trends reflects not only branding's status as the perceived economic cure-all, but also a corresponding devaluation of the production process and of producers in general. Branding, in other words, has been hogging all the "value-added."

When the actual manufacturing process is so devalued, it stands to reason that the people doing the work of production are likely to be treated like detritus – the stuff left behind. The idea has a certain symmetry: ever since mass production created the need for branding in the first place, its role has slowly been expanding in importance until, more than a century and a half after the Industrial Revolution, it occurred to these companies that maybe branding could replace production entirely. As tennis pro Andre Agassi said in a 1992 Canon camera commercial, "Image is everything."

Agassi may have been pitching for Canon at the time but he is first and foremost a member of Team Nike, the company that pioneered the business philosophy of no-limits spending on branding, coupled with a near-total divestment of the contract workers that make its shoes in tucked-away factories. As Phil Knight has said, "There is no value in making things any more. The value is added by careful research, by innovation and by marketing."[6] For Phil Knight, production is not the building block of his branded empire, but is instead a tedious, marginal chore.

Which is why many companies now bypass production completely. Instead of making the products themselves, in their own factories, they "source" them, much as corporations in the natural-resource industries source uranium, copper or logs. They close existing factories, shifting to contracted-out, mostly offshore, manufacturing. And as the old jobs fly offshore, something else is flying away with them: the old-fashioned idea that a manufacturer is responsible for its own workforce. Disney spokesman Ken

Green gave an indication of the depth of this shift when he became publicly frustrated that his company was being taken to task for the desperate conditions in a Haitian factory that produces Disney clothes. "We don't employ anyone in Haiti," he said, referring to the fact that the factory is owned by a contractor. "With the newsprint you use, do you have any idea of the labour conditions involved to produce it?" Green demanded of Cathy Majtenyi of the *Catholic Register*.[7]

From El Paso to Beijing, San Francisco to Jakarta, Munich to Tijuana, the global brands are sloughing the responsibility of production onto their contractors; they just tell them to make the damn thing, and make it cheap, so there's lots of money left over for branding. Make it *really* cheap.

Exporting the Nike Model

Nike, which began as an import/export scheme of made-in-Japan running shoes and does not own any of its factories, has become a prototype for the product-free brand. Inspired by the swoosh's staggering success, many more traditionally run companies ("vertically integrated," as the phrase goes) are busy imitating Nike's model, not only copying the company's marketing approach, as we saw in "No Space," but also its on-the-cheap outsourced production structure. In the mid-nineties, for instance, the Vans running-shoe company pulled up stakes in the old-fashioned realm of manufacturing and converted to the Nike way. In a prospectus for an initial public stock offering, the company lays out how it "recently repositioned itself from a domestic manufacturer to a market-driven company" by sponsoring hundreds of athletes as well as high-profile extreme sporting events such as the Vans Warped Tour. The company's "expenditure of significant funds to create consumer demand" was financed by closing an existing factory in California and contracting production in South Korea to "third party manufacturers."[8]

Adidas followed a similar trajectory, turning over its operation in 1993 to Robert Louis-Dreyfus, formerly a chief executive at advertising giant Saatchi & Saatchi. Announcing that he wanted to capture the heart of the "global teenager," Louis-Dreyfus promptly shut down the company-owned factories in Germany, and moved to contracting-out in Asia.[9] Freed from the chains of production, the company had newfound time and money to create a Nike-style brand image. "We closed down everything," Adidas spokesperson Peter Csanadi says proudly. "We only kept one small factory which is our global technology centre and makes about 1 percent of total output."[10] (See Table 9.1, Appendix, page 473.)

Though they don't draw the headlines they once did, more factory closures are announced in North America and Europe each week – 45,000 U.S. apparel workers lost their jobs in 1997 alone.[11] That sector's job-flight patterns have been equally dramatic around the globe. (See Table 9.2, Appendix, page 475.) Though plant closures themselves have barely slowed down since the darkest days of the late-eighties/early-nineties recession, there has been a marked shift in the reason given for these "reorganizations." Mass layoffs were previously presented as an unfortunate necessity, tied to disappointing company performance. Today they are simply savvy shifts in corporate strategy, a "strategic redirection," to use the Vans term. More and more, these layoffs are announced in conjunction with pledges to increase revenue through advertising spending, with executives vowing to refocus on the needs of their brands, as opposed to the needs of their workers.

Consider the case of Sara Lee Corp., an old-style conglomerate that encompasses not only its frozen-food namesake but also such "unintegrated" brands as Hanes underwear, Wonderbra, Coach leather goods, Champion sports apparel, Kiwi shoe polish and Ball Park Franks. Despite the fact that Sara Lee enjoyed solid growth, healthy profits, good stock return and no debt, by the mid-nineties Wall Street had become disenchanted with the company and was undervaluing its stock. Its profits had risen 10 percent in the 1996–97 fiscal year, hitting $1 billion, but Wall Street, as we have seen, is guided by spiritual goals as well as economic ones.[12] And Sara Lee, driven by the corporeal stuff of real-world products, as opposed to the sleek ideas of brand identity, was simply out of economic fashion. "Lumpy-object purveyors," as Tom Peters might say.[13]

To correct the situation, in September 1997 the company announced a $1.6 billion restructuring plan to get out of the "stuff" business by purging its manufacturing base. Thirteen of its factories, beginning with yarn and textile plants, would be sold to contractors who would become Sara Lee's suppliers. The company would be able to dip into the money saved to double its ad spending. "It's passé for us to be as vertically integrated as we were," explained Sara Lee CEO John H. Bryan.[14] Wall Street and the business press loved the new marketing-driven Sara Lee, rewarding the company with a 15 percent jump in stock price and flattering profiles of its bold and imaginative CEO. "Bryan's shift away from manufacturing to focus on brand marketing recognizes that the future belongs to companies – like Coca-Cola Co. – that own little but sell much," enthused one article in *Business Week.*[15] Even more telling was the analogy chosen by *Crain's Chicago Business*: "Sara Lee's goal is to become more like Oregon-based Nike Inc., which outsources its manufacturing and focuses primarily on product development and brand management."[16]

In November 1997, Levi Strauss announced a similarly motivated shake-up. Company revenue had dropped between 1996 and 1997, from $7.1 billion to $6.8 billion. But a 4 percent dip hardly seems to explain the company's decision to shut eleven plants. The closures resulted in 6,395 workers being laid off, one-third of its already downsized North American workforce. In this process, the company shut down three of its four factories in El Paso, Texas, a city where Levi's was the single largest private employer. Still unsatisfied with the results, the following year Levi's announced another round of closures in Europe and North America. Eleven more of its North American factories would be shut down and the total toll of laid-off workers rose to 16,310 in only two years.[17]

John Ermatinger, president of Levi's Americas division, had a familiar explanation. "Our strategic plan in North America is to focus intensely on brand management, marketing and product design as a means to meet the casual clothing wants and needs of consumers," he said.[18] Levi's chairman, Robert Haas, who on the same day received an award from the UN for making life better for his employees, told *The Wall Street Journal* that the closures reflected not just "overcapacity" but also "our own desire to refocus marketing, to inject more quality and distinctiveness into the brand."[19] In 1997, this quality and distinctiveness came in the form of a particularly funky international ad campaign rumored to have cost $90 million, Levi's most expensive campaign ever, and more than the company spent advertising the brand in all of 1996.

"This Is Not a Job-Flight Story"

In explaining the plant closures as a decision to turn Levi's into "a marketing company," Robert Haas was careful to tell the press that the jobs that were eliminated were not "leaving," they were just sort of evaporating. "This is not a job-flight story," he said after the first round of layoffs. The statement is technically true. Seeing Levi's as a job-flight story would miss the more fundamental – and more damaging – shift that the closures represent. As far as the company is concerned, those 16,310 jobs are off the payrolls for good, replaced, according to Ermatinger, by "contractors throughout the world." Those contractors will perform the same tasks as the old Levi's-owned factories – but the workers inside will never be employed by Levi Strauss.

For some companies a plant closure is still a straightforward decision to move the same facility to a cheaper locale. But for others – particularly those with strong brand identities like Levi Strauss and Hanes – layoffs are only the most visible manifestation of a much more fundamental shift: one that is less about where to produce than how. Unlike factories that hop from one place to another, these factories will never rematerialize. Mid-flight, they morph into something else entirely: "orders" to be placed with a contractor, who may well turn over those orders to as many as ten subcontractors, who – particularly in the garment sector – may in turn pass a portion of the subcontracts on to a network of home workers who will complete the jobs in basements and living rooms. Sure enough, only five months after the first round of plant closures was announced, Levi's made another public statement: it would resume manufacturing in China. The company had pulled out of China in 1993, citing concerns about human-rights violations. Now it has returned, not to build its own factories, but to place orders with three contractors that the company vows to closely monitor for violations of labor law.[20]

This shift in attitude toward production is so profound that where a previous era of consumer goods corporations displayed their logos on the façades of their factories, many of today's brand-based multinationals now maintain that the location of their production operations is a "trade secret," to be guarded at all costs. When asked by human-rights groups in April 1999 to disclose the names and addresses of its contract factories, Peggy Carter, a vice president at Champion clothing, replied: "We have no interest in our competition learning where we are located and taking advantage of what has taken us years to build."[21]

Increasingly, brand-name multinationals – Levi's, Nike, Champion, Wal-Mart, Reebok, the Gap, IBM and General Motors – insist that they are just like any one of us: bargain hunters in search of the best deal in the global mall. They are very picky customers, with specific instructions about made-to-order design, materials, delivery dates and, most important, the need for rock-bottom prices. But what they are *not* interested in is the burdensome logistics of how those prices fall so low; building factories, buying machinery and budgeting for labor have all been lobbed squarely into somebody else's court.

And the real job-flight story is that a growing number of the most high-profile and profitable corporations in the world are fleeing the jobs business altogether.

The Unbearable Lightness of Cavite: Inside the Free-Trade Zones

Despite the conceptual brilliance of the "brands, not products" strategy, production has a pesky way of never quite being transcended entirely: *somebody* has to get down and dirty and make the products the global brands will hang their meaning on. And that's where the free-trade zones come in. In Indonesia, China, Mexico, Vietnam, the Philippines and elsewhere, export processing zones (as these areas are also called) are emerging as leading producers of garments, toys, shoes, electronics, machinery, even cars.

If Nike Town and the other superstores are the glittering new gateways to the branded dreamworlds, then the Cavite Export Processing Zone, located ninety miles south of Manila in the town of Rosario, is the branding broom closet. After a month visiting similar industrial areas in Indonesia, I arrived in Rosario in early September 1997, at the tail end of monsoon season and the beginning of the Asian economic storm. I'd come to spend a week in Cavite because it is the largest free-trade zone in the Philippines, a 682-acre walled-in industrial area housing 207 factories that produce goods strictly for the export market. Rosario's population of 60,000 all seemed to be on the move; the town's busy, sweltering streets were packed with army jeeps converted into minibuses and with motorcycle taxis with precarious sidecars, its sidewalks lined with stalls selling fried rice, Coke and soap. Most of this commercial activity serves the 50,000 workers who rush through Rosario on their way to and from work in the zone, whose gated entrance is located smack in the middle of town.

Inside the gates, factory workers assemble the finished products of our branded world: Nike running shoes, Gap pajamas, IBM computer screens, Old Navy jeans. But despite the presence of such illustrious multinationals, Cavite – and the exploding number of export processing zones like it throughout the developing world – could well be the only places left on earth where the superbrands actually keep a low profile. Indeed, they are positively self-effacing. Their names and logos aren't splashed on the façades of the factories in the industrial zone. And here, competing labels aren't segregated each in its own superstore; they are often produced side by side in the same factories, glued by the very same workers, stitched and soldered on the very same machines. It was in Cavite that I finally found a piece of unswooshed space, and I found it, oddly enough, in a Nike shoe factory.

I was only permitted one visit inside the zone's gates to interview officials – individual factories, I was told, are off limits to anyone but potential importers or exporters. But a few days later, with the help of an eighteen-year-old worker who had been laid off from his job in an electronics factory, I managed to sneak back to get the unofficial tour. In the rows of virtually identical giant shed-like structures, one factory stood out: the name on the white rectangular building said "Philips," but through its surrounding fence I could see mountains of Nike shoes piled high. It seems that in Cavite, production has been banished to our age's most worthless status: its factories are unbrandable, unswooshworthy; producers are the industrial untouchables. Is this what Phil Knight meant, I wondered, when he said his company wasn't about the sneakers?

Manufacturing is concentrated and isolated inside the zone as if it were toxic waste: pure, 100 percent production at low, low prices. Cavite, like the rest of the zones that compete with it, presents itself as the buy-in-bulk Price Club for multinationals on the lookout for bargains – grab a really big shopping cart. Inside, it's obvious that the row of factories, each with its own gate and guard, has been carefully planned to squeeze the maximum amount of production out of this swath of land. Windowless workshops made of cheap plastic and aluminum siding are crammed in next to each other, only feet apart. Racks of time cards bake in the sun, making sure the maximum amount of work is extracted from each worker, the maximum number of working hours extracted from each day. The streets in the zone are eerily empty, and open doors – the ventilation system for most factories – reveal lines of young women hunched in silence over clamoring machines.

In other parts of the world, workers live inside the economic zones, but not in Cavite: this is a place of pure work. All the bustle and color of Rosario abruptly stops at the gates, where workers must show their ID cards to armed guards in order to get inside. Visitors are rarely permitted in the zone and little or no internal commerce takes place on its orderly streets, not even candy and drink vending. Buses and taxicabs must drop their speed and silence their horns when they get into the zone – a marked change from the boisterous streets of Rosario. If all of this makes Cavite feel as if it's in a different country, that's because, in a way, it is. The zone is a tax-free economy, sealed off from the local government of both town and province – a miniature military state inside a democracy.

As a concept, free-trade zones are as old as commerce itself, and were all the more relevant in ancient times when the transportation of goods required multiple holdovers and rest stops. Pre-Roman Empire city-states, including Tyre, Carthage and Utica, encouraged trade by declaring themselves "free cities," where goods in transit could be stored without tax, and merchants would be protected from harm. These tax-free areas developed further economic significance during colonial times, when entire cities – including Hong Kong, Singapore and Gibraltar – were designated as "free ports" from which the loot of colonialism could be safely shipped back to England, Europe or America with low import tariffs.[22] Today, the globe is dotted with variations on these tax-free pockets, from duty-free shops in airports and the free banking zones of the Cayman Islands to bonded warehouses and ports where goods in transit are held, sorted and packaged.

Though it has plenty in common with these other tax havens, the export processing zone is really in a class of its own. Less holding tank than sovereign territory, the EPZ is an area where goods don't just pass through but are actually manufactured, an area, furthermore, where there are no import and export duties, and often no income or property taxes either. The idea that EPZs could help Third World economies first gained currency in 1964 when the United Nations Economic and Social Council adopted a resolution endorsing the zones as a means of promoting trade with developing nations. The idea didn't really get off the ground, however, until the early eighties, when India introduced a five-year tax break for companies manufacturing in its low-wage zones.

Since then, the free-trade-zone industry has exploded. There are fifty-two economic zones in the Philippines alone, employing 459,000 people – that's up from only 23,000 zone workers in 1986 and 229,000 as recently as 1994. The largest zone economy is China, where by conservative estimates there are 18 million people in 124 export processing zones.[23] In total, the International Labor Organization says that there are at least 850 EPZs in the world, but that number is likely much closer to 1,000, spread through seventy countries and employing roughly 27 million workers.[24] The World Trade Organization estimates that between $200 and $250 billion worth of trade flows through the zones.[25] The number of individual factories housed inside these industrial parks is also expanding. In fact, the free-trade factories along the U.S.–Mexico border – in Spanish, *maquiladoras* (from *maquillar*, "to make up, or assemble") – are probably the only structures that proliferate as quickly as Wal-Mart outlets: there were 789 maquiladoras in 1985. In 1995, there were 2,747. By 1997, there were 3,508 employing about 900,000 workers.[26]

Regardless of where the EPZs are located, the workers' stories have a certain mesmerizing sameness: the workday is long – fourteen hours in Sri Lanka, twelve hours in Indonesia, sixteen in Southern China, twelve in the Philippines. The vast majority of the workers are women, always young, always working for contractors or subcontractors from Korea, Taiwan or Hong Kong. The contractors are usually filling orders for companies based in the U.S., Britain, Japan, Germany or Canada. The management is military-style, the supervisors often abusive, the wages below subsistence and the work low-skill and tedious. As an economic model, today's export processing zones have more in common with fast-food franchises than sustainable developments, so removed are they from the countries that host them. These pockets of pure industry hide behind a cloak of transience: the contracts come and go with little notice; the workers are predominantly migrants, far from home and with little connection to the city or province where zones are located; the work itself is short-term, often not renewed.

As I walk along the blank streets of Cavite, I can feel the threatening impermanence, the underlying instability of the zone. The shed-like factories are connected so tenuously to the surrounding country, to the adjacent town, to the very earth they are perched upon, that it feels as if the jobs that flew here from the North could fly away again just as quickly. The factories are cheaply constructed and tossed together on land that is rented, not owned. When I climb up the water tower on the edge of the zone and look down at the hundreds of factories, it seems as if the whole cardboard complex could lift up and blow away, like Dorothy's house in *The Wizard of Oz*. No wonder the EPZ factories in Guatemala are called "swallows."

Fear pervades the zones. The governments are afraid of losing their foreign factories; the factories are afraid of losing their brand-name buyers; and the workers are afraid of losing their unstable jobs. These are factories built not on land but on air.

"It Should Have Been a Different Rosario"

The air the export processing zones are built upon is the promise of industrialization. The theory behind EPZs is that they will attract foreign investors, who, if all goes well, will decide to stay in the country, and the zones' segregated assembly lines will turn into lasting development: technology transfers and domestic industries. To lure the swallows into this clever trap, the governments of poor countries offer tax breaks, lax regulations and the services of a military willing and able to crush labor unrest. To sweeten the pot further, they put their own people on the auction block, falling over each other to offer up the lowest minimum wage, allowing workers to be paid less than the real cost of living.

In Cavite, the economic zone is designed as a fantasyland for foreign investors. Golf courses, executive clubs and private schools have been built on the outskirts of Rosario to ease the discomforts of Third World life. Rent for factories is dirt cheap: 11 pesos per square foot – less than a cent. For the first five years of their stay, corporations are treated to an all-expenses-paid "tax holiday" during which they pay no income tax and no property tax. It's a good deal, no doubt, but it's nothing compared to Sri Lanka, where EPZ investors stay for ten years before having to pay any tax.[27]

The phrase "tax holiday" is oddly fitting. For the investors, free-trade zones are a sort of corporate Club Med, where the hotel pays for everything and the guests live free, and where integration with the local culture and economy is kept to a bare minimum. As one International Labor Organization report puts it, the EPZ "is to the inexperienced foreign investor what the package holiday is to the cautious tourist." Zero-risk globalization. Companies just ship in the pieces of cloth or computer parts – free of import tax – and the cheap, non-union workforce assembles it for them. Then the finished garments or electronics are shipped back out, with no export tax.

The rationale goes something like this: *of course* companies must pay taxes and strictly abide by national laws, but just in this one case, on this one specific piece of land, for just a little while, an exception will be made – for the cause of future prosperity. The EPZs, therefore, exist within a kind of legal and economic set of brackets, apart from the rest of their countries – the Cavite zone, for example, is under the sole jurisdiction of the Philippines' federal Department of Trade and Industry; the local police and municipal government have no right even to cross the threshold. The layers of blockades serve a dual purpose: to keep the hordes away from the costly goods being manufactured inside the zone, but also, and perhaps more important, to shield the country from what is going on inside the zone.

Because such sweet deals have been laid out to entice the swallows, the barriers around the zone serve to reinforce the idea that what is happening inside is only temporary, or is not really happening at all. This collective denial is particularly important in Communist countries where zones house the most Wild West forms of capitalism this side of Moscow: this is *definitely* not really happening, *certainly* not here where the government in power maintains that capital is the devil and workers reign supreme. In her book *Losing Control?*, Saskia Sassen writes that the zones are a part of a process of carving up nations so that "an actual piece of land becomes denationalized...."[28] Never mind that the boundaries of these only-temporary, not-really-happening, denationalized spaces keep expanding to engulf more and more of their actual nations. Twenty-seven million people worldwide are now living and working in brackets, and the brackets, instead of being slowly removed, just keep getting wider.

It is one of the zones' many cruel ironies that every incentive the governments throw in to attract the multinationals only reinforces the sense that the companies are economic tourists rather than long-term investors. It's a classic vicious cycle: in an attempt to alleviate poverty, the governments offer more and more incentives; but then the EPZs must be cordoned off like leper colonies, and the more they are cordoned off, the more the factories appear to exist in a world entirely separate from the host country, and outside the zone the poverty only grows more desperate. In Cavite, the zone is a kind of futuristic industrial suburbia where everything is ordered; the workers are uniformed, the grass manicured, the factories regimented. There are cute signs all around the grounds instructing workers to "Keep Our Zone Clean" and "Promote Peace and Progress of the Philippines." But walk out of the gate and the bubble bursts. Aside from the swarms of workers at the start and end of shifts, you'd never know that the town of Rosario is home to more than two hundred factories. The roads are a mess, running water is scarce and garbage is overflowing.

Many of the workers live in shantytowns on the outskirts of town and in neighboring villages. Others, particularly the youngest workers, live in the dormitories, a hodgepodge of concrete bunkers separated from the zone enclave by only a thick wall. The structure is actually a converted farm, and some rooms, the workers tell me, are really pigpens with roofs slapped on them.

The Philippines' experience of "industrialization in brackets" is by no means unique. The current mania for the EPZ model is based on the successes of the so-called Asian Tiger economies, in particular the economies of South Korea and Taiwan. When only a few countries had the zones, including South Korea and Taiwan, wages rose steadily, technology transfers occurred and taxes were gradually introduced. But as critics of EPZs are quick to point out, the global economy has become much more competitive since those countries made the transition from low-wage industries to higher-skill ones. Today, with seventy countries competing for the export-processing-zone dollar, the incentives to lure investors are increasing and the wages and standards are being held hostage to the threat of departure. The upshot is that entire countries are being turned into industrial slums and low-wage labor ghettos, with no end in sight. As Cuban president Fidel Castro thundered to the assembled world leaders at the World Trade Organization's fiftieth-birthday celebration in May 1998, "What are we going to live on?... What industrial production will be left for us? Only low-tech, labor-intensive and highly contaminating ones? Do they perhaps want to turn a large part of the Third World into a huge free trade zone full of assembly plants which don't even pay taxes?"[29]

As bad as the situation is in Cavite, it doesn't begin to compare with Sri Lanka, where extended tax holidays mean that towns can't even provide public transportation for EPZ workers. The roads they walk to and from the factories are dark and dangerous, since there is no money for streetlights. Dormitory rooms are so overcrowded that they have white lines painted on the floor to mark where each worker sleeps – they "look like car parks," as one journalist observed.[30]

Jose Ricafrente has the dubious honor of being mayor of Rosario. I met with him in his small office, while a lineup of needy people waited outside. A once-modest fishing village, his town today has the highest per capita investment in all of the Philippines – thanks to the Cavite zone – but it lacks even the basic resources to clean up the mess that the factories create in the community. Rosario has all the problems of industrialization – pollution, an exploding population of migrant workers, increased crime, rivers of sewage – without any of the benefits. The federal government estimates that only 30 of the zone's 207 factories pay any taxes at all, but everybody else questions even that low figure. The mayor says that many companies are granted extensions of their tax holiday, or they close and reopen under another name, then take the free ride all over again. "They fold up before the tax holiday expires, then they incorporate to another company, just to avoid payment of taxes. They don't pay anything to the government, so we're in a dilemma right now," Ricafrente told me. A small man with a deep and powerful voice, Ricafrente is loved by his constituents for the outspoken positions he took on human rights and democracy during Ferdinand Marcos's brutal rule. But the day I met him, the mayor seemed exhausted, worn down by his powerlessness to affect the situation in his own backyard.[31] "We cannot even provide the basic services that our people expect from us," he said, with a sort of matter-of-fact rage. "We need water, we need roads, we need medical services, education. They expect us to deliver all of them at the same time, expecting that we've got money from taxes from the places inside the zone."

The mayor is convinced that there will always be a country – whether Vietnam, China, Sri Lanka or Mexico – that is willing to bid lower. And in the process, towns like Rosario will have sold out their people, compromised their education system and polluted their natural resources. "It should be a symbiotic relationship," Ricafrente says of foreign investment. "They derive income from us, so the government should also derive income from them.... It should have been a different Rosario."

Working in Brackets

So, if it's clear by now that the factories don't bring in taxes or create local infrastructures, and that the goods produced are all exported, why do countries like the Philippines still bend over backward to lure them inside their borders? The official reason is a trickle-down theory: these zones are job-creation programs and the income the workers earn will eventually fuel sustainable growth in the local economy.

The problem with this theory is that the zone wages are so low that workers spend most of their pay on shared dorm rooms and transportation; the rest goes to noodles and fried rice from vendors lined up outside the gate. Zone workers certainly cannot dream of affording the consumer goods they produce. These low wages are partly a result of the fierce competition for factories coming from other developing countries. But, above all, the government is extremely reluctant to enforce its own labor laws for fear of scaring away the swallows. So labor rights are under such severe assault inside the zones that there is little chance of workers earning enough to adequately feed themselves, let alone stimulate the local economy.

The Philippine government denies this, of course. It says that the zones are subject to the same labor standards as the rest of Philippine society: workers must be paid the minimum wage, receive social security benefits, have some measure of job security, be dismissed only with just cause and be paid extra for overtime, and they have the right to form independent trade unions. But in reality, the government views working conditions in the export factories as a matter of foreign trade policy, not a labor-rights issue. And since the government attracted the foreign investors with promises of a cheap and docile workforce, it intends to deliver. For this reason, labor department officials turn a blind eye to violations in the zone or even facilitate them.

Many of the zone factories are run according to iron-fist rules that systematically break Philippine labor law. Some employers, for instance, keep bathrooms padlocked except during two fifteen-minute breaks, during which time all the workers have to sign in and out so management can keep track of their nonproductive time. Seamstresses at a factory sewing garments for the Gap, Guess and Old Navy told me that they sometimes have to resort to urinating in plastic bags under their machines. There are rules against talking, and at the Ju Young electronics factory, a rule against smiling. One factory shames those who disobey by posting a list of "The Most Talkative Workers."

Factories regularly cheat on their workers' social security payments and gather illegal "donations" from workers for everything from cleaning materials to factory Christmas parties. At a factory that makes IBM computer screens, the "bonus" for working hours of overtime isn't a higher hourly wage but doughnuts and a pen. Some owners expect workers to pull weeds from the ground on their way into the factory; others must clean the floors and the washrooms after their shifts end. Ventilation is poor and protective gear scarce.

Then there is the matter of wages. In the Cavite zone, the minimum wage is regarded more as a loose guideline than as a rigid law. If $6 a day is too onerous, investors can apply to the government for a waiver on that too. So while some zone workers earn the minimum wage, most – thanks to the waivers – earn less.[32]

Not Low Enough: Squeezing Wages in China

Part of the reason the threat of factory flight is so tangible in Cavite is that compared with China, Filipino wages are very high. In fact, everyone's wages are high compared with China. But what is truly remarkable about that is that the most egregious wage cheating goes on inside China itself.

Labor groups agree that a living wage for an assembly-line worker in China would be approximately US87 cents an hour. In the United States and Germany, where multinationals have closed down hundreds of domestic textile factories to move to zone production, garment workers are paid an average of US$10 and $18.50 an hour, respectively.[33] Yet even with these massive savings in labor costs, those who manufacture for the most prominent and richest brands in the world are still refusing to pay workers in China the 87 cents that would cover their cost of living, stave off illness and even allow them to send a little money home to their families. A 1998 study of brand-name manufacturing in the Chinese special economic zones found that Wal-Mart, Ralph Lauren, Ann Taylor, Esprit, Liz Claiborne, Kmart, Nike, Adidas, J.C. Penney and the Limited were only paying a fraction of that miserable 87 cents – some were paying as little as 13 cents an hour. (See Table 9.3, Appendix, page 474.)

The only way to understand how rich and supposedly law-abiding multinational corporations could regress to nineteenth-century levels of exploitation (and get caught repeatedly) is through the mechanics of subcontracting itself: at every layer of contracting, subcontracting and homework, the manufacturers bid against each other to drive down the price, and at every level the contractor and subcontractor exact their small profit. At the end of this bid-down, contract-out chain is the worker – often three or four times removed from the company that placed the original order – with a paycheck that has been trimmed at every turn. "When the multinationals squeeze the subcontractors, the subcontractors squeeze the workers," explains a 1997 report on Nike's and Reebok's Chinese shoe factories.[34]

"No Union, No Strike"

A large sign is posted at a central intersection in the Cavite Export Processing Zone: "DO NOT LISTEN TO AGITATORS AND TROUBLE MAKERS." The words are in English, painted in bright red capital letters and everyone knows what they mean. Although trade unions are technically legal in the Philippines, there is a widely understood – if unwritten – "no union, no strike" policy inside the zones. As the sign suggests, workers who do attempt to organize unions in their factories are viewed as troublemakers, and often face threats and intimidation.

One of the reasons I went to Cavite is that I had heard this zone was a hotbed of "troublemaking," thanks to a newly formed organization called the Workers' Assistance Center. Attached to Rosario's Catholic church only a few blocks from the zone's entrance, the center is trying to break through the wall of fear that surrounds free-trade zones in the Philippines. Slowly, they have been collecting information about working conditions inside the zone. Nida Barcenas, one of the organizers at the center, told me, "At first, I used to have to follow workers home and beg them to talk to me. They were so scared – their families said I was a troublemaker." But after the center had been up and running for a year, the zone workers flocked there after their shifts – to hang out, eat dinner and attend seminars. I had heard about the center back in Toronto, told by several international labor experts that the research and organizing on free-trade zones coming out of this little bare-bones operation is among the most advanced being done anywhere in Asia.

The Workers' Assistance Center, known as WAC, was founded to support the factory workers' constitutional right to fight for better conditions – zone or no zone. Zernan Toledo is the center's most intense and radical organizer, and though he is only twenty-five and looks like a college student, he runs the center's affairs with all the discipline of a revolutionary cell. "Outside the zone, workers are free to organize a union, but inside they cannot stage pickets or have demonstrations," Toledo told me in my two-hour "orientation session" at the center. "Group discussions in the factories are prohibited and we cannot enter the zone," he said, pointing to a diagram of the zone layout hanging on the wall.[35] This catch-22 exists throughout the quasi-private zones. As the International Confederation of Free Trade Unions report puts it: "The workers are effectively living in 'lawless' territory where to defend their rights and interests they are constantly forced to take 'illegal' action themselves."[36]

In the Philippines, the zone's culture of incentives and exceptions, which was intended to be phased out as the foreign companies joined the national economy, has had the opposite effect. Not only have new swallows landed, but unionized factories already in the country have shut themselves down and reopened inside the Cavite Export Processing Zone in order to take advantage of all the incentives. For instance, Marks & Spencer goods used to be manufactured in a unionized factory north of Manila. "It only took ten trucks to bring Marks & Spencer to Cavite," a labor organizer in the area told me. "The union was eliminated."

Cavite is by no means exceptional in this regard. Union organizing is a source of great fear throughout the zones, where a successful drive can have dire consequences for both organizers and workers. That was the lesson learned in December 1998, when the American shirtmaker Phillips-Van Heusen closed down the only unionized export apparel factory in all of Guatemala, laying off five hundred workers. The Camisas Modernas plant was unionized in 1997, after a long and bitter organizing drive and significant pressure placed on the company by U.S. human-rights groups. With the union, wages went up from US$56 a week to $71 and the previously squalid factory was cleaned up. Jay Mazur, president of the Union of Needletrades, Industrial and Textile Employees (UNITE) – America's largest apparel union – called the contract "a beacon of hope for more than 80,000 maquiladora workers in Guatemala."[37] When the factory closed, however, the beacon of hope turned into a flashing red danger signal, reinforcing the familiar warning: no union, no strike.

Patriotism and national duty are bound up in the exploitation of the export zones, with young people – mostly women – sent off to sweatshop factories the way a previous generation of young men were sent off to war. No questioning of authority is expected or permitted. In some Central American and Asian EPZs, strikes are officially illegal; in Sri Lanka, it is illegal to do anything at all that might jeopardize the country's export earnings, including publishing and distributing critical material.[38] In 1993, a Sri Lankan zone worker by the name of Ranjith Mudiyanselage was killed for appearing to challenge this policy. After complaining about a faulty machine that had sliced off a co-worker's finger, Mudiyanselage was abducted on his way out of an inquiry into the incident. His body was found beaten and burning on a pile of old tires outside a local church. The man's legal adviser, who had accompanied him to the inquiry, was murdered in the same way.[39]

Despite the constant threat of retaliation, the Workers' Assistance Center has made some modest attempts to organize unions inside the Cavite zone factories, with varying degrees of success. For instance, when a drive was undertaken at the All Asia garment factory, the organizers came up against a very challenging obstacle: worker exhaustion. The biggest complaint among the All Asia seamstresses who stitch clothes for Ellen Tracy and Sassoon is forced overtime. Regular shifts last from 7 a.m. to 10 p.m., but on a few nights a week employees must work "late" – until 2 a.m. During peak periods, it is not uncommon to work two 2 a.m. shifts in a row, leaving many women only a couple of hours of sleep before they have to start their commute back to the factory. But that also means most All Asia workers spend their precious thirty-minute breaks at the factory napping, not talking about unions. "I have a hard time talking with the workers because the workers are always very sleepy," a mother of four tells me, explaining why she has had no luck in her attempts to bring a union to the All Asia factory. She has been with the company for four years and still lacks basic job security and health insurance.

Work in the zone is characterized by this brutal combination of tremendous intensity and nonexistent job security. Everyone works six or seven days a week, and when a big order is due to be shipped out, employees work until it is done. Most workers want some overtime hours because they need the money, but the overnight shifts are widely considered a burden. Refusing to stay, however, is not an option. For instance, according to the official rule book of the Philips factory (a contractor that has filled orders for both Nike and Reebok), "Refusal to render overtime work when so required" is an offense "punishable with dismissal." The same is true at all the factories I encountered, and there are many reports of workers asking to leave early – before 2 a.m., for instance – and being told not to return to work the next day.

Overtime horror stories pour out of the export processing zones, regardless of location: in China, there are documented cases of three-day shifts, when workers are forced to sleep under their machines. Contractors often face heavy financial penalties if they fail to deliver on time, no matter how unreasonable the deadline. In Honduras, when filling out a particularly large order on a tight deadline, factory managers have been reported injecting workers with amphetamines to keep them going on forty-eight-hour marathons.[40]

What Happened to Carmelita...

In Cavite, you can't talk about overtime without the conversation turning to Carmelita Alonzo, who died, according to her co-workers, "of overwork." Alonzo, I was told again and again – by groups of workers gathered at the Workers' Assistance Center and by individual workers in one-on-one interviews – was a seamstress at the V.T. Fashions factory, stitching clothes for the Gap and Liz Claiborne, among many other labels. All of the workers I spoke with urgently wanted me to know how this tragedy happened so that I could explain it to "the people in Canada who buy these products." Carmelita Alonzo's death occurred following a long stretch of overnight shifts during a particularly heavy peak season. "There were a lot of products for ship-out and no one was allowed to go home," recalls Josie, whose denim factory is owned by the same firm as Carmelita's, and who also faced large orders at that time. "In February, the line leader had overnights almost every night for one week." Not only had Alonzo been working those shifts, but she had a two-hour commute to get back to her family. Suffering from pneumonia – a common illness in factories that are suffocatingly hot during the day but fill with condensation at night – she asked her manager for time off to recover. She was denied. Alonzo was eventually admitted to hospital, where she died on March 8, 1997 – International Women's Day.

I asked a group of workers gathered late one evening around the long table at the center how they felt about what happened to Carmelita. The answers were confused at first. "Feel? But Carmelita is us." But then Salvador, a sweet-faced twenty-two-year-old from a toy factory, said something that made all of his co-workers nod in vigorous agreement. "Carmelita died because of working overtime. It is possible to happen to any one of us," he explained, the words oddly incongruous with his pale blue *Beverly Hills 90210* T-shirt.

Much of the overtime stress could be alleviated if the factories would just hire more workers and create two shorter shifts. But why should they? The government official appointed to oversee the zone isn't interested in taking on the factory owners and managers about the overtime violations. Raymondo Nagrampa, the zone administrator, acknowledged that it would certainly be better if the factories hired more people for fewer hours, but, he told me, "I think I will leave that. I think this is more of a management decision."

For their part, the factory owners are in no rush to expand the size of their workforce, because after a big order is filled there could be a dry spell and they don't want to be stuck with more employees than work. Since following Philippine labor law is "a management decision," most decide that it is more convenient for management to have one pool of workers who are simply forced to work more hours when there is more work and fewer when there is less of it. And this is the flip side of the overtime equation: when a factory is experiencing a lull in orders or a shipment of supplies has been delayed, workers are sent home without pay, sometimes for a week at a time. The group of workers gathered around the table at the Workers' Assistance Center burst out laughing when I asked them about job security or a guaranteed number of working hours. "No work, no pay!" the young men and women exclaim in unison.

The "no work, no pay" rule applies to all workers, contract or "regular." Contracts, when they exist, last only five months or less, after which time workers have to "recontract." Many of the factory workers in Cavite are actually hired through an employment agency, inside the zone walls, that collects their checks and takes a cut – a temp agency for factory workers, in other words, and one more level in the multiple-level system that lives off their labor. Management uses a variety of tricks in the different zones to keep employees from achieving permanent status and collecting the accompanying rights and benefits. In the Central American maquiladoras, it is a common practice for factories to fire workers at the end of the year and rehire them a few weeks later so that they don't have to grant them permanent status; in the Thai zones, the same practice is known as "hire and fire."[41] In China, many workers in the zones have no contracts at all, which leaves them without any rights or recourse whatsoever.[42]

It is in this casual new relationship to factory employment that the EPZ system breaks down completely. In principle, the zones are an ingenious mechanism for global wealth redistribution. Yes, they lure jobs from the North, but few fair-minded observers would deny the proposition that as industrialized nations shift to higher-tech economies, it is only a matter of global justice that the jobs upon which our middle classes were built should be shared with countries still enslaved by poverty. The problem is that the workers in Cavite, and in zones throughout Asia and Latin America, are not inheriting "our" jobs at all. Gerard Greenfield, former research director of the Asian Monitoring and Resource Centre in Hong Kong, says, "One of the myths of relocation is that those jobs that seemed to be transferred from the so-called North to the South are perceived as similar jobs to what was already being done before." They are not. Just as company-owned manufacturing turned – somewhere over the Pacific Ocean – into "orders" to be placed with third-party contractors, so did full-time employment undergo a mid-flight transformation into "contracts." "The biggest challenge to those in Asia," says Greenfield, "is that the new employment created by Western and Asian multinationals investing in Asia is temporary and short-term employment."[43]

In fact, zone workers in many parts of Asia, the Caribbean and Central America have more in common with office-temp workers in North America and Europe than they do with factory workers in those Northern countries. What is happening in the EPZs is a radical alteration in the very nature of factory work. That was the conclusion of a 1996 study conducted by the International Labor Organization, which stated that the dramatic relocation of production in the garment and shoe industries "has been accompanied by a parallel shift of production from the formal to the informal sector in many countries, with generally negative consequences on wage levels and conditions of work." Employment in these sectors, the study went on, has shifted from "full-time in-plant jobs to part-time and temporary jobs and, especially in clothing and footwear, increasing resort to homework and small shops."[44]

Indeed, this is not simply a job-flight story.

A Floating Workforce

On my last night in Cavite, I met a group of six teenage girls in the workers' dormitories who shared a six-by-eight-foot concrete room: four slept on the makeshift bunk bed (two to a bed), the other two on mats spread on the floor. The girls who made Aztek, Apple and IBM CD-ROM drives shared the top bunk; the ones who sewed Gap clothing, the bottom. All were the children of farmers, away from their families for the first time.

Their jam-packed shoebox of a home had the air of an apocalyptic slumber party – part prison cell, part *Sixteen Candles*. It may have been a converted pigsty, but these were sixteen-year-old girls, and like teenage girls the world over they had covered the gray, stained walls with pictures: of fluffy animals, Filipino action-movie stars, and glossy magazine ads of women modeling lacy bras and underwear. After a little while, serious talk of working conditions erupted into fits of giggles and hiding under bedcovers. It seems that my questions reminded two of the girls of a crush they had on a labor organizer who had recently given a seminar at the Workers' Assistance Center on the risks of infertility from working with hazardous chemicals.

Were they worried about infertility?

"Oh, yes. Very worried now."

All through the Asian zones, the roads are lined with teenage girls in blue shirts, holding hands with their friends and carrying umbrellas to shield them from the sun. They look like students coming home from school. In Cavite, as elsewhere, the vast majority of workers are unmarried women between the ages of seventeen and twenty-five. Like the girls in the dorms, roughly 80 percent of the workers have migrated from other provinces of the Philippines to work in the factories – a mere 5 percent are native to the town of Rosario. Like the swallow factories, they too are only tenuously connected to this place.

Raymondo Nagrampa, the zone administrator, says migrants are recruited for the zone to compensate for something innate in "the Cavite character," something that makes local people unfit to work in the factories situated near their homes. "I don't mean any offense to the Cavite personality," he explained, in his spacious air-conditioned office. "But from what I gather, this particular character is not suited for the factory life – they'd rather go into something quickly. They do not have the patience to be right there in the factory line." Nagrampa attributes this to the fact that Rosario is so close to Manila "and so we can say that the Cavitenians are not running scared with regard to getting some income for their daily subsistence....

"But in the case of those from the provinces, from the lower areas, they are not exposed to the big-city lifestyle. They feel more comfortable just working in the factory line, for, after all, this is a marked improvement from the farm work that they've been accustomed to, where they were exposed to the sun. To them, for the lowly province rural worker, working inside an enclosed factory is better off than being outside."

I asked dozens of zone workers – all of them migrants from rural areas – about what Raymondo Nagrampa had said. Every one of them responded with outrage.

"It's not human!" exclaimed Rosalie, a teenager whose job is installing the "backlights" in IBM computer screens. "Our rights are being trampled and Mr. Nagrampa says that because he has not experienced working in a factory and the conditions inside."

Salvador, in his *90210* T-shirt, was beside himself: "Mr. Nagrampa earns a lot of money and he has an air-conditioned room and his own car, so of course he would say that we prefer this work – it is beneficial to him, but not to us.... Working on the farm is difficult, yes, but there we have our family and friends and instead of always eating dried fish, we have fresh food to eat."

His words clearly struck a chord with a homesick Rosalie: "I want to be together with my family in the province," she said quietly, looking even younger than her nineteen years. "It's better there because when I get sick, my parents are there, but here there is no one to take care of me."

Many other rural workers told me that they would have stayed home if they could, but the choice was made for them: most of their families had lost their farms, displaced by golf courses, botched land-reform laws and more export processing zones. Others said that the only reason they came to Cavite was that when the zone recruiters came to their villages, they promised that workers would earn enough in the factories to send money home to their impoverished families. The same inducement had been offered to other girls their age, they told me, to go to Manila to work in the sex trade.

Several more young women wanted to tell me about those promises, too. The problem, they said, is that no matter how long they work in the zone, there is never more than a few pesos left over to send home. "If we had land we would just stay there to cultivate the land for our needs," Raquel, a teenage girl from one of the garment factories, told me. "But we are landless, so we have no choice but to work in the economic zone even though it is very hard and the situation here is very unfair. The recruiters said we would get a high income, but in my experience, instead of sending my parents money, I cannot maintain even my own expenses."

So the workers in Cavite have lost on all counts: they are penniless *and* homeless. It's a potent combination. In the dormitories, sleep deprivation, malnutrition and homesickness mingle to create an atmosphere of deep disorientation. "We are alien in the factories. We are also alien in the boardinghouse because we all come from faraway provinces," Liza, an electronics worker, told me. "We are strangers here."

Cecille Tuico, one of the organizers at the Workers' Assistance Center, was listening in on the conversation. After the workers left to make their way through Rosario's dark streets and back to the dormitories, she pointed out that the alienation the workers so poignantly describe is precisely what the employers look for when they seek out migrants instead of locals to work in the zone. With the same muted, matter-of-fact anger I have come to recognize in so many Filipino human-rights activists, Tuico said that the factory managers prefer young women who are far from home and have not finished high school, because "they are scared and uneducated about their rights."

The Zones' Other Product: A New Kind of Factory Worker

Their naiveté and insecurity undoubtedly make discipline easier for factory managers, but younger workers are preferred for other reasons, too. Women are often fired from their zone jobs in their mid-twenties, told by supervisors that they are "too old," and that their fingers are no longer sufficiently nimble. This practice is a highly effective way of minimizing the number of mothers on the company payroll.

In Cavite, the workers tell me stories about pregnant women forced to work until 2 a.m., even after pleading with the supervisor; of women who work in the ironing section giving birth to babies with burns on their skin; of women who mold the plastic for cordless phones giving birth to stillborn infants. The evidence I hear in Cavite is anecdotal, told to me quietly and urgently by women with the same terrified expression I saw when conversation turned to Carmelita Alonzo. Some of the stories are certainly apocryphal – fear-fueled zone legends – but the abuse of pregnant women in export processing zones is also well documented and the problem reaches far beyond Cavite.

Because most zone employers want to avoid paying benefits, assigning workers to a predictable schedule or offering any job security, motherhood has become the scourge of these pink-collar zones. A study by Human Rights Watch that has become the basis for a grievance under the NAFTA side agreement on labor found that women applying for jobs in the Mexican maquiladoras routinely had to undergo pregnancy tests. The study, which implicates such investors in the zones as Zenith, Panasonic, General Electric, General Motors and Fruit of the Loom, found that "pregnant women are denied hiring. Moreover, maquiladora employers sometimes mistreat and discharge pregnant employees."[45] The researchers uncovered mistreatment designed to encourage workers to resign: pregnant women were required to work the night shift, or to take on exceptionally long hours of unpaid overtime and physically strenuous tasks. They were also refused time off work to go to the doctor, a practice that has led to on-the-job miscarriages. "In this way," the study reports, "a pregnant worker is forced to choose between having a healthy, full-term pregnancy and keeping her job."[46]

Other methods of sidestepping the costs and responsibilities of employing workers with children are reported on a more haphazard basis throughout the zones. In Honduras and El Salvador the garbage dumps in the zones are littered with empty packets of contraceptive pills that are reportedly passed out on the factory floor. In the Honduran zones there have been reports of management forcing workers to have abortions. At some Mexican maquiladoras, women are required to prove they are menstruating through such humiliating practices as monthly sanitary-pad checks. Employees are kept on twenty-eight-day contracts – the length of the average menstrual cycle – making it easy, as soon as a pregnancy comes to light, for the worker to be dismissed.[47] In a Sri Lankan zone, one worker was reported to be so terrified of losing her job after giving birth that she drowned her newborn baby in a toilet.[48]

The widespread assault on women's reproductive freedoms in the zones is the most brutal expression of the failure on the part of many consumer-goods corporations to live up to their traditional role as mass employers. Today's "new deal" with workers is a non-deal; one-time manufacturers, turned marketing mavens, are so resolutely intent on evading any and all commitments that they are creating a workforce of childless women, a system of footloose factories employing footloose workers. In a letter to Human Rights Watch explaining why it discriminated against pregnant women in the maquiladoras, General Motors stated plainly that it "will not hire female job applicants found to be pregnant" in an effort to avoid "substantial financial liabilities imposed by the Mexican social security system."[49] Since the critical report was published, GM has changed the policy. It remains, however, a stark contrast to the days when the company made it a banner policy that the adult men working in its auto plants should earn enough not only to support a family of four but to drive them around in a GM car or truck. General Motors has cut about 82,000 jobs in the U.S. since 1991 and expects to cut another 40,000 by the year 2003, moving production to the maquiladoras and their clones around the globe.[50] A far cry from those days when it proudly proclaimed, "What's good for General Motors is good for the country."

Migrant Factories

Within this reengineered system, the workers aren't the only ones on a day pass. The swallow factories that employ them have been built to maximize flexibility: to follow the tax breaks and incentives, to bend with the currency devaluations and benefit by the strict rule of dictators. In North America and Europe, job flight is a threat with which workers have become all too familiar. A study commissioned by the NAFTA labor commission found that in the United States, between 1993 and 1995, "employers threatened to close the plant in 50 percent of all union certification elections.... Specific, unambiguous threats ranged from attaching shipping labels to equipment throughout the plant with a Mexican address, to posting maps of North America with an arrow pointing from the current plant site to Mexico." The study found that the employers followed through on the threats, shutting down all or part of newly unionized plants, in 15 percent of these cases – triple the closing rate of the pre-NAFTA 1980s.[51] In China, Indonesia, India and the Philippines the threat of plant closure and job flight is even more powerful. Since the industries are quick to flee escalating wages, environmental regulation and taxes, factories are made to be mobile. Some of these swallow factories may well be on their third or even fourth flight, and as the history of subcontracting makes clear, they touch down more lightly at each new stop.

When the flying multinationals first landed in Taiwan, Korea and Japan, many of their factories were owned and operated by local contractors. In Pusan, South Korea, for instance – known during the eighties as "the sneaker capital of the world" – Korean entrepreneurs ran factories for Reebok, L.A. Gear and Nike. But when, in the late eighties, Korean workers began to rebel against their dollar-a-day wages and formed trade unions to fight for better conditions, the swallows once again took flight. Between 1987 and 1992, 30,000 factory jobs were lost in Korea's export processing zones, and in less than three years one-third of the shoe jobs had disappeared. The story is much the same in Taiwan. The migration patterns have been clearly documented with Reebok's manufacturers. In 1985, Reebok produced almost all its sneakers in South Korea and Taiwan and none in Indonesia and China. By 1995, nearly all those factories had flown out of Korea and Taiwan and 60 percent of Reebok's contracts had landed in Indonesia and China.[52]

But on this new leg of the journey, the factories were not owned by local Indonesian and Chinese contractors. Instead they were owned and run by the same Korean and Taiwanese companies that ran them before the move. When the multinationals pulled their orders from Korea and Taiwan, their contractors followed, closing up shop in their home countries and building the new factories in countries where labor was still cheap: China, Indonesia, Thailand and the Philippines. One of these contractors – the largest single supplier for Reebok, Adidas and Nike – is a Taiwanese-owned company called Yue Yuen. Yue Yuen has closed most of its factories in its homeland of

Taiwan and chased the low wages to China, where it employs 54,000 people in a single factory complex. For Chi Neng Tsai, one of the company's owners, it simply makes good business sense to go where the workers are hungry: "Thirty years ago, when Taiwan was hungry, we also were more productive," he says.[53]

Taiwanese and Korean bosses are uniquely positioned to exploit this hunger: they can tell workers from personal experience what happens when unions come in and wages go up. And maintaining contractors who have had the rug pulled out from under them once before is a stroke of management genius on the part of the Western multinationals. What better way to keep costs down than to make yesterday's casualties today's wardens?

It is a system that doesn't do much for the sense of stability in Cavite, or for the Philippine economy in general, which is already unusually vulnerable to global forces, since the majority of its companies are owned by foreign investors. As Filipino economist Antonio Tujan told me, "The contractors have displaced the Filipino middleman."[54] In fact, Tujan, the director of a Manila-based think tank highly critical of Philippine economic policy, corrects me when I refer to the buildings I saw inside the Cavite Export Processing Zone as "factories." They aren't factories, he says, "they are labor warehouses."

He explains that since all the materials are imported, nothing is actually manufactured in the factories, only assembled. (The components are manufactured in yet another country, where the workers are more highly skilled, though still cheaper than U.S. or European workers.) It's true, now that Tujan mentions it, that when I climbed up the water tower and looked down on the zone, part of what contributed to the unbearable lightness of Cavite was that apart from one incinerator, there were no smokestacks. That's a bonus for the air quality in Rosario but odd for an industrial park of Cavite's size. Neither was there any local rhyme or reason to what was being produced. When I walked the zone's freshly paved streets, I was surprised by the variety of manufacturing going on. Like most people, I had thought that Asian export zones were mostly filled with garment and electronics producers, but not Cavite: a factory making car seats sat next to one making sneakers, across the way from a factory with dozens of aluminum speedboats piled up by its gate. On another street, the open doors of a factory revealed racks of dresses and jackets, right next to the plant where Salvador made novelty key chains and other small toys. "You see?" says Antonio Tujan. "We have a country whose industry is so deformed, so unbelievably mishmash, that it cannot exist by itself. It's all a myth, you know. They talk about industrialization in the context of globalization, but it's all a myth."

No wonder the promise of industrialization in Cavite feels more like a threat. The place is a development mirage.

The Shoppers Take Flight

The fear that the flighty multinationals will once again pull their orders and migrate to more favorable conditions underlies everything that takes place in the zones. It makes for an odd dissonance: despite the fact that they have no local physical holdings – they don't own the buildings, land or equipment – brands like Nike, the Gap and IBM are omnipresent, invisibly pulling all the strings. They are so powerful as buyers that the hands-on involvement owning the factories would entail has come to look, from their perspective, like needless micromanagement. And because the actual owners and factory managers are completely dependent on their large contracts to make the machines run, workers are left in a uniquely weak bargaining position: you can't sit down and bargain with an order form. So even the classic Marxist division between workers and owners doesn't quite work in the zone, since the brand-name multinationals have divested the "means of production," to use Marx's phrase, unwilling to encumber themselves with the responsibilities of actually owning and managing the factories, and employing a labor force.

If anything, the multinationals have more power over production by not owning the factories. Like most committed shoppers, they see no need to concern themselves with how their bargains were produced – they simply pounce on them, keeping the suppliers on their toes by taking bids from slews of other contractors. One contractor, Young Il Kim of Guatemala, whose Sam Lucas factory produces clothing for Wal-Mart and J.C. Penney, says of his big-brand clients, "They're interested in a high-quality garment, fast delivery, and cheap sewing charges – and that's all."[55] In this cutthroat context, each contractor swears he could deliver the goods cheaper if the brands would only start producing in Africa, Vietnam or Bangladesh, or if they would shift to homeworkers.

More blatantly, the power of the brands may occasionally be invoked to affect public policy in the countries where export zones are located. Companies or their emissaries may make public statements about how a raise in the legal minimum wage could price a certain Asian country "out of the market," as Nike's and Reebok's contractors have been quick to tell the Indonesian government whenever strikes get out of hand.[56] Calling a strike at a Nike factory "intolerable," Anton Supit, chairman of the Indonesian Footwear Association, which represents contractors for Nike, Reebok and Adidas, called on the Indonesian military to intervene. "If the authorities don't handle strikes, especially ones leading to violence and brutality, we will lose our foreign buyers. The government's income from exports will decrease and unemployment will worsen."[57] The corporate shoppers may also help draft international trade agreements to reduce quotas and tariffs, or even lobby a government directly to loosen regulations. In describing the conditions under which Nike decided to begin "sourcing" its shoes in China, for instance, company vice president David Chang explained that "one of the first things we told the Chinese was that their prices had to be more competitive with our other Far East sources because the cost of doing business in China was so enormous.... The hope is for a 20 percent price advantage over Korea."[58] After all, what price-conscious consumer doesn't comparison shop? And if a shift to a more "competitive" country causes mass layoffs somewhere else in the world, that is somebody else's blood on somebody else's hands. As Levi's CEO Robert Haas said, "This is not a job-flight story."

Multinational corporations have vehemently defended themselves against the accusation that they are orchestrating a "race to the bottom" by claiming that their presence has helped to raise the standard of living in underdeveloped countries. As Nike CEO Phil Knight said in 1996, "For the past 25 years, Nike has provided good jobs, improved labor practices and raised standards of living wherever we operate."[59] Confronted with the starvation wages in Haiti, a Disney spokesperson told *The Globe and Mail*, "It's a process all developing countries go through, like Japan and Korea, who were at this stage decades ago."[60] And there is no shortage of economists to spin the mounting revelations of corporate abuse, claiming that sweatshops are not a sign of eroded rights but a signal that prosperity is just around the corner. "My concern," said famed Harvard economist Jeffrey D. Sachs, "is not that there are too many sweatshops but that there are too few... those are precisely the jobs that were the stepping stones for Singapore and Hong Kong and those are the jobs that have to come to Africa to get them out of back-breaking rural poverty."[61] Sachs's colleague Paul Krugman concurred, arguing that in the developing world the choice is not between bad jobs and good jobs but between bad jobs and no jobs. "The overwhelming mainstream view among economists is that the growth of this kind of employment is tremendous good news for the world's poor."[62]

The no-pain-no-gain defense of sweatshops, however, took a severe beating when the currencies of those very countries supposedly benefiting most from this development model began crashing like cheap plates. First in Mexico, then Thailand, South Korea, the Philippines and Indonesia, workers were, and in many cases still are, bringing home minimum-wage paychecks worth less than when the "economic miracle" first came to bless their nations years ago. Nike's public-relations director, Vada Manager, used to claim that "the job opportunities that we have provided to women and men in developing economies like Vietnam and Indonesia have provided a bridge of opportunity for these individuals to have a much better quality of life,"[63] but by the winter of 1998, nobody knew better than Nike that that bridge had collapsed. With currency devaluation and soaring inflation, real wages in Nike's Indonesian factories fell by 45 percent in 1998.[64] In July of that year, Indonesian president B.J. Habibie urged his 200 million citizens to do their part to conserve the country's dwindling rice supply by fasting for two days out of each week, from dawn until dusk. Development built on starvation wages, far from kick-starting a steady improvement in conditions, has proved to be a case of one step forward, three steps back. And by early 1998 there were no more shining Asian Tigers to point to, and those corporations and economists that had mounted such a singular defense of sweatshops had had their arguments entirely discredited.

The fear of flying has been looming large in Cavite of late. The currency began its downward spiral a few weeks before I arrived, and since then conditions have only worsened. By early 1999, the price of basic commodities like cooking oil, sugar, chicken and soap had increased by as much as 36 percent from the year before. Paychecks that barely made ends meet now no longer accomplish even that. Workers who had begun to find the courage to stand up to management are now living not only under the threat of mass layoffs and factory flight but with the reality. In 1998, 3,072 businesses in the Philippines either closed down or scaled back operation – a 166 percent increase over the year before.[65] For its part, Nike has laid off 268 workers at the Philips factory, where I had seen, through the surrounding fence, the shoes lying in great piles. A few months later, in February 1999, Nike pulled out of two other Philippine factories as well, these ones located in the nearby Bataan export zone; 1,505 workers were affected by the closures.[66] But Phil Knight didn't have to do the dirty work himself – he just cut the orders and left the rest to the contractors. Like the factories themselves, these job losses went unswooshed.

The transience woven into the fabric of free-trade zones is an extreme manifestation of the corporate divestment of the world of work, which is taking place at all levels of industry. Cavite may be capitalism's dream vacation, but casualization is a game that can be played at home, and contracting out, as *Business Week* reporter Aaron Bernstein has written, is trickling up. "While outsourcing started in manufacturing in the early 1980s, it has expanded through virtually every industry as companies rush to shed staff in everything from human resources to computer systems."[67] The same impetus that lies behind the brands-versus-products and contracts-versus-jobs conflict is fueling the move to temp, part-time, freelance and homework in North America and Europe, as we will see in the next chapter.

This is not a job-flight story. It is a flight-from-jobs story.

Top: The quintessential free agent. *Bottom:* Based on a “culture jam” from *Adbusters*.

CHAPTER TEN

THREATS AND TEMPS

From Working for Nothing to "Free Agent Nation"

A sense of impermanence is blowing through the labor force, destabilizing everyone from office temps to high-tech independent contractors to restaurant and retail clerks. Factory jobs are being outsourced, garment jobs are morphing into homework, and in every industry, temporary contracts are replacing full, secure employment. In a growing number of instances, even CEOs are opting for shorter stints at one corporation after another, breezing in and out of different corner offices and purging half the employees as they come and go.

Almost every major labor battle of the decade has focused not on wage issues but on enforced casualization, from the United Parcel Service workers' stand against "part-time America" to the unionized Australian dockworkers fighting their replacement by contract workers, to the Canadian autoworkers at Ford and Chrysler striking against the outsourcing of their jobs to non-union factories. All these stories are about different industries doing variations on the same thing: finding ways to cut ties to their workforce and travel light. The underbelly of the shiny "brands, not products" revelation can be seen increasingly in every workplace around the globe. Every corporation wants a fluid reserve of part-timers, temps and freelancers to help it keep overheads down and ride the twists and turns in the market. As British management consultant Charles Handy says, savvy companies prefer to see themselves as "organizers" of collections of contractors, as opposed to "employment organizations."[1] One thing is certain: offering employment – the steady kind, with benefits, holiday pay, a measure of security and maybe even union representation – has fallen out of economic fashion.

Branded Work: Hobbies, Not Jobs

Though an entire class of consumer-goods companies has transcended the need to produce what it sells, so far not even the most weightless multinational has been able to free itself entirely from the burden of employees. Production may be relegated to contractors, but clerks are still needed to sell the brand-name goods at the point of purchase, especially given the growth of branded retail. In the service industry, however, big-brand employers have become artful at dodging most commitments to their employees, expertly fostering the notion that their clerks are somehow not quite legitimate workers, and thus do not really need or deserve job security, livable wages and benefits.

Most of the large employers in the service sector manage their workforce as if their clerks didn't depend on their paychecks for anything essential, such as rent or child support. Instead, retail and service employers tend to view their employees as children: students looking for summer jobs, spending money or a quick stopover on the road to a more fulfilling and better-paying career. These are great jobs, in other words, for people who don't really need them. And so the mall and the superstore have given birth to a ballooning subcategory of joke jobs – the frozen-yogurt jerk, the Orange Julius juicer, the Gap greeter, the Prozac-happy Wal-Mart "sales associate" – that are notoriously unstable, low-paying and overwhelmingly part-time. (See Table 10.1, Appendix, page 475.)

What is distressing about this trend is that over the past two decades, the relative importance of the service sector as a source of jobs has soared. The decline in manufacturing, as well as the waves of downsizing and cutbacks in the public sector, have been met by dramatic growth in the numbers of service-sector jobs to the extent that services and retail now account for 75 percent of total U.S. employment.[2] (See Table 10.2, page 235.) Today, there are four and a half times as many Americans selling clothes in specialty and department stores as there are workers stitching and weaving them, and Wal-Mart isn't just the biggest retailer in the world, it is also the largest private employer in the United States.

And yet despite these shifts in employment patterns, most brand-name retail, service and restaurant chains have opted to put on economic blinders, insisting that they are still offering hobby jobs for kids. Never mind that the service sector is now filled with workers who have multiple university degrees, immigrants unable to find manufacturing jobs, laid-off nurses and teachers, and downsized middle managers. Never mind, too, that the students who do work in retail and fast food – as many of them do – are facing higher tuition costs, less financial assistance from parents and government and more years in school. Never mind that the food service workforce has been steadily aging over the last decade so that more than half are now over twenty-five years old. (See Table 10.3, Appendix, page 475.) Or that a 1997 study found that 25 percent of non-management Canadian retail workers had been with the same company for eleven years or more and that 39 percent had been there for between four and ten years.[3] That's a lot longer than "Chainsaw" Al Dunlap lasted as CEO of Sunbeam Corp. But never mind all that. Everyone knows that a job in the service sector is a hobby, and retail is a place where people go for "experience," not a livelihood.

Nowhere has this message been more successfully absorbed than at the cash register and the takeout counter, where many workers say they feel as if they are just passing through even after logging a decade in the McWork sector. Brenda Hilbrich, who works at Borders Books and Music in Manhattan, explains how difficult it is to reconcile the quality of her employment with a sense of personal success: "You're stuck with this dichotomy of 'I'm supposed to do better but yet I can't because I can't find another job.' So you tell yourself, 'I'm only here temporarily because I'm going to find something better.'"[4] This internalized state of perpetual transience has been convenient for service-sector employers who have been free to let wages stagnate and to provide little room for upward mobility, since there is no urgent need to improve the conditions of jobs that everyone agrees are only temporary. Borders clerk Jason Chappell says that the retail chains work hard to reinforce feelings of transience in their workers in order to protect this highly profitable formula. "So much of the company propaganda is convincing you that you're not workers, that it's something else, that you're not working class.... Everyone thinks they are middle class even when they're making $13,000 a year."[5]

TRACY
Mc
d's

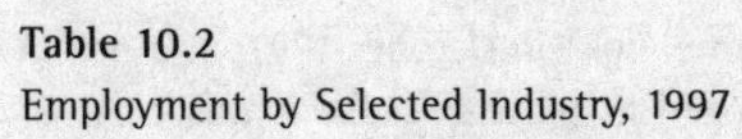

Table 10.2

Employment by Selected Industry, 1997

U.S.

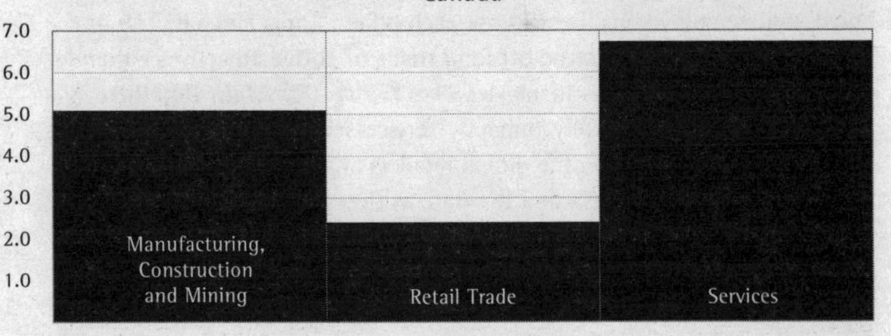

Canada

millions of people employed
7.0
6.0
5.0
4.0
3.0
2.0
1.0
Manufacturing, Construction and Mining
Retail Trade
Services

U.K.

Source for U.S. figures: "Employment and Earnings," Bureau of Labor Statistics. Source for Canadian figures: "Annual Estimates of Employment, Earnings and Hours 1985-1997," Statistics Canada. Source for U.K. figures: Office for National Statistics.

I met with Chappell and Hilbrich late one night in October 1997, at a deli in Manhattan's financial district. We chose this place because it was close to the Borders outlet at the base of the World Trade Center where they both work. I had heard about the pair because of their successful efforts to bring a union to Borders, part of a flurry of labor organizing inside the large chains since the mid-nineties: at Starbucks, Barnes & Noble, Wal-Mart, Kentucky Fried Chicken, McDonald's. It seems as if more and more of the twenty-something-going-on-thirty-something clerks working for the super-brands are looking around – at the counters in front of them where they serve Sumatran coffee, and at the best-selling books, and made-in-China sweaters – and are acknowledging that, for better or worse, some of them aren't going anywhere fast. Laurie Bonang, who works at Starbucks in Vancouver, British Columbia, told me that "people our age are finally realizing that we get out of university, we're a zillion dollars in debt, and we're working in Starbucks. This isn't how we want to spend the rest of our lives, but for right now the dream job isn't waiting for us anymore.... I was hoping that Starbucks would be a stepping stone to bigger and better things, but unfortunately it's a stepping stone to a big sinkhole."[6]

As Bonang told her story, she was painfully aware that she is living out one of the most hackneyed pop-culture clichés of our branded age: this is the stuff of *Saturday Night Live*'s "Gap Girls" skit, circa 1993, in which bored, underemployed mall chicks ask each other: "Didja cinch it?" Or of the Starbucks "baristas" who rattle off long trains of coffee adjectives – grande-decaf-low-fat-moccacino – in movies like *You've Got Mail.* But there is a reason why the most vocally unhappy service-sector workers are the ones working for the highest-profile global retailers and restaurants. Large chains such as Wal-Mart, Starbucks and the Gap, as they have proliferated since the mid-eighties, have been lowering workplace standards in the service sector, fueling their marketing budgets, imperialistic expansion and high-concept "retail experiences" by lowballing their clerks on wages and hours. Most of the big-name brands in the service sector pay the legal minimum wage or slightly more, even though the average wage for retail workers is several dollars higher.[7] Wal-Mart clerks in the U.S., for instance, earn an average of $7.50 an hour and since Wal-Mart classifies "full time" as twenty-eight hours a week, the average annual income is $10,920 – significantly less than the industry average. (See Table 10.4, Appendix, page 476.)[8] Kmart wages are also low and the benefits are considered so substandard that when a 172,000-square-foot Super Kmart opened in San Jose, California, in October 1997, the local city council voted to endorse a boycott of the retailer. Council member Margie Fernandes said that the low wages, minimal health benefits and part-time hours are far below those provided by other area retailers, and that these are not the kind of jobs the community needs. "San Jose is a very, very expensive place to live and we need to make sure the people who work here can afford to live here," Fernandes explained.[9]

McDonald's and Starbucks staff, meanwhile, frequently earn less than the employees of single-outlet restaurants and cafés, which explains why McDonald's is widely credited for pioneering the throwaway "McJob" that the entire fast-food industry has since moved to emulate. At Britain's McLibel Trial, in which the company contested claims made by two Greenpeace activists about its employment practices, international trade unionist Dan Gallin defined a McJob as "a low skill, low pay, high stress, exhausting and unstable job."[10] Though the activists on trial for libel were found guilty on several counts, in his verdict Chief Justice Rodger Bell ruled that in the matter of McJobs the defendants had a point. The chain has had a negative impact on food-service wages as a whole, he wrote, and the allegation that McDonald's "pays its workers low wages, helping to depress wages for workers in the catering trade in Britain has been proved to be true. It is justified."[11]

As we have seen in Cavite, the brand-name multinationals have freed themselves of the burden of providing employees with a living wage. In the malls of North America and England, on the high street, in the food court and at the superstore, they have managed a similar trick. In some cases, particularly in the garment sector, these retailers are the very same companies that are doing business in the export processing zones, meaning that their responsibilities as employers have been sharply reduced at both the production and service ends of the economic cycle. Wal-Mart and the Gap, for instance, contract out their production to EPZs dotting the Southern Hemisphere, where goods are produced mostly by women in their teens and twenties who earn minimum wage or less and live in cramped dorm rooms. Those goods – sweatshirts, baby clothes, toys and Walkmans – are then sold by another workforce, concentrated in the North, which is also largely filled with young people earning approximately minimum wage, most in their teens and early twenties.

Though in many ways it is indecent to compare the relative privilege of retail workers at the mall with the abuse and exploitation suffered by zone workers, there is an undeniable pattern at work. In general, the corporations in question have ensured that they do not have to confront the possibility that adults with families are depending on the wages that they pay, whether at the mall or in the zone. Just as factory jobs that once supported families have been reconfigured in the Third World as jobs for teenagers, so have the brand-name clothing companies and restaurant chains given legitimacy to the idea that fast-food and retail-sector jobs are disposable, and unfit for adults.

As in the zones, the youthfulness of the sector is far from accidental. It reflects a distinct preference on the part of service-sector employers, achieved through a series of overt and covert management actions. Young workers are consistently hired over older ones, and workers who have been on staff for a few years – building up higher wages and seniority – often report losing precious shifts to new batches of younger and cheaper clerks. Other anti-adult tactics have included the targeting of older workers for harassment – the issue that served as the catalyst for the first strike at a McDonald's outlet. In April 1998, after witnessing a verbally abusive supervisor reduce an elderly co-worker to tears, the teenage workers at the Golden Arches in Macedonia, Ohio, walked off the job in protest. They didn't return until management agreed to undergo "people skills" training. "We get verbally harassed, and physically too. Not me, but basically just the elderly woman," teen striker Bryan Drapp said on *Good Morning America*. Drapp was fired two months later.[12]

Brenda Hilbrich of Borders contends that justifying low wages on the grounds that young workers are just passing through is a handy self-fulfilling prophecy – particularly in her field, bookselling. "It doesn't have to have a high turnover," she says. "If the conditions are good and you're making a nice salary, people actually like working in the service industry.

They like working with books. A lot of people who have left have said, 'This was my favorite job, but I had to go because I can't make enough money to live.'"[13]

The fact is that the economy needs steady jobs that adults can live on. And it's clear that many people would stay in retail if it paid adult rates, the proof being that when the sector does pay decently, it attracts older workers, and the rate of staff turnover falls in line with the rest of the economy. But at the large chains, which seem at least for now to have bottomless resources to build superstores and to sink millions into expanding and synergizing their brands, the idea of paying a living wage is rarely considered. At Borders, where most clerks earn wages in line with other bookstore chains but below the retail average, company president Richard L. Flanagan wrote a letter to all his clerks, addressing the question of whether Borders could pay a "living wage" as opposed to what it reportedly pays now – between US$6.63 and $9.27 an hour. "While the concept is romantically appealing," he wrote, "it ignores the practicalities and realities of our business environment."[14]

Much of what makes paying a living wage seem so "romantic" has to do with the rapid expansion described in Part II, "No Choice." For companies whose business plans depend upon becoming dominant in their market before their nearest competitor beats them to it, new outlets come before workers – even when those workers are a key part of the chain's image. "They expect us to look like a Gap ad, professional, clean and neat all the time, and I can't even pay to do laundry," says Laurie Bonang of Starbucks. "You can buy two grande mocha cappuccinos with my hourly salary." Like millions of her demographic coevals on the payrolls of all-star brands like the Gap, Nike and Barnes & Noble, Bonang is living inside a stunning corporate success story – though you'd never know it from the resignation and anger in her voice. All the brand-name retail workers I spoke with expressed their frustration at helping their stores rake in, to them, unimaginable profits, and then having to watch that profit get funneled into compulsive expansion. Employee wages, meanwhile, stagnate or even decline. At Starbucks in British Columbia new workers faced an actual wage decrease – from Can$7.50 to $7 an hour – during a period when the chain was doubling its profits and opening 350 new stores a year. "I do the banking. I know how much the store pulls in a week," Laurie Bonang says. "They just take all that revenue and open up new stores."[15]

Borders clerks also maintain that wages have suffered as a result of rapid growth. They say that their chain used to be a more equitable place to work before the neck-and-neck race with Barnes & Noble took over corporate priorities; there was a profit-sharing program and a biannual 5 percent raise for all workers. "Then came expansion and corresponding cuts," reads a statement from disgruntled employees at a downtown Philadelphia outlet of Borders. "Profit sharing was dropped, raises were cut..."[16]

In sharp contrast to the days when corporate employees took pride in their company's growth, seeing it as the result of a successful group effort, many clerks have come to see themselves as being in direct competition with their employers' expansion dreams. "If Borders opened thirty-eight new stores a year instead of forty," reasoned Jason Chappell, sitting next to Brenda Hilbrich on the vinyl seats of our deli booth, "they could afford to give us a nice wage increase. On average it costs $7 million to open a superstore. That's Borders' own figures...."

"But," Brenda interrupted, "if you say that directly to them, they say, 'Well, that's two markets we don't get into.'"

"We have to saturate markets," Chappell said, nodding.

"Yeah," Brenda added. "We have to compete with Barnes & Noble."

The retail clerks employed by the superchains are only too familiar with the manic logic of expansion.

Busting the McUnion

The need to prevent workers from weighing too heavily on the bottom line is the main reason that the branded chains have fought off the recent wave of unionization with such ferocity. McDonald's, for instance, has been embroiled in bribery scandals during German union drives, and over the course of a 1994 union drive in France, ten McDonald's managers were arrested for violating labor laws and trade-union rights.[17] In June 1998, the company fired the two young workers who organized the strike in Macedonia, Ohio.[18] In 1997, when the employees at a Windsor, Ontario, Wal-Mart were about to hold an election on joining a union, a series of not-so-subtle management hints led many workers to believe that if they voted yes their store would be shut down. The Ontario Labour Relations Board reviewed the process and found that the behavior of Wal-Mart managers and supervisors before the vote amounted to "a subtle but extremely effective threat," which caused "the average reasonable employee to conclude that the store would close if the union got in."[19]

Other chains have not hesitated to make good on the threat to close. In 1997, Starbucks decided to shut down its Vancouver distribution plant after workers unionized. In February 1998, just as a union certification for a Montreal-area outlet of McDonald's was being reviewed by the Quebec Labour Commission, the franchise owner closed down the outlet. Shortly after the closure, the labor commission accredited the union – cold comfort, since no one works there anymore. Six months later, another McDonald's restaurant was successfully unionized, this one a busy outlet in Squamish, British Columbia, near the Whistler ski resort. The organizers were two teenage girls, one sixteen, the other seventeen. It wasn't about wages, they said – they were just tired of being scolded like children in front of the customers. The outlet remains open, making it the only unionized McDonald's in North America, but at the time of writing, the company was on the verge of having the union decertified. Fighting the battle on the public-relations front, in mid-1999 the fast-food chain launched an international television campaign featuring McDonald's workers serving up shakes and fries under the captions "future lawyer," "future engineer" and so on. Here was the true McDonald's workforce, the company seemed to be saying: happy, contented and just passing through.

During the late 1990s, the process of turning the service sector into a low-wage ghetto advanced rapidly in Germany. The German unemployment rate reached 12.6 percent in 1998, primarily because the economy could not absorb the massive layoffs in the manufacturing sector that occurred after reunification – four out of five East German factory jobs were lost. To make up for the shortfall, the service sector was touted by the business press and the political right wing as the economic panacea. There was just one catch: before the mall could step in to save the German economy, the minimum wage would have to be substantially lowered and benefits such as long holidays for all workers would have to be dismantled. In other words, good jobs with security and a living wage would have to be turned into bad jobs. Then Germany too would enjoy the benefits of a service-based economic recovery.

It is one of the paradoxes of service-sector employment that the more prominent a role it plays in the labor landscape, the more casual service-sector companies became in their attitude toward providing job security. Nowhere is this more in evidence than in the industry's increasing reliance on part-timers. (See Table 10.5, Appendix, page 476.) Starbucks, for instance, staffs its outlets almost exclusively with part-timers while only one-third of Kmart's workforce is full-time. Workers at the ill-fated Montreal-area McDonald's cited as their principal reason for unionization the fact that they often couldn't get shifts longer than three hours.

In the U.S. the number of part-timers has tripled since 1968, while in Canada, between 1975 and 1997, the growth rate of part-time jobs was nearly three times the rate of full-time jobs.[20] But the problem is not the part-time nature of work per se. In Canada, only one-third of part-timers want but cannot find full-time jobs (which is an increase from one-fifth in the late eighties). In the U.S., only one-quarter want full-time jobs but can't find them. The vast majority of part-timers are students and women, many of whom are juggling childcare and paid work. (See Table 10.6, Appendix, page 476.)

But while many workers are indeed drawn to flexible work arrangements, their definition of what constitutes "flexibility" is dramatically different from the one favored by service-sector bosses. For instance, while studies have shown that working mothers define flexibility as "having the ability to work less than full-time hours at decent wages and benefits, while still working a regular schedule,"[21] the service sector has a different view of part-time work, and a different agenda. A handful of brand-name chains, including Starbucks and Borders, bolster low wages by offering health and dental benefits to their part-timers. For other employers, however, part-time positions are used as a loophole to keep wages down and to avoid benefits and overtime; "flexibility" becomes a code for "no promises," making the juggling of other commitments – both financial and parental – more challenging, not less. At some retail outlets I've researched, the allotment of hours is so random that the ritual of posting next week's schedule prompts the staff to gather around anxiously, craning their necks and hopping up and down as if they are checking to see who got the lead in the high-school musical.

Furthermore, the "part-time" classification is often more a technicality than a reality, with retail employers keeping their part-timers just below the forty-hour legal cutoff for full-time – Laurie Bonang, for instance, clocks between thirty-five and thirty-nine hours a week at Starbucks. For all intents and purposes, she has the duties of a full-time employee, but under forty hours the company does not have to pay overtime or guarantee full-time hours. Other chains are equally creative. Borders instituted a company-wide thirty-seven-and-a-half-hour work week for all employees, and Wal-Mart caps its work week at thirty-three hours, defining base "full time" as twenty-eight hours. What all of this means in the lives of workers is a scheduling roller coaster that in many ways is more demanding than the traditional forty-hour week. For instance, the Gap – which defines full-time as thirty hours a week – has a system of keeping clerks "on call" for certain shifts during which time they aren't scheduled or paid to work but must be available to come in if the manager calls. (One worker joked to me that she had to buy a beeper in case a folding crisis flared up in Gap Kids.)

Starbucks has been the most innovative in the modern art of supple scheduling. The company has created a software program called Star Labor that allows head office maximum control over the schedules of its clerks down to the minute. With Star Labor, gone is anything as blunt and imprecise as a day or evening shift. The software measures exactly when each latte is sold and by whom, then tailor-makes shifts – often only a few hours long – to maximize coffee-selling efficiency. As Laurie Bonang explains, "They give you an arbitrary skill number from one to nine and they plug in when you're available, how long you've been there, when customers come in and when we need more staff, and the computer spits out your schedule based on that."[22] While Starbucks' breakthrough in "just-in-time" frothing looks great on a spreadsheet, for Steve Emery it meant hauling himself out of bed to start work at 5 a.m., only to leave at 9:30 a.m. after the morning rush had peaked and, according to Star Labor, he was no longer working at maximum efficiency. Wal-Mart has introduced a similar centralized scheduling system, effectively reducing employee hours by pinning them precisely to in-store traffic. "It's done just like we order merchandise," says Wal-Mart CEO David Glass.[23]

The vast gulf between employee and employer definitions of "flexibility" was the central issue of the United Parcel Service strike in the summer of 1997, the largest U.S. job action in fourteen years. Despite profits of $1 billion in 1996, UPS had kept 58 percent of its workers classified as part-time and was rapidly moving toward an even more "flexible" workforce. Of the 43,000 jobs UPS had created since 1992, only 8,000 were full time. The system worked well for the courier company, since it was able to ride the peaks and valleys of the delivery cycle that sees heavy pickups and deliveries in the morning and evening but lulls during the day. "There's too much downtime in between to hire full-time workers," explained UPS spokesperson Susan Rosenberg.[24]

Building up a part-time workforce had other cost-saving benefits. Before the strike, the company paid its part-timers roughly half the hourly wage of its full-timers for performing the same tasks.[25] Furthermore, the union claimed that 10,000 of the company's so-called part-timers were, like Laurie Bonang at Starbucks, actually working between thirty-five and thirty-nine hours a week—just under the cutoff that would require overtime pay, full benefits and the higher wage scale.

Some service-sector companies have made much of the fact that they offer stock options or "profit-sharing" to low-level employees, among them Wal-Mart, which calls its clerks "sales associates"; Borders, which refers to them as "co-owners"; and Starbucks, which prefers the term "partners." Many employees do appreciate these gestures, but others claim that while the workplace democracy schemes sparkle on a corporate Web site, they rarely translate into much of substance. Most part-time workers at Starbucks, for instance, can't afford to buy into the employee stock-option program since their salaries barely cover their expenses. And where profit-sharing schemes are automatic, as at Wal-Mart, workers say their "share" of the $118 billion of annual sales their company hauls in is laughable. Clerks in the Windsor, Ontario, outlet of Wal-Mart, for example, say they only saw an extra $70 during the first three years that their store was open. "Never mind that from the viewpoint of the boardroom, the pension plan's best feature was that it kept 28 million more shares in firm control of company executives," writes *The Wall Street Journal*'s Bob Ortega of the Wal-Mart plan. "Most workers *perceived* that they could cash in, so the cost of the plan paid off in spades by helping keep the unions out and the wages low" (italics his).[26]

Free Work: More Fake Jobs, Courtesy of the Superbrands

One thing you can say about the retail and service industries: at least they pay their workers a little something for their trouble. Not so for some other industries that have liberated themselves from the chains of social-security forms with such free-market gusto that many young workers receive no pay from them at all. Perhaps predictably, the culture industry has led the way in the blossoming of unpaid work, blithely turning a blind eye to the unglamorous fact that many people under thirty are saddled with the mundane responsibility of actually having to support themselves.

Writing about his former job, which involved hiring unpaid interns to send faxes and run errands for *Men's Journal* magazine, Jim Frederick notes that many of his applicants had already worked for nothing at *Interview*, CBS News, MTV, *The Village Voice* and so on. "'Very impressive,' I would say. By my quick calculations they had contributed, conservatively, five or six thousand dollars' worth of uncompensated work to various media conglomerates."[27] Of course, the media conglomerates – the broadcasters, magazines and book publishers – insist that they are generously offering young people precious experience in a hard employment market – a foot in the door on the old-fashioned "apprenticeship" model. Besides, they say, sounding suspiciously like McDonald's managers the world over, the interns are just kids – they don't really *need* the money.

And getting two "unreal" jobs for the price of one, most interns subsidize their unpaid day job by working in the service industry at night and on weekends, as well as by living at home to a later age. But in the U.S. – where it has become commonplace to hop from one unpaid culture job to the next for a year or two – a disproportionate number of interns, as Frederick observes, appear to be living off trust funds, seemingly without any immediate concerns about earning a living. But just as the service-sector employers will not admit that the youthfulness of their workforce might have something to do with the wages they pay and the security they fail to offer, you will never catch a television network or a publisher confessing that the absence of remuneration for internships might also have something to do with the relative privilege of those applying for these positions at their companies. This racket is not only exploitative in the classic sense, it also has some very real implications for the future of cultural production: today's interns are tomorrow's managers, producers and editors and, as Frederick writes, "If you can't get a job unless you've had an internship, and you can't take an internship unless you can get supported by daddy for a couple of months, then the system guarantees an applicant pool that is decidedly privileged."[28]

Music video stations such as MTV have been among the more liberal users of the unpaid internship system. When it was first introduced, the music video channel represented a managerial coup in low-cost, high-profit broadcasting since the stations primarily play videos that are produced out of house and supplied by record labels. While some stations, including Canada's MuchMusic, now play licensing and royalty fees to broadcast videos, these pale in comparison to the production costs of the videos in a single top 30 countdown. Inside the stations, on air-hosts, producers and technicians work alongside unpaid, mostly student, interns who sometimes are rewarded with jobs and sometimes stay at the station for many months, hoping for their big break. Which is where the legendary success stories come in – the famous V.J. who started off answering phones, or the greatest success story of them all: the tale of Rick the Temp. In 1996, Rick won the annual "Be a Temp at MuchMusic Contest" and was welcomed to the station with cross promotional fanfare and branded giveaways. One year later, Rick was on the air in his new job as V.J., but the kicker was that even after he became a big star, he kept the moniker Rick the Temp. There was Rick on TV, interviewing the Backstreet Boys, and although he was always paid for his work, for would-be interns, his success served a daily advertisement for the glory and glamour that awaits if you donate your labor as a gift to a major media company.

Temps: The Rented Worker

Rick the Temp isn't just the Great White Hope for unpaid interns. He also represents the pinnacle of another subcategory of New Age workers: the temps. And temps, it must be said, need all the hope they can get. The use of temp labor in the U.S. has increased by 400 percent since 1982 and that growth has been steady.[29] Annual industry revenue among American temp firms has increased by about 20 percent every year since 1992, with the firms pulling in revenues of $58.7 billion in 1998.[30] The mammoth international temp agency Manpower Temporary Services rivals Wal-Mart as the largest private employer in the U.S.[31] According to a 1997 study, 83 percent of the fastest-growing American companies are now outsourcing jobs they once hired people to perform – compared with 64 percent just three years before.[32] In Canada, the Association of Canadian Search, Employment & Staffing Services estimates that more than 75 percent of businesses use the services of the $2 billion Canadian temp industry.

The most dramatic growth, however, is taking place not in North America but in Western Europe, where temp agencies are among Europe's fastest-growing companies.[33] In France, Spain, the Netherlands and Germany, hiring workers on long-term temporary contracts has become a well-trampled back entranceway to the labor market, allowing employers to sidestep tough laws that provide generous employee benefits and make firing without just cause far more difficult than in the United States. France, for instance, has become the second-largest temp-services market after the U.S., making up 30 percent of worldwide temp revenue. And though temping accounts for only 2 percent of all the country's jobs, according to France's labor minister, Martine Aubry, "86 per cent of new hires are on short-term contracts."[34] Manpower Europe, an outpost of the U.S.-based temp firm, saw its revenue in Spain jump a staggering 719 percent in just one year, from $6.1 million in 1996 to $50 million in 1997. Italy didn't legalize temp agencies until 1997, but when it did, Manpower Europe rushed in to open thirty-five offices in 1998.[35]

These companies all have the formula. They don't take you on full time. They don't pay benefits. Then their profits go through the roof.

– Laura Pisciotti, UPS worker, on strike, August 1997

Every day, 4.5 million workers are assigned to jobs through temp agencies in Europe and the U.S., but since only 12.5 percent of temps are placed on any given day, the real number of total temporary employees in Europe and the U.S. is closer to 36 million people.[36] More significant than soaring numbers, however, is a major shift under way in the nature of the temporary work industry. Temp agencies are no longer strictly in the business of farming out rent-a-receptionists when the secretary calls in sick. For starters, temps are no longer all that temporary: in the U.S., 29 percent stay at the same posting for a year or more.[37] Their agencies, meanwhile, have become full-service human resource departments for all your no-commitment staffing needs, including accounting, filing, manufacturing and computer services. And according to Bruce Steinberg, director of research at the U.S.-based National Association of Temporary and Staffing Services, "a quiet evolution is taking place throughout the staffing services industry" – rather than renting out workers, the agencies are "providing a complete service solution."[38] What that means is that more companies are contracting out entire functions and divisions – work previously performed in-house – to outside agencies charged not only with staffing but, like the contract factories in the export processing zones, administration and maintenance of the task as well. For instance, in 1993 American Airlines outsourced the ticket counters at twenty-eight U.S. airports to outside agencies. Around 550 ticketing-agent jobs went temp and, in some cases, workers who had earned $40,000 were offered their same jobs back for $16,000.[39] A similar reshuffling took place when UPS decided to turn over its customer-service centers to outside contractors – 5,000 employees earning $10 to $12 an hour were replaced with temps earning between $6.50 and $8.[40]

As Tom Peters says, "You're a damn fool if you own it!"[41] Bruce Steinberg concurs: by amputating whole divisions and sloughing them off on "managed services arrangements, the business can concentrate its time, energy and resources on core business while staffing service practices its core competency of managing workers."[42] Hiring and managing workers, in other words, is not the base of a healthy company but a specialized task – somebody else's "core competency" that is better left to the experts, while the real business is tended to by an ever-shrinking number of workers, as the next chapter will show.

Yes, but ... Won't Bill Gates Save Us?

Any discussion of the plight of corporate temps, UPS couriers, outsourced GM workers, Gap greeters, MTV interns and Starbucks "baristas" leads inevitably to the same place: Yes, but ... what about all the great new jobs in the growing high-tech world? For my generation of workers, the legendary riches awaiting technology workers in Seattle and Silicon Valley are the "yes, but" answer to any and all grievances about employment exclusions. Standing in contrast to all the downer stories about layoffs and McJobs is this shimmering digital mecca where fifteen-year-olds design video games for Sega, where AT&T hires hackers just to keep an eye on them and where scores of young workers become millionaires from their lavish stock options. Yes, but ... Bill Gates will make it all okay, won't he?

It was Microsoft, with its famous employee stock-option plan, that developed and fostered the mythology of Silicon Gold, but it is also Microsoft that has done the most to dismantle it. The golden era of the geeks has come and gone, and today's high-tech jobs are as unstable as any other. Part-timers, temps and contractors are rampant in Silicon Valley – a recent labor study of the region estimates that between 27 and 40 percent of the Valley's employees are "contingency workers," and the use of temps there is increasing at twice the rate of the rest of the country. The percentage of Silicon Valley workers employed by temp agencies is nearly three times the national average.[43]

And Microsoft, the largest of the software firms, didn't just lead the way to this part-time promised land, it wrote the operating manual. For more than a decade, the company has been busily closing ranks around the programmers who got there first, and banishing as many other employees as it can from that sacred inner circle. Through extensive use of independent contractors, temps and "full-service employment solutions" Microsoft is well on its way to engineering the perfect employee-less corporation, a jigsaw puzzle of outsourced divisions, contract factories and freelance employees. Gates has already converted one-third of his general workforce into temps, and in the Interactive Media Division, where CD-ROMs and Internet products are developed, about half the workers are officially employed by outside "payroll agencies," who deliver tax-free workers like printer cartridges.[44]

Microsoft's two-tier workforce is a microcosm of the job market's New Age new deal. At the center is the high-tech dream: permanent, full-time employees, with benefits and generous stock options, working and playing on the youthful corporate "campus." These Microserfs are cultishly loyal to their corporation, its soaring stock price and its staggering 51 percent operating profit margin ("Show me the money!" they roared at the annual staff meeting in Seattle's Kingdome Stadium in fall 1997).[45] And why shouldn't they be loyal? They earn an average of $220,000 a year, and that's not even factoring in the top five superrich executives.

Orbiting around this starry-eyed core are between 4,000 and 5,750 temporary workers.[46] The temps work side by side with members of the core group – as technicians, designers and programmers – and perform many of the same jobs. About 1,500 have been with the company for so long they have taken to calling themselves "permatemps." The only way to tell the temps from the "real" Microserfs is by the color of their badges: blue for perms, orange for permatemps.

Like the fleet of part-timers who give UPS the "flexibility" to employ workers only during peak hours, and the contract workers in Cavite who provide their factory owners with the "flexibility" to send them home during dry spells, what thousands of temps means for Microsoft is the freedom to expand and contract its workforce at will. "We use them," says Microsoft personnel officer Doug McKenna, "to provide us with flexibility and to deal with uncertainty."[47]

Trouble began in 1990 when the Internal Revenue Service challenged Microsoft's classification of orange badges as independent contractors, ruling that these people were actually employees of Microsoft and the company should be paying their payroll tax. Based in part on this finding, in 1993 a group of employees classified by Microsoft as contractors launched a lawsuit against the company, claiming they were regular workers and deserved the same benefits and stock options as their permanent colleagues. In July 1997, Microsoft lost the landmark case when an eleven-judge Court of Appeals panel ruled that the freelancers were "common law" employees and had the right to the company's benefits program, to its pension and to its stock-purchasing plan.[48]

Microsoft's response to this setback, however, has not been to add freelancers to its payroll but simply to work more assiduously to marginalize the temps. To this end, the company has moved away from hiring "independent contractors" directly. Instead, after employees have been scouted, interviewed and selected by Microsoft, they are instructed to register with one of five payroll agencies that have special arrangements with the company. MicroTemps are then hired through an agency that acts as the official employer: cutting paychecks, withholding income taxes and sometimes providing bare-bones benefits. Laird Post, a principal with management consultant Towers Perrin in Seattle, Washington, explains the legalities of this new arrangement. "It's hard to rationalize legally that the person is not an employee unless they are an employee of someone else" – in Microsoft's case, that someone else is the payroll agency.[49] To make sure that the temps will never again be confused with actual Microsoft workers, they are barred from all extracurricular company functions, including taking part in late-night pizza meals and after-hours parties. And in June 1998 the company introduced a new policy requiring temps who have been on an assignment with the company for a year or more to take a thirty-one-day break before they can take another "temporary" post.[50] As Sharon Decker, Microsoft's director of contingency staffing, explains, "We are refocusing a lot of policies we had in place so everyone understands how a temp should be treated and what is appropriate."[51]

In addition to staffing its campus with permatemps, in 1997 Microsoft initiated a series of moves to disentangle itself from other earthly and cumbersome aspects of running a multibillion-dollar company. "Don't get caught with useless fixed assets," Bob Herbold, Microsoft's chief operating officer, says, explaining his staffing philosophy to a group of shareholders.[52] According to Herbold, pretty much everything but the core functions of programming and product development fall into the "useless fixed assets" category – including the company's sixty-three receptionists, who were laid off, losing benefits and stock options, and told to reapply through the Tascor temp agency. "We were overpaying them," Herbold said.[53]

In the same stroke, Microsoft sliced and diced its Redmond campus and parceled out the pieces (along with employees who wanted to hold on to their jobs) to outside "vendors": Pitney Bowes took over the mail room; the print and copy center is now operated by Xerox personnel; the CD-ROM factory was sold to KAO Information Systems; even the company store was outsourced to Benussen Deutsch & Associates. In this latest round of restructuring, 680 jobs were cut from the payroll and $500 million slashed from the operating budget.[54] With all these contractors on the campus, Herbold noted, "just managing the outsourcers is quite a task" – and there was no reason for Microsoft to get saddled with that useless fixed asset. In a stroke of divestment genius, Microsoft contracted out the task of managing the contractors to Johnson Controls, which also takes care of the campus facilities. "Our revenue has gone up 91 percent and our head count has actually decreased 19 percent," Bob Herbold says proudly. And what did Microsoft do with the savings? "We're plowing them into R&D and we're plowing them into profit, obviously."[55]

"Free Agent Nation"

It must be said that many of Microsoft's high-tech freelancers are hardly defenseless victims of Bill Gates's payroll concoctions, but are freelancers by choice. Like many contractors, the "software gypsies," as high-tech freelancers are sometimes called, have made a conscious decision to put independence and mobility before institutional loyalty and security. Some of them are even what Tom Peters likes to call a "Brand Called You."

Tom Peters's latest management-guru idea is that just as companies must reach branding nirvana by learning to let go of manufacturing and employment, so must individual workers empower themselves by abandoning the idea of being employees. According to this logic, if we are to be successful in the new economy, all of us must self-incorporate into our very own brand – a Brand Called You. Success in the job market will only come when we retrofit ourselves as consultants and service providers, identify our own Brand You equities and lease ourselves out to targeted projects that will in turn increase our individual portfolio of "braggables." "I call the approach Me Inc.," Peters writes. "You're Chairperson/CEO/Entrepreneur-in-Chief of your own professional service firm."[56] Faith Popcorn, the management guru who came to prominence with her 1991 best-seller, *The Popcorn Report*, goes so far as to recommend that we change our names to better "click" with our carefully designed and marketed brand image. *She* did – her name used to be Faith Plotkin.

Even more than Popcorn or Peters, however, it is a man named Daniel H. Pink who is the dean at Brand You U. Pink has seen the growth in temporary and contract work, as well as the rise in self-employment, and has declared the arrival of "Free Agent Nation." Not only is he writing a book by that title, but Pink himself is a proud patriot of the nation. After quitting a prestigious White House job as Al Gore's chief speechwriter, Pink went on a journey in search of fellow "free agents": people who had chosen a life of contracts and freelance gigs over bosses and benefits. What he found, as he relayed in a cover article in *Fast Company*, was the sixties. The citizens of Pink's nation are marketing consultants, headhunters, copywriters and software designers who are all striving to achieve a Zen-like balance of work and personal life. They practice their yoga positions and play with their dogs in their wired home offices, while earning more money – by jumping from one contract to the next – than they did when they were tied to one company and paid a fixed salary. "This is the summer of love revisited, man!" we hear from Bo Rinald, an agent representing a thousand freelance software developers in Silicon Valley.[57] For Pink's free agents, the end of jobs is the baby-boomer dream come true: free-market capitalism without neckties; dropped out of the corporate world in body but plugged-in in spirit. Everyone knows that you can't be a cog in the machine if you work from your living room....

A younger – and, of course, hipper – version of Free Agent Nation was articulated in a special work issue of *Details* magazine. For Gen-Xers with MBAs, the future of work is apparently filled with stunningly profitable snowboarding businesses, video-game companies and cool-hunting firms. "Opportunity Rocks!" crowed the headline of an article that laid out the future of work as a nonstop party of extreme self-employment: "Life without jobs, work without bosses, money without salaries, lives without limits."[58] According to the writer, Rob Lieber, "The time of considering yourself an 'employee' has passed. Now it's time to start thinking of yourself as a service provider, hiring out your skills and services to the highest, or most interesting, bidder."[59]

I admit to being lured by the sirens of free agency myself. About four years ago, I quit my job as a magazine editor to go freelance, and like Pink I've never looked back. Of course I love the fact that no one boss controls my every working hour (that privilege is now spread around to dozens of people), that I'm not subject to the arbitrary edicts of petty managers and, most important, that I can work in my pajamas if I feel like it. I know from first-hand experience that freelance life can indeed mean freedom, just as part time, for others, can live up to its promise of genuine flexibility. Pink has a point when he says of free agency, "This is a legitimate way to work – it isn't some poor laid-off slob struggling to find his way back to the corporate bosom."[60] However, there's a problem when it's people like Pink – or other freelance writers overly euphoric about working in their pajamas – who hold themselves up as living proof that divestment from corporate employment is a win-win formula. And it does seem as if most of the major articles about the joys of freelancing have been written by successful freelance writers under the impression that they themselves represent the millions of contractors, temps, freelancers, part-timers and the self-employed. But writing, because of its solitary nature and low overhead, is one of the very few professions that are genuinely compatible with homework, and study after study shows that it is absurd to equate the experience of being a freelance journalist, or having your own advertising company, with that of being a temp secretary at Microsoft or a contract factory worker in Cavite. On the whole, casualization pans out as the worst of both worlds: monotonous work at lower wages, with no benefits or security, and even less control over scheduling.

The bottom line is that the advantages and drawbacks of contract and contingency work have a simple correlation to the class of the individuals doing the work: the higher up they are on the income scale, the more chance they have to leverage their comings and goings. The further down they are, the more vulnerable they are to being yanked around and bargained even lower. The top 20 percent of wage earners tend to more or less maintain their high wages whether they are in full-time jobs or on freelance contracts. But according to a 1997 U.S. study, 52 percent of women in nonstandard work arrangements are being paid "poverty-level wages" – compared with only 27.6 percent in the full-time female worker population being paid those low wages. In other words, most nonstandard workers aren't members of Free Agent Nation. According to the study, "58.2 per cent are in the lowest quality work arrangements – jobs with substantial pay penalties and few benefits relative to full-time standard workers." (See Table 10.7, Appendix, page 477.)[61] Furthermore, the real wages of temp workers in the U.S. actually went down, on average, by 14.7 percent between 1989 and 1994.[62] In Canada, nonpermanent jobs pay one-third less than permanent jobs, and 30 percent of nonpermanent employees work irregular hours.[63] Clearly, temping puts the most vulnerable workforce further at risk, and no matter what *Details* says, it doesn't rock.

Moreover, there is a direct cause-and-effect relationship between the free agents skipping and hopping on the top rungs of the corporate ladder, and the agents hanging off the bottom who have been "freed" of such pesky burdens as security and benefits. Nobody is more liberated, after all, than the CEOs themselves, who, like Nike's cabal of Über-athletes, have formed their own Dream Team to be traded back and forth between companies whenever some star power is needed to boost Wall Street morale. Temp CEOs, as writer Clive Thompson calls them, now shuttle from multinational to multinational, staying for an average term of only five years, collecting multimillion-dollar incentive packages on the way in, and multimillion-dollar golden handshakes on the way out.[64] "Companies are changing executives like baseball managers," says John Challenger, executive vice president of the outplacement firm Challenger, Gray & Christmas. "The replacement will typically arrive like a SWAT team and sweep out the old and restaff with his or her own people."[65] When "Chainsaw" Al Dunlap was appointed CEO of Sunbeam in July 1996, Scott Graham, an analyst at Oppenheimer & Co., commented, "This is like the Lakers signing Shaquille O'Neal."[66]

The two extreme poles of workplace transience – represented by the contractor in Cavite afraid of flying factories, and the temp CEO unveiling restructuring plans in New York – work together like a global seesaw. Since the CEO superstars earn their reputation on Wall Street through such kamikaze missions as auctioning off their company's entire manufacturing base or initiating a grandiose merger that will save millions of dollars in job duplication, the more mobile the CEOs become, the more unstable the position of the broader workforce will be. As Daniel Pink points out, the word "free-lance" is derived from the age when mercenary soldiers rented themselves – and their lances – out for battle. "The free lancers roamed from assignment to assignment – killing people for money."[67] Granted it's a little dramatic, but it's not a half-bad job description for today's free-agent executives. In fact, it is the precise reason CEO salaries skyrocketed during the years that layoffs were at their most ruthless. Ira T. Kay, author of *CEO Pay and Shareholder Value*, knows why. Writing in *The Wall Street Journal*, Kay points out that the exorbitant salaries American companies have taken to paying their CEOs is a "crucial factor making the U.S. economy the most competitive in the world" because without juicy bonuses company heads would have "no economic incentive to face up to difficult management decisions, such as layoffs." In other words, as satirist Wayne Grytting retorted, we are "supporting those executive bonuses so we can get... fired."[68]

It's a fair enough equation, particularly in the U.S. According to the AFL-CIO, "the CEOs of the 30 companies with the largest announced layoffs saw their salaries, bonuses, and long-term compensation increase by 67.3 per cent."[69] The man responsible for the most layoffs in 1997 – Eastman Kodak CEO George Fisher, who cut 20,100 jobs – received an options grant that same year estimated to be worth $60 million.[70] And the highest-paid man in the world in 1997 was Sanford Wiell, who earned $230 million as head of the Travelers Group. The first thing Wiell did in 1998 was announce that Travelers would merge with Citicorp, a move that, while sending stock prices soaring, is expected to throw thousands out of work. In the same spirit, John Smith, the General Motors chairman implementing those 82,000 job cuts discussed in the last chapter, received a $2.54 million bonus in 1997 that was tied to the company's record earnings.[71]

There are many others in the business community who, unlike Ira T. Kay, are appalled by the amounts executives have been paying themselves in recent years. In *Business Week*, Jennifer Reingold writes with some disgust, "Good, bad, or indifferent, virtually anyone who spent time in the corner office of a large public company in 1997 saw his or her net worth rise by at least several million."[72] For Reingold, the injustice lies in the fact that CEOs are able to collect raises and bonuses even when their company's stock price drops and shareholders take a hit. For instance, Ray Irani, CEO of Occidental Petroleum, collected $101 million in compensation in 1997, the same year that the company lost $390 million.

This camp of market watchers has been pushing for CEO remuneration to be directly linked to stock performance; in other words, "You make us rich, you get a healthy cut. But if we take a hit, then you take one too." Though this system protects stockholders from the greed of ineffective executives, it actually puts ordinary workers at even greater risk, by creating direct incentives for the quick and dirty layoffs that are always sure to rally stock prices and bring on the bonuses. For instance, at Caterpillar – the model of the incentive-driven corporation – executives get paid in stocks that have consistently been inflated by massive plant closures and worker wage rollbacks. What is emerging out of this growing trend of tying executive pay to stock performance is a corporate culture so damaged that workers must often be fired or shortchanged for the boss to get paid.

This last point raises the most interesting question of all, I think, about the long-term effect of the brand-name multinationals' divestment of the jobs business. From Starbucks to Microsoft, from Caterpillar to Citibank, the correlation between profit and job growth is in the process of being severed. As Buzz Hargrove, president of the Canadian Auto Workers, says, "Workers can work harder, their employers can be more successful, but – and downsizing and outsourcing are only one example – the link between overall economic success and the guaranteed sharing in that success is weaker than ever before."[73] We know what this means in the short term: record profits, giddy shareholders and no seats left in business class. But what does it mean in the slightly longer term? What of the workers who fell off the payroll, whose bosses are voices on the phone at employment agencies, who lost their reason to take pride in their company's good fortune? Is it possible that the corporate sector, by fleeing from jobs, is unwittingly pouring fuel on the fire of its own opposition movement?

Top: Bill Gates, Microsoft President and CEO, gets pied. *Bottom:* Biotic Baking Brigade strikes again. Economist Milton Friedman, architect of the global corporate takeover, gets his just deserts.

CHAPTER ELEVEN

BREEDING DISLOYALTY

What Goes Around, Comes Around

In our manufacturing, administrative, and distribution facilities, we have a specific philosophy – cameras keep honest people honest.

– Leo Myers, safety and security systems engineer for Mattel, explains the company's enthusiastic use of video surveillance on its global workforce, 1990

When I dropped out of university in 1993, I could count on the fingers of one hand the number of my friends who had jobs. "The Recession," we repeated to one another over and over again, through years of jobless summers, through listless decisions to slog it out in grad school, through periods of cutbacks to our universities, through miserable stretches when parents were out of work. Just as we would later blame El Niño for everything from droughts to floods, the Recession was an economic bad weather system that had sucked up all the jobs as if they were Missouri trailer parks.

When the jobs disappeared, we understood that it was a result of the tough economic times that seemed to be affecting everyone (though perhaps not everyone equally) from company presidents facing bankruptcy to ax-wielding politicians – everybody, men and women, old and young, in all walks of life and work, right on down to me and my middle-class friends and our halfhearted job searches. The shift from the Recession to the cutthroat global economy happened so suddenly I feel as if I was sick that day and missed the whole thing – as with Grade 10 algebra, I will forever be playing catch-up. All I know is that one minute we were all in the Recession together. The next, a new strain of business leader was rising like a phoenix from the ashes – suit freshly pressed, enthusiasm pumped – announcing the arrival of a new golden age. But as we have seen in the last two chapters, when the jobs came back (*if* the jobs came back), they came back changed. For the workers in the contract factories of the export processing zones, and for the legions of temps, part-timers, contract and service-sector workers in industrialized countries, the modern employer has begun to look like a one-night stand who has the audacity to expect monogamy after a meaningless encounter. And many of them even got it for a while. Running scared from years of layoffs and gloomy economic projections, most of us did swallow the rhetoric that we should be happy picking up whatever pay stubs were scattered our way. There is mounting evidence, however, that workplace transience is finally eroding our collective faith, not only in individual corporations but in the very principle of trickle-down economics.

Soaring profits and growth rates, as well as the mind-boggling salaries and bonuses that CEOs of large corporations pay themselves, have radically changed the conditions under which workers originally came to accept lower wages and diminished security, leaving many feeling that they've been had. Nowhere was this shift in attitude more apparent than in the public's sympathy for the striking United Parcel Service workers in 1997. Though Americans are notorious for their lack of sympathy for labor strikes, the plight of UPS part-timers struck a chord. Polls found that 55 percent of Americans supported the UPS workers, and only 27 percent sided with the company. Keffo, the editor of a bitter zine for temporary workers, summed up the public sentiment: "Day after day, [people] read and heard how great the economy is and it doesn't take a rocket scientist to realize well, duh, if UPS is doing so well, why can't they pay their workers more, or hire part timers as full timers, or keep their grubby fingers out of their workers' pension fund. So in a hilarious twist of fate, all the 'good' economic news works against UPS in favour of the Teamsters."[1]

Realizing that it had become a lightning rod for a broader malaise, UPS agreed to convert 10,000 part-time jobs to full-time jobs at twice the hourly pay, and increased pay for part-timers by 35 percent over five years. In explaining the concessions, UPS vice chairman John W. Alden said the company never foresaw its workers becoming symbols of the rage against the New Economy. "If I had known that it was going to go from negotiating for UPS to negotiating for part-time America, we would've approached it differently."[2]

From Job Creators to Wealth Creators

As we have seen, it has only been in the past three or four years that corporations have stopped hiding layoffs and restructuring behind the rhetoric of necessity and begun to speak openly and unapologetically about their aversion to hiring people and, in extreme cases, their total exodus from the employment business. Multinationals that once boasted of their role as "engines of job growth" – and used it as leverage to extract all kinds of government support – now prefer to identify themselves as engines of "economic growth." It's a subtle difference, but not if you happen to be looking for work. Corporations are indeed "growing" the economy, but they are doing it, as we have seen, through layoffs, mergers, consolidation and outsourcing – in other words, through job debasement and job loss. And as the economy grows, the percentage of people directly employed by the world's largest corporations is actually decreasing. Transnational corporations, which control more than 33 percent of the world's productive assets, account for only 5 percent of the world's direct employment.[3] And although the total assets of the world's one hundred largest corporations increased by 288 percent between 1990 and 1997, the number of people those corporations employed grew by less than 9 percent during that same period of tremendous growth.[4]

The most striking figure is the most recent: in 1998, despite the stellar performance of the U.S. economy and despite the record low unemployment rate, U.S. corporations eliminated 677,000 permanent jobs – more job cuts than in any other year this decade. One in nine of those cuts came in the aftermath of mergers; many others came from the manufacturing sector. As the low U.S unemployment rate suggests, two-thirds of the companies that eliminated jobs created new ones and laid-off workers found alternative employment relatively quickly.[5] But what those dramatic job cuts demonstrate is that a stable, reliable relationship between workers and their corporate employers has little or nothing to do with either the unemployment rate or the relative health of the economy. People are experiencing less stability even in the very best of economic times – in fact, these good economic times may be flowing, at least in part, from that loss of stability.

Job creation as part of the corporate mission, particularly the creation of full-time, decently paid, stable jobs, appears to have taken a back seat in many major corporations, regardless of company profits. (See related tables on page 264.) Rather than being one component of a healthy operation, labor is increasingly treated by the corporate sector as an unavoidable burden, like paying income tax; or an expensive nuisance, like not being allowed to dump toxic waste into lakes. Politicians may say that jobs are their priority, but the stock market responds cheerfully every time mass layoffs are announced, and sinks gloomily whenever it looks as if workers might get a raise. Whatever bizarre route we took to get here, an unmistakable message now emanates from our free markets: good jobs are bad for business, bad for "the economy" and should be avoided at all cost. Although this equation has undeniably reaped record profits in the short term, it may well prove to be a strategic miscalculation on the part of our captains of industry. By discarding their self-identification as job creators, companies leave themselves open to a kind of backlash that can come only from a population that knows that the smooth sailing of the economy is of little demonstrable benefit to them. (See Tables 11.1–11.4, pages 264–65.)

According to the 1997 report of the United Nations Conference on Trade and Development (UNCTAD), "Rising inequalities pose a serious threat of a political backlash against globalization, one that is as likely to come from the North as well as from the South.... The 1920s and 1930s provide a stark, and disturbing, reminder of just how quickly faith in markets and economic openness can be overwhelmed by political events."[6] With the effects of the Asian and Russian economic crises in full swing, a UN report on "human development" issued the following year was even more severe: noting the growing disparities between rich and poor, James Gustave Speth, administrator of the United Nations Development Program, said, "The numbers are shockingly high, amid the affluence. Progress must be more evenly distributed."[7]

You hear this kind of talk more and more these days. Ominous warnings about a simmering antiglobalization backlash cast a shadow over the usual euphoria of the annual gathering of corporate and political leaders in Davos, Switzerland. The business press is littered with more uneasy forecasts, such as the one in *Business Week* that noted, "The sight of bulging corporate coffers co-existing with a continuous stagnation in Americans' living standards could become politically untenable."[8] And that's America, which has a record low unemployment rate. The situation becomes even less comfortable in Canada, where unemployment is at 8.3 percent, and in European Union countries that are stuck with an average unemployment rate of 11.5 percent. (See Table 11.5, Appendix, page 477.)

At a speech delivered to the Business Council on National Issues, Ted Newall, chief executive officer of Nova Corp. in Calgary, Alberta, called the fact that more than 20 percent of Canadians live below the poverty line a "time bomb that is just waiting to go off." Indeed, a little side industry has developed of CEOs falling over each other to proclaim themselves ethical clairvoyants: they write books about the new "stockholder society," publicly berate their peers at luncheon addresses for their lack of scruples and announce that the time has come for corporate leaders to address the growing economic disparities. Trouble is, they can't agree on who is going to go first.

The fear that the poor will storm the barricades is as old as the castle moat, particularly during periods of great economic prosperity accompanied by inequitable distribution of wealth. Bertrand Russell writes that the Victorian élite in England were so consumed by paranoia that the working class would revolt against their "appalling poverty" that "at the time of Peterloo many large country houses kept artillery in readiness, lest they should be attacked by the mob. My maternal grandfather, who died in 1869, while wandering in his mind during his last illness, heard a loud noise in the street and thought it was the revolution breaking out, showing that at least unconsciously, the thought of revolution had remained with him throughout long prosperous years."[9]

A friend of mine whose family lives in India says her Punjabi aunt is so afraid of an insurrection of her own household staff that she keeps the kitchen knives locked up, leaving the servants to chop vegetables with sharpened sticks. It's not so different from the growing numbers of Americans moving into gated communities because the suburbs no longer provide adequate protection from the perceived urban threat.

Despite the widening gulf between rich and poor consistently reported by the UN and despite the much-discussed disappearance of the middle class in

Table 11.1
Total Assets of Top 100 Transnational Corporations, 1980 and 1995

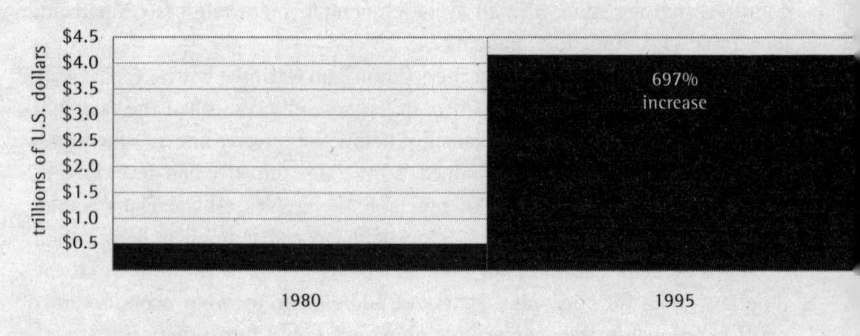

Source: Transnational Corporations in World Development: Third Survey (UN: 1983); Transnational Corporations in World Development: Trends and Prospects (UN: 1988); World Investment Reports (UN: 1993, 1994, 1997).

Table 11.2
Direct Employment in Top 100 Transnational Corporations, 1980 and 1995

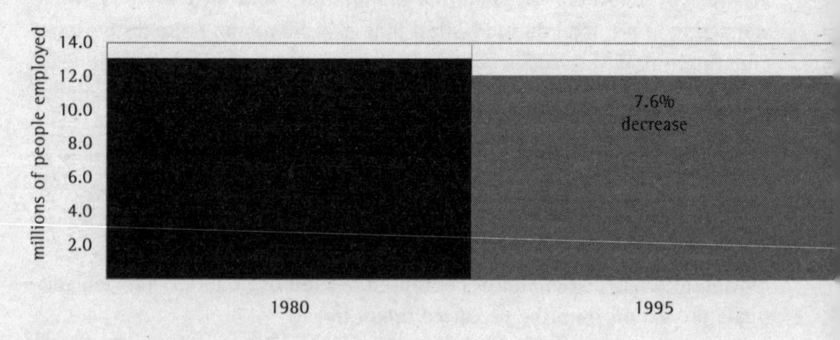

Source: Transnational Corporations in World Development: Third Survey (UN: 1983); Transnational Corporations in World Development: Trends and Prospects (UN: 1988); World Investment Reports (UN: 1993, 1994, 1997).

Table 11.3

Growth of Employment through Temp Agencies in Europe and U.S., 1988 and 1996

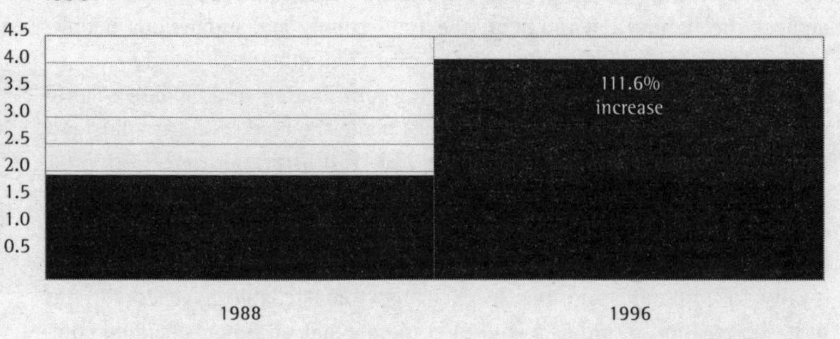

Source: International Confederation of Temporary Work Businesses (CIETT); Countries included: U.K., France, Netherlands, Germany, Spain, Belgium, Denmark and the U.S.

Table 11.4

Average Number of People Employed Daily through U.S. Temp Agencies, 1970 and 1998

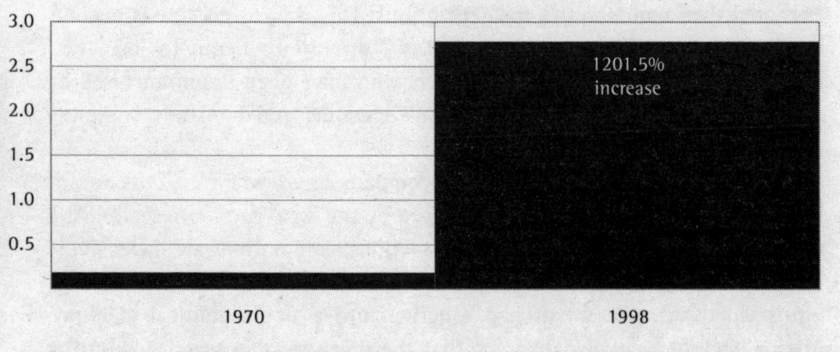

Source: Bruce Steinberg, "Temporary Help Annual Update for 1997," *Contemporary Times*, Spring 1998; Timothy W. Brogan, "Staffing Services Annual Update" (1999), National Association of Temporary and Staffing Services.

the West, the attack on jobs and income levels is probably not the most serious corporate offense we face as global citizens: it is, in theory, not irreversible. Far worse, in the long term, are the crimes committed by corporations against the natural environment, the food supply and indigenous peoples and cultures. Nevertheless, the erosion of a commitment to steady employment is the single most significant factor contributing to a climate of anti-corporate militancy and it is this that has made the markets most vulnerable to widespread "social unrest," to quote *The Wall Street Journal.*[10]

When corporations are perceived as functioning vehicles of wealth distribution – effectively trickling down jobs and tax revenue – they at least provide the bedrock for the often Faustian bargains by which citizens offer loyalty to corporate priorities in exchange for a reliable paycheck. In the past, job creation served as a kind of corporate suit of armor, shielding companies from the wrath that might otherwise have been directed their way as a result of environmental or human-rights abuses.

Nowhere was this armor more protective than in the "jobs vs. the environment" debates of the late eighties and early nineties, when progressive movements were sharply divided, for example, between those who supported the rights of loggers and those who wanted to protect old-growth forests. In British Columbia, activists were people who came in by bus from the city while loggers loyally stood by the multinational corporations that had anchored their communities for generations. This kind of division is becoming less clear for many participants, as corporations begin to lose their natural allies among blue-collar workers who have been disenfranchised by callously executed layoffs, sudden mill closures and constant company threats to move offshore.

Today, it's hard to find a contented company town, where citizens do not feel they have in some way been betrayed by the local corporate sector. And rather than dividing communities into factions, corporations are increasingly serving as the common thread by which labor, environmental and human-rights violations can be stitched together into a single political ideology. After a while it becomes apparent that the unsustainable search for profits that, for example, leads to the clear-cutting of old-growth forests is the same philosophy that devastates logging towns by moving the mills to

Indonesia. John Jordan, a British anarchist environmentalist, puts it this way: "Transnationals are affecting democracy, work, communities, culture and the biosphere. Inadvertently, they have helped us see the whole problem as one system, to connect every issue to every other issue, to not look at one problem in isolation."

This simmering backlash is about more than personal grievances. Even if you happen to be one of the lucky ones who has landed a good job and has never been laid off, everyone has heard the warnings – if not for themselves, then for their children or their parents or their friends. We live in a culture of job insecurity, and the messages of self-sufficiency have reached every one of us. In North America, the back end of an eighteen-wheeler heading for Mexico, workers weeping at the factory gate, the boarded-up windows of a hollowed-out factory town and people sleeping in doorways and on sidewalks have been among the most powerful economic images of our time: metaphors, seared into the collective consciousness, for an economy that consistently and unapologetically puts profits before people.

That message has perhaps been received most vividly by the generation that came of age since the recession hit in the early nineties. Almost without exception, they mapped out their life plan while listening to a chorus of voices telling them to lower their expectations, to rely on no one for their success. If they wanted a job with General Motors, Nike or General Electric, or indeed anywhere in the corporate sector, the message was the same: count on no one. Just in case they weren't paying attention, it was reinforced by high-school guidance counselors holding seminars on how to become "Me Inc.," by nightly newscasts filled with stories about how pension funds will soon be empty and by companies like Prudential Insurance urging us all to "Be your own rock." At university campuses across North America, orientation-week events – the time when students are first introduced to campus life – are now sponsored by mutual-fund companies, which use the opportunity to prod incoming students to start saving for their retirement before they've even picked a major.

All this has had its effects. According to the bible of demographic marketing, *The Yankelovich Report*, the belief in the need to be self-reliant has increased by one-third with every generation – from the "Matures" (born

1909–1945), to "Boomers" (born 1946–1964) to "Xers" (defined loosely and somewhat inaccurately as everyone born between 1965 and the present). "Over two-thirds of Xers agree that, 'I have to take whatever I can get in this world because no one is going to give me anything.' Far fewer Boomers and Matures agree – only half and one-third, respectively," the report states.[11] The New York advertising firm DMB&B found similar attitudes in its study of global teens. "From a lengthy battery of attitudinal items, the one that teens most agree with worldwide is: 'It's up to me to get what I want out of life.'" Nine out of ten young Americans polled agreed with this sentiment of total self-reliance.[12]

This shift in attitude has translated into a serious boom for the mutual-fund industry. Young people, it seems, are buying more RSPs than ever before. "Why is Generation X more focused on the need to save?" wonders a reporter in *Business Week*. "Much of it has to do with self-reliance. They believe they'll succeed only on their own initiative and have little confidence that either Social Security or traditional employer pensions will be around to support them in retirement."[13] In fact, if you believe the business press, the only impact this spirit of self-reliance will have is the spearheading of a new wave of cutthroat entrepreneurial initiatives as the kids who can't count on anyone look out for Number One.

There is no question that many young people have compensated for the fact that they don't trust politicians or corporations by adopting the social-Darwinist values of the system that engendered their insecurity: they will be greedier, tougher, more focused. They will Just Do It. But what of those who didn't go the MBA route, who don't want to be the next Bill Gates or Richard Branson? Why should they stay invested in the economic goals of corporations that have so actively divested them? What is the incentive to be loyal to a sector that has bombarded them, for their entire adult life, with a single message: Don't count on us?

This issue is not only about unemployment per se. It would be a grave mistake to assume that any old paycheck will buy the level of loyalty and protection to which many corporations – sometimes rightly – were once accustomed. Casual, part-time and low-wage work does not bring about the same identification with one's employer as the lifelong contracts of yesterday. Go to any mall fifteen minutes after the stores close and you'll see the new employment relationship in action: all the minimum-wage clerks are lined up, their purses and backpacks open for "bag check." It's standard practice, retail workers will tell you, for managers to search them daily for stolen goods. And according to an annual industry survey conducted by the University of Florida's Security Research Project, there is reason for suspicion: the study shows that employee theft accounted for 42.7 percent of the total amount of goods stolen from U.S. retailers in 1998, the highest rate ever recorded by the survey. Starbucks clerk Steve Emery likes to quote a line he got from a sympathetic customer: "You pay peanuts, so you get monkeys." When he told me that, it reminded me of something I had heard only two months earlier from a group of Nike workers in Indonesia. Sitting cross-legged in a circle at one of the dorms, they told me that, deep down, they hoped their factory would burn to the ground. Understandably, the factory workers' sentiments were much more extreme than the resentments expressed by McWorkers in the West – then again, the guards doing "bag check" at the gated entrance to the Nike factory in Indonesia were armed with revolvers.

But it is in the ranks of the millions of temp workers that the true breeding grounds of the anticorporate backlash will most likely be found. Since most temps don't stay at one post long enough for anyone to keep track of the value of their labor, the merit principle – once a sacred capitalist tenet – is becoming moot. And the situation can be intensely demoralizing. "Pretty soon, I'll run out of places to work in this city," writes Debbie Goad, a temp with twenty years of secretarial experience. "I'm registered at fifteen temporary agencies. It's like playing the slots in Vegas. They constantly call me, sounding like used-car salesmen. 'I know I'll get you the perfect job soon.'"[14]

She wrote those words in *Temp Slave*, a little publication out of Madison, Wisconsin, devoted to tapping a seemingly bottomless well of worker resentment. In it, workers who have been branded as disposable vent their anger at the corporations that rent them like pieces of equipment, then return them, used, to the agency. Temps traditionally have had no one to talk to about these issues – the nature of the work keeps them isolated from each other and also, inside their temporary workplaces, from their salaried co-workers.

So it's no surprise that *Temp Slave*, and Web sites like Temp 24–7, boil with repressed hostility, offering helpful tips on how to sabotage your employer's computer system, as well as essays with titles such as "Everybody hates temps. The feeling is mutual!" and "The boredom, the sheer boredom of office life for temps."

Just as temp workforces mess with the merit principle, so does the growing practice of swapping CEOs like pro ballplayers. Temp CEOs are a major assault on the capitalist folklore of the mail-room boy who works his way up to becoming president of the company. Today's executives, since they just seem to trade the top spot with one another, appear to be born into their self-enclosed stratospheres like kings. In such a context, there is less room for the dream of making it up from the mail room – especially since the mail room has probably been outsourced to Pitney Bowes and staffed with permatemps.

That is the situation at Microsoft, and it is part of the reason why temp rage seethes there like nowhere else. Another is that Microsoft openly admits that its reserve of temps exists to protect the core of permanent workers from the ravages of the free market. When a product line is discontinued, or costs are cut in ingenious new ways, it's the temps that absorb the blows. If you ask the agencies, they say that their clients don't mind being treated like outdated software – after all, Bill Gates never promised them a thing. "When people know it's a temporary arrangement, some day, when the assignment ends, there's not a sense of a broken trust," explains Peg Cheirett, president of Wasser Group, one of the agencies that supplies Microsoft with temps.[15]

There's no doubt Gates has devised a means of downsizing that avoids those high-pitched wails of betrayal that IBM bosses faced in the late eighties when they eliminated 37,000 jobs, shocking employees who were under the impression they had secured jobs for life. Microsoft's temps have no basis to expect anything of Bill Gates – that much is true – but while that fact may keep pickets from blocking the entrance to the Microsoft Campus, it does little to protect the company from getting hacked from inside its own computer system. (As it did throughout 1998, when the hacker cabal Cult of the Dead Cow released a made-for-Microsoft hacking program called

Back Orifice. It was downloaded from the Internet 300,000 times.) Microsoft's permatemps brush up against the hyperactive capitalist dream of Silicon Gold every day, and yet they – more than anyone else – know that it's an invitation-only affair. So while Microsoft's permanent employees are renowned for their corporate cultishness, Microsoft permatemps are almost unparalleled in their rancor. Asked by journalists what they think of their employer, they offer up such choice comments as: "They treat you like pond scum"[16] or "It's a system of having two classes of people, and instilling fear and inferiority and loathing."[17]

Divestment: A Two-Way Transaction

Commenting on this shift, Charles Handy, author of *The Hungry Spirit*, writes that "it is clear that the psychological contract between employers and employed has changed. The smart jargon now talks of guaranteeing 'employability' not 'employment,' which, being interpreted, means don't count on us, count on yourself, but we'll try to help if we can."[18]

But for some – particularly younger workers – there is a silver lining. Because young people tend not to see the place where they work as an extension of their souls, they have, in some cases, found freedom in knowing they will never suffer the kind of heart-wrenching betrayals their parents did. For almost everyone who has entered the job market in the past decade, unemployment is a known quantity, as is self-generated and erratic work. In addition, losing one's job is much less frightening when getting it seemed an accident in the first place. Such familiarity with unemployment creates its own kind of worker divestment – divestment of the very notion of total dependency on stable work. We may begin to wonder whether we should even want the same job for our whole lives, and, more important, why we should depend on the twists and turns of large institutions for our sense of self.

This slow divestment by corporate culture has implications that reach far beyond the psychology of the individual: a population of skilled workers who don't see themselves as corporate lifers could lead to a renaissance in creativity and a revitalization of civic life, two very hopeful prospects. One thing is certain: it is already leading to a new kind of anticorporate politics.

KRAFT
FRI 13

Table 11.6

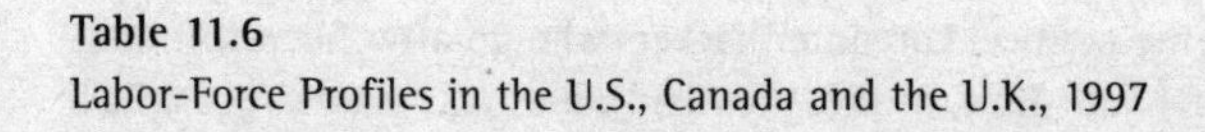
Labor-Force Profiles in the U.S., Canada and the U.K., 1997

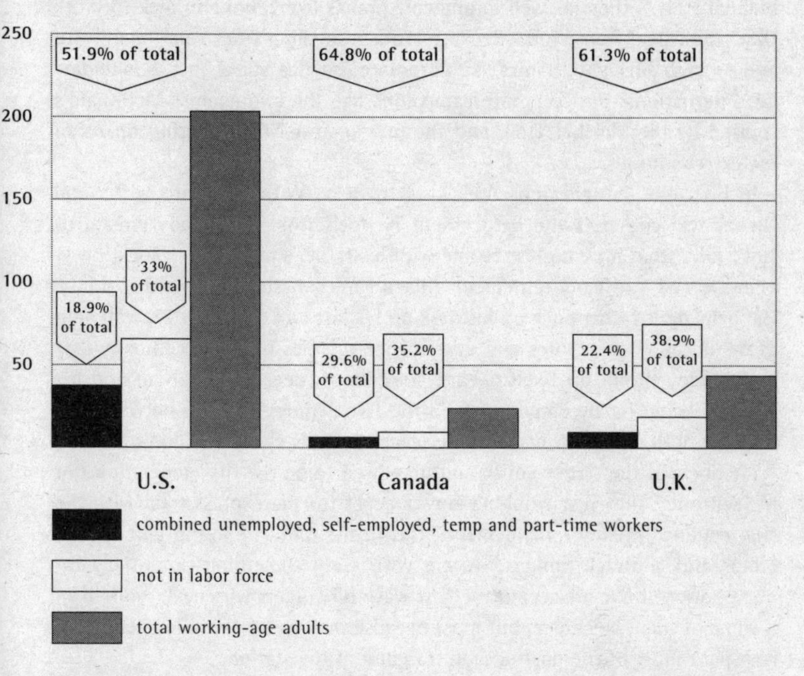

Source: Bureau of Labor Statistics; Statistics Canada; Office for National Statistics (Labor Force Survey); International Confederation of Temporary Work Businesses (CIETT). For Canadian and U.S. statistics, overlap of temp and part-time workers is accounted for. In the U.K., overlap between part-time and self-employed is accounted for.

[Taking the U.S. statistics as an example: the unemployed, part-time, temporary and replacement workers make up close to 40 percent of people actively working or looking for work. However, if you factor in the 67 million working-age Americans who are not included in the unemployment figures because they are not actively looking for work, the percentage of adults holding down full-time permanent jobs slips into the minority.]

You can see it in the political computer hackers who go after Microsoft and, as the next chapter will show, in the guerrilla "adbusters" who target urban billboards. It is there as well in anarchic pranks like "Phone in Sick to Work Day," the "Steal from Work! Because Work Is Stealing from You!" manifesto and on Web sites with names like Corporate America Sucks, just as it underlies international anticorporate campaigns like the one against McDonald's spurred by the McLibel Trial, and the one against Nike, focusing on Asian factory conditions.

In his essay "Stupid Jobs Are Good to Relax With," Toronto writer Hal Niedzviecki contrasts the detachment he feels from the steady stream of "joke jobs" that junk up his résumé with his father's profound dislocation at being forced into early retirement after a career of steady upward mobility. Hal helped his father pack up his desk on his last day at the office, watching as he nicked Post-it Notes and other office supplies from the company that had employed him for twelve years. "Despite his decades of labour and my years of being barely employed (and the five degrees we have between us), we have both ended up in the same place. He feels cheated. I don't."[19]

Members of the sixties youth culture vowed to be the first generation not to "sell out": they just wouldn't buy a ticket for the express train with the sign reading "lifelong employment." But in the ranks of young part-timers, temps and contract workers, we are witnessing something potentially far more powerful. We are seeing the first wave of workers who never bought in – some of them by choice, but most because that lifelong-employment train has spent most of the past decade standing in the station.

The extent of this shift cannot be overstated. Among the total number of working-age adults in the U.S., Canada and the U.K., those with full-time, permanent jobs working for someone other than themselves are in the minority. Temps, part-timers, the unemployed and those who have opted out of the labor force entirely – some because they don't want to work but many more because they have given up looking for jobs – now make up more than half of the working-age population. (See Table 11.6, page 273.)

In other words, the people who don't have access to a corporation to which they can offer lifelong loyalty are the majority. And for young workers, consistently overrepresented among the unemployed, part-time and temporary sectors, the relationship to the work world is even more tenuous. (See Table 11.7, Appendix, page 478.)

From No Jobs to No Logo...

It should come as no surprise that the companies that increasingly find themselves at the wrong end of a bottle of spray paint, a computer hack or an international anticorporate campaign are the ones with the most cutting-edge ads, the most intuitive market researchers and the most aggressive in-school outreach programs. With the dictates of branding forcing companies to sever their traditional ties to steady job creation, it is no exaggeration to say that the "strongest" brands are the ones generating the worst jobs, whether in the export processing zones, in Silicon Valley or at the mall. Furthermore, the companies that advertise aggressively on MTV, Channel One and in *Details*, selling sneakers, jeans, fast food and Walkmans, are the very ones that pioneered the McJob sector and led the production exodus to cheap labor enclaves like Cavite. After pumping young people up with go-get-'em messages – the "Just Do It" sneakers, "No Fear" T-shirts and "No Excuses" jeans – these companies have responded to job requests with a resounding "Who, me?" The workers in Cavite may be unswooshworthy, but Nike's and Levi's core consumers have received another message from the brands' global shuffle: they are unjobworthy.

To add insult to injury, as we saw in Part I, "No Space," this abandonment by brand-name corporations is occurring at the very moment when youth culture is being sought out for more aggressive branding than ever before. Youth style and attitude are among the most effective wealth generators in our entertainment economy, but real live youth are being used around the world to pioneer a new kind of disposable workforce. It is in this volatile context, as the final section will show, that the branding economy is becoming the political equivalent of a sign hanging on the back of the body corporate that says "Kick Me."

NO LOGO

Top: A call to Depression-era ad jammers from *The Ballyhoo*. *Bottom:* Two tobacco ad parodies by Ron English.

CHAPTER TWELVE

CULTURE JAMMING

Ads Under Attack

Advertising men are indeed very unhappy these days, very nervous, with a kind of apocalyptic expectancy. Often when I have lunched with an agency friend, a half dozen worried copy writers and art directors have accompanied us. Invariably they want to know when the revolution is coming, and where will they get off if it does come.

– Ex-adman James Rorty, *Our Master's Voice*, 1934

It's Sunday morning on the edge of New York's Alphabet City and Jorge Rodriguez de Gerada is perched at the top of a high ladder, ripping the paper off a cigarette billboard. Moments before, the billboard at the corner of Houston and Attorney sported a fun-loving Newport couple jostling over a pretzel. Now it showcases the haunting face of a child, which Rodriguez de Gerada has painted in rust. To finish it off, he pastes up a few hand-torn strips of the old Newport ad, which form a fluorescent green frame around the child's face.

When it's done, the installation looks as the thirty-one-year-old artist had intended: as if years of cigarette, beer and car ads had been scraped away to reveal the rusted backing of the billboard. Burned into the metal is the real commodity of the advertising transaction. "After the ads are taken down," he says, "what is left is the impact on the children in the area, staring at these images."[1]

Unlike some of the growing legion of New York guerrilla artists, Rodriguez de Gerada refuses to slink around at night like a vandal, choosing instead to make his statements in broad daylight. For that matter, he doesn't much like the phrase "guerrilla art," preferring "citizen art" instead. He wants the dialogue he has been having with the city's billboards for more than ten years to be seen as a normal mode of discourse in a democratic society – not as some edgy vanguard act. While he paints and pastes, he wants kids to stop and watch – as they do on this sunny day, just as an old man offers to help support the ladder.

Rodriguez de Gerada even claims to have talked cops out of arresting him on three different occasions. "I say, 'Look, look what's around here, look what's happening. Let me explain to you why I do it.'" He tells the police officer about how poor neighborhoods have a disproportionately high number of billboards selling tobacco and hard liquor products. He talks about how these ads always feature models sailing, skiing or playing golf, making the addictive products they promote particularly glamorous to kids stuck in the ghetto, longing for escape. Unlike the advertisers who pitch and run, he wants his work to be part of a community discussion about the politics of public space.

Rodriguez de Gerada is widely recognized as one of the most skilled and creative founders of culture jamming, the practice of parodying advertisements and hijacking billboards in order to drastically alter their messages. Streets are public spaces, adbusters argue, and since most residents can't afford to counter corporate messages by purchasing their own ads, they should have the right to talk back to images they never asked to see. In recent years, this argument has been bolstered by advertising's mounting aggressiveness in the public domain – the ads discussed in "No Space," painted and projected onto sidewalks; reaching around entire buildings and buses; into schools; onto basketball courts and on the Internet. At the same time, as discussed in "No Choice," the proliferation of the quasi-public "town squares" of malls and superstores has created more and more spaces where commercial messages are the only ones permitted. Adding even greater urgency to their cause is the belief among many jammers that concentration of media ownership has successfully devalued the right to free speech by severing it from the right to be heard.

All at once, these forces are coalescing to create a climate of semiotic Robin Hoodism. A growing number of activists believe the time has come for the public to stop asking that some space be left unsponsored, and to begin seizing it back. Culture jamming baldly rejects the idea that marketing – because it buys its way into our public spaces – must be passively accepted as a one-way information flow.

The most sophisticated culture jams are not stand-alone ad parodies but interceptions – counter-messages that hack into a corporation's own method of communication to send a message starkly at odds with the one that was intended. The process forces the company to foot the bill for its own subversion, either literally, because the company is the one that paid for the billboard, or figuratively, because anytime people mess with a logo, they are tapping into the vast resources spent to make that logo meaningful. Kalle Lasn, editor of Vancouver-based *Adbusters* magazine, uses the martial art of jujitsu as a precise metaphor to explain the mechanics of the jam. "In one simple deft move you slap the giant on its back. We use the momentum of the enemy." It's an image borrowed from Saul Alinsky who, in his activist bible, *Rules for Radicals*, defines "mass political jujitsu" as "utilizing the power of one part of the power structure against another part ... the superior strength of the Haves become their own undoing."[2] So, by rappelling off the side of a thirty-by-ninety-foot Levi's billboard (the largest in San Francisco) and pasting the face of serial killer Charles Manson over the image, a group of jammers attempts to leave a disruptive message about the labor practices employed to make Levi's jeans. In the statement it left on the scene, the Billboard Liberation Front said they chose Manson's face because the jeans were "Assembled by prisoners in China, sold to penal institutions in the Americas."

The term "culture jamming" was coined in 1984 by the San Francisco audio-collage band Negativland. "The skillfully reworked billboard ... directs the public viewer to a consideration of the original corporate strategy," a band member states on the album *Jamcon '84*. The jujitsu metaphor isn't as apt for jammers who insist that they aren't inverting ad messages but are rather improving, editing, augmenting or unmasking them. "This is extreme truth in advertising," one billboard artist tells me.[3] A good jam, in other words, is an X-ray of the subconscious of a campaign, uncovering not an opposite meaning but the deeper truth hiding beneath the layers of advertising euphemisms. So, according to these principles, with a slight turn of the imagery knob, the now-retired Joe Camel turns into Joe Chemo, hooked up to an IV machine. That's what's in his future, isn't it? Or Joe is shown about fifteen years younger than his usual swinger self (see image, page 278). Like Baby Smurf, the "Cancer Kid" is cute and cuddly and playing with building blocks instead of sports cars and pool cues. And why not? Before R.J. Reynolds reached a $206 billion settlement with forty-six states, the American government accused the tobacco company of using the cartoon camel to entice children to start smoking – why not go further, the culture jammers ask, and reach out to even younger would-be smokers? Apple computers' "Think Different" campaign of famous figures both living and dead has been the subject of numerous simple hacks: a photograph of Stalin appears with the altered slogan "Think Really Different"; the caption for the ad featuring the Dalai Lama is changed to "Think Disillusioned" and the rainbow Apple logo is morphed into a skull (see image on page 344). My favorite truth-in-advertising campaign is a simple jam on Exxon that appeared just after the 1989 Valdez spill: "Shit Happens. New Exxon," two towering billboards announced to millions of San Francisco commuters.

Attempting to pinpoint the roots of culture jamming is next to impossible, largely because the practice is itself a cutting and pasting of graffiti, modern art, do-it-yourself punk philosophy and age-old pranksterism. And using billboards as an activist canvas isn't a new revolutionary tactic either. San Francisco's Billboard Liberation Front (responsible for the Exxon and Levi's jams) has been altering ads for twenty years, while Australia's Billboard Utilizing Graffitists Against Unhealthy Promotions (BUG-UP) reached its peak in 1983, causing an unprecedented $1 million worth of damage to tobacco billboards in and around Sydney.

It was Guy Debord and the Situationists, the muses and theorists of the theatrical student uprising of Paris, May 1968, who first articulated the power of a simple *détournement*, defined as an image, message or artifact lifted out of its context to create a new meaning. But though culture jammers borrow liberally from the avant-garde art movements of the past – from Dada and Surrealism to Conceptualism and Situationism – the canvas these art revolutionaries were attacking tended to be the art world and its passive culture of spectatorship, as well as the anti-pleasure ethos of mainstream capitalist society. For many French students in the late sixties, the enemy was the rigidity and conformity of the Company Man; the company itself proved markedly less engaging. So where Situationist Asger Jorn hurled paint at pastoral paintings bought at flea markets, today's culture jammers prefer to hack into corporate advertising and other avenues of corporate speech. And if the culture jammers' messages are more pointedly political than their predecessors', that may be because what were indeed subversive messages in the sixties – "Never Work," "It Is Forbidden to Forbid," "Take Your Desires for Reality" – now sound more like Sprite or Nike slogans: Just Feel It. And the "situations" or "happenings" staged by the political pranksters in 1968, though genuinely shocking and disruptive at the time, are the Absolut Vodka ad of 1998 – the one featuring purple-clad art school students storming bars and restaurants banging on bottles.

In 1993, Mark Dery wrote "Culture Jamming: Hacking, Slashing and Sniping in the Empire of Signs," a booklet published by the Open Magazine Pamphlet Series. For Dery, jamming incorporates such eclectic combinations of theater and activism as the Guerrilla Girls, who highlighted the art world's exclusion of female artists by holding demonstrations outside the Whitney Museum in gorilla masks; Joey Skagg, who has pulled off countless successful media hoaxes; and Artfux's execution-in-effigy of arch-Republican Jesse Helms on Capitol Hill. For Dery, culture jamming is anything, essentially, that mixes art, media, parody and the outsider stance. But within these subcultures, there has always been a tension between the forces of the merry prankster and the hard-core revolutionary. Nagging questions re-emerge: are play and pleasure themselves revolutionary acts, as the Situationists might argue? Is screwing up the culture's information flows inherently subversive, as Skagg would hold? Or is the mix of art and politics just a matter of making sure, to paraphrase Emma Goldman, that somebody has hooked up a good sound system at the revolution?

Though culture jamming is an undercurrent that never dries up entirely, there is no doubt that for the last five years it has been in the midst of a revival, and one focused more on politics than on pranksterism. For a growing number of young activists, adbusting has presented itself as the perfect tool with which to register disapproval of the multinational corporations that have so aggressively stalked them as shoppers, and so unceremoniously dumped them as workers. Influenced by media theorists such as Noam Chomsky, Edward Herman, Mark Crispin Miller, Robert McChesney and Ben Bagdikian, all of whom have explored ideas about corporate control over information flows, the adbusters are writing theory on the streets, literally deconstructing corporate culture with a waterproof magic marker and a bucket of wheatpaste.

Jammers span a significant range of backgrounds, from purer-than-thou Marxist-anarchists who refuse interviews with "the corporate press" to those like Rodriguez de Gerada who work in the advertising industry by day (his paying job, ironically, is putting up commercial signs and superstore window displays) and long to use their skills to send messages they consider constructive. Besides a fair bit of animosity between these camps, the only ideology bridging the spectrum of culture jamming is the belief that free speech is meaningless if the commercial cacophony has risen to the point that no one can hear you. "I think everyone should have their own billboard, but they don't," says Jack Napier (a pseudonym) of the Billboard Liberation Front.[4]

On the more radical end of the spectrum, a network of "media collectives" has emerged, decentralized and anarchic, that combine adbusting with zine publishing, pirate radio, activist video, Internet development and community activism. Chapters of the collective have popped up in Tallahassee, Boston, Seattle, Montreal and Winnipeg – often splintering off into other organizations. In London, where adbusting is called "subvertising," a new group has been formed, called the UK Subs after the seventies punk group of the same name. And in the past two years, the real-world jammers have been joined by a global network of on-line "hacktivists" who carry out their raids on the Internet, mostly by breaking into corporate Web sites and leaving their own messages behind.

More mainstream groups have also been getting in on the action. The U.S. Teamsters have taken quite a shine to the ad jam, using it to build up support for striking workers in several recent labor disputes. For instance, Miller Brewing found itself on the receiving end of a similar jam when it laid off workers at a St. Louis plant. The Teamsters purchased a billboard that parodied a then current Miller campaign; as *Business Week* reported, "Instead of two bottles of beer in a snowbank with the tagline 'Two Cold,' the ad showed two frozen workers in a snowbank labeled 'Too Cold: Miller canned 88 St. Louis workers.'"[5] As organizer Ron Carver says, "When you're doing this, you're threatening multimillion-dollar ad campaigns."[6]

One high-profile culture jam arrived in the fall of 1997 when the New York antitobacco lobby purchased hundreds of rooftop taxi ads to hawk "Virginia Slime" and "Cancer Country" brand cigarettes. All over Manhattan, as yellow cabs got stuck in gridlock, the jammed ads jostled with the real ones.

"Mutiny on the Corporate Sponsor Ship" – Paper Tiger, 1997 slogan

The rebirth of culture jamming has much to do with newly accessible technologies that have made both the creation and the circulation of ad parodies immeasurably easier. The Internet may be bogged down with brave new forms of branding, as we have seen, but it is also crawling with sites that offer links to culture jammers in cities across North America and Europe, ad parodies for instant downloading and digital versions of original ads, which can be imported directly onto personal desktops or jammed on site. For Rodriguez de Gerada, the true revolution has been in the impact desktop publishing has had on the techniques available to ad hackers. Over the course of the last decade, he says, culture jamming has shifted "from low-tech to medium-tech to high-tech," with scanners and software programs like Photoshop now enabling activists to match colors, fonts and materials precisely. "I know so many different techniques that make it look like the whole ad was reprinted with its new message, as opposed to somebody coming at it with a spray-paint can."

This is a crucial distinction. Where graffiti traditionally seek to leave dissonant tags on the slick face of advertising (or the "pimple on the face of the retouched cover photo of America," to use a Negativland image), Rodriguez de Gerada's messages are designed to mesh with their targets, borrowing visual legitimacy from advertising itself. Many of his "edits" have been so successfully integrated that the altered billboards look like originals, though with a message that takes viewers by surprise. Even the child's face he put up in Alphabet City – not a traditional parody jam – was digitally output on the same kind of adhesive vinyl that advertisers use to seamlessly cover buses and buildings with corporate logos. "The technology allows us to use Madison Avenue's aesthetics against itself," he says. "That is the most important aspect of this new wave of people using this guerrilla tactic, because that's what the MTV generation has become accustomed to – everything's flashy, everything's bright and clean. If you spend time to make it cleaner it will not be dismissed."

But others hold that jamming need not be so high tech. The Toronto performance artist Jubal Brown spread the visual virus for Canada's largest billboard-busting blitz with nothing more than a magic marker. He taught his friends how to distort the already hollowed out faces of fashion models by using a marker to black out their eyes and draw a zipper over their mouths – presto! Instant skull. For the women jammers in particular, "skulling" fitted in neatly with the "truth in advertising" theory: if emaciation is the beauty ideal, why not go all the way with zombie chic – give the advertisers a few supermodels from beyond the grave? For Brown, more nihilist than feminist, skulling was simply a détournement to highlight the cultural poverty of the sponsored life. ("Buy Buy Buy! Die Die Die!" reads Brown's statement displayed in a local Toronto art gallery.) On April Fool's Day, 1997, dozens of people went out on skulling missions, hitting hundreds of billboards on busy Toronto streets (see image, page 344). Their handiwork was reprinted in *Adbusters*, helping to spread skulling to cities across North America.

And nobody is riding the culture-jamming wave as high as *Adbusters*, the self-described "house-organ" of the culture-jamming scene. Editor Kalle Lasn, who speaks exclusively in the magazine's enviro-pop lingo, likes to say that we are a culture "addicted to toxins" that are poisoning our bodies, our "mental environment" and our planet. He believes that adbusting will eventually spark a "paradigm shift" in public consciousness. Published by the Vancouver-based Media Foundation, the magazine started in 1989 with 5,000 copies. It now has a circulation of 35,000 – at least 20,000 copies of which go to the United States. The foundation also produces "uncommercials" for television that accuse the beauty industry of causing eating disorders, attack North American overconsumption, and urge everyone to trade their cars in for bikes. Most television stations in Canada and the U.S. have refused to air the spots, which gives the Media Foundation the perfect excuse to take them to court and use the trials to attract press attention to their vision of more democratic, publicly accessible media.

Culture jamming is enjoying a resurgence, in part because of technological advancements, but also more pertinently, because of the good old rules of supply and demand. Something not far from the surface of the public psyche is delighted to see the icons of corporate power subverted and mocked. There is, in short, a market for it. With commercialism able to overpower the traditional authority of religion, politics and schools, corporations have emerged as the natural targets for all sorts of free-floating rage and rebellion. The new ethos that culture jamming taps into is go-for-the-corporate-jugular. "States have fallen back and corporations have become the new institutions," says Jaggi Singh, a Montreal-based anticorporate activist.[7] "People are just reacting to the iconography of our time." American labor rights activist Trim Bissell goes further, explaining that the thirsty expansion of chains like Starbucks and the aggressive branding of companies like Nike have created a climate ripe for anticorporate attacks. "There are certain corporations which market themselves so aggressively, which are so intent on stamping their image on everybody and every street, that they build up a reservoir of resentment among thinking people," he says. "People resent the destruction of culture and its replacement with these mass-produced corporate logos and slogans. It represents a kind of cultural fascism."[8]

Most of the superbrands are of course well aware that the very imagery that has generated billions for them in sales is likely to create other, unintended, waves within the culture. Well before the anti-Nike campaign began in earnest, CEO Phil Knight presciently observed that "there's a flip side to the emotions we generate and the tremendous well of emotions we live off of. Somehow, emotions imply their opposites and at the level we operate, the reaction is much more than a passing thought."[9] The reaction is also more than the fickle flight of fashion that makes a particular style of hip sneaker suddenly look absurd, or a played-to-death pop song become, overnight, intolerable. At its best, culture jamming homes in on the flip side of those branded emotions, and refocuses them, so that they aren't replaced with a craving for the next fashion or pop sensation but turn, slowly, on the process of branding itself.

It's hard to say how spooked the advertisers are about getting busted. Although the U.S. Association of National Advertisers has no qualms about lobbying police on behalf of its members to crack down on adbusters, they are generally loath to let the charges go to trial. This is probably wise. Even though ad companies try to paint jammers as "vigilante censors" in the media,[10] they know it wouldn't take much for the public to decide that the advertisers are the ones censoring the jammers' creative expressions.

So while most big brand names rush to sue for alleged trademark violations and readily take each other to court for parodying slogans or products (as Nike did when Candies shoes adopted the slogan "Just Screw It"), multinationals are proving markedly less eager to enter into legal battles that will clearly be fought less on legal than on political grounds. "No one wants to be in the limelight because they are the target of community protests or boycotts," one advertising executive told *Advertising Age.*[11] Furthermore, corporations rightly see jammers as rabid attention seekers and have learned to avoid anything that could garner media coverage for their stunts. A case in point came in 1992 when Absolut Vodka threatened to sue *Adbusters* over its "Absolut Nonsense" parody. The company immediately backed down when the magazine went to the press and challenged the distiller to a public debate on the harmful effects of alcohol.

And much to Negativland's surprise, Pepsi's lawyers even refrained from responding to the band's 1997 release, *Dispepsi* – an anti-pop album consisting of hacked, jammed, distorted and disfigured Pepsi jingles. One song mimics the ads by juxtaposing the product's name with a laundry list of random unpleasant images: "I got fired by my boss. Pepsi/ I nailed Jesus to the cross. Pepsi/ ... The ghastly stench of puppy mills. Pepsi" and so on.[12] When asked by *Entertainment Weekly* magazine for its response to the album, the soft-drink giant claimed to think it was "a pretty good listen."[13]

Identity Politics Goes Interactive

There is a connection between the ad fatigue expressed by the jammers and the fierce salvos against media sexism, racism and homophobia that were so much in vogue when I was an undergraduate in the late eighties and early nineties. This connection is perhaps best traced through the evolving relationships that feminists have had with the ad world, particularly since the movement deserves credit for laying the groundwork for many of the current ad critiques. As Susan Douglas notes in *Where the Girls Are*, "Of all the social movements of the 1960s and '70s, none was more explicitly anti-consumerist than the women's movement. Feminists had attacked the ad campaigns for products like Pristeen and Silva Thins, and by rejecting makeup, fashion and the need for spotless floors, repudiated the very need to buy certain products at all."[14] Furthermore, when *Ms.* magazine was relaunched in 1990, the editors took advertiser interference so seriously that they made the unprecedented move of banishing lucrative advertisements from their pages entirely. And the "No Comment" section – a back-page gallery of sexist ads reprinted from other publications – remains one of the highest-profile forums for adbusting.

Many female culture jammers say they first became interested in the machinations of marketing via a "Feminism 101" critique of the beauty industry. Maybe they started by scrawling "feed me" on Calvin Klein ads in bus shelters, as the skateboarding members of the all-high-school Bitch Brigade did. Or maybe they got their hands on a copy of Nomy Lamm's zine, *I'm So Fucking Beautiful*, or they stumbled onto the "Feed the Super Model" interactive game on the official RiotGrrrl Web site. Or maybe, like Toronto's Carly Stasko, they got started through grrrly self-publishing. Twenty-one-year-old Stasko is a one-woman alternative-image factory: her pocket and backpack overflow with ad-jammed stickers, copies of her latest zine and handwritten flyers on the virtues of "guerrilla gardening." And when Stasko is not studying semiotics at the University of Toronto, planting sunflower seeds in abandoned urban lots or making her own media, she's teaching courses at local alternative schools where she shows classes of fourteen-year-olds how they too can cut and paste their own culture jams.

Stasko's interest in marketing began when she realized the degree to which contemporary definitions of female beauty – articulated largely through the media and advertisements – were making her and her peers feel insecure and inadequate. But unlike my generation of young feminists who had dealt with similar revelations largely by calling for censorship and re-education programs, she caught the mid-nineties self-publishing craze. Still in her teens, Stasko began publishing *Uncool*, a photocopied zine crammed with collages of sliced-and-diced quizzes from women's magazines, jammed ads for tampons, manifestos on culture jamming and, in one issue, a full-page ad for Philosophy Barbie. "What came first?" Stasko's Barbie wonders. "The beauty or the myth?" and "If I break a nail, but I'm asleep, is it still a crisis?"

She says that the process of making her own media, adopting the voice of the promoter and hacking into the surface of the ad culture began to weaken advertising's effect on her. "I realized that I can use the same tools the media does to promote my ideas. It took the sting out of the media for me because I saw how easy it was."[15]

Although he is more than ten years older than Stasko, the road that led Rodriguez de Gerada to culture jamming shares some of the same twists. A founding member of the political art troop Artfux, he began adbusting coincident with a wave of black and Latino community organizing against cigarette and alcohol advertising. In 1990, thirty years after the National Association for the Advancement of Colored People first lobbied cigarette companies to use more black models in their ads, a church-based movement began in several American cities that accused these same companies of exploiting black poverty by target-marketing inner cities for their lethal product. In a clear sign of the times, attention had shifted from who was in the ads to the products they sold. Reverend Calvin O. Butts of the Abyssinian Baptist Church in Harlem took his parishioners on billboard-busting blitzes during which they would paint over the cigarette and alcohol advertisements around their church. Other preachers took up the fight in Chicago, Detroit and Dallas.[16]

Reverend Butts's adbusting consisted of reaching up to offending billboards with long-handled paint rollers and whitewashing the ads. It was functional, but Rodriguez de Gerada decided to be more creative: to replace the companies' consumption messages with more persuasive political messages of his own. As a skilled artist, he carefully morphed the faces of cigarette models so they looked rancid and diseased. He replaced the standard Surgeon General's Warning with his own messages: "Struggle General's Warning: Blacks and Latinos are the prime scapegoats for illegal drugs, and the prime targets for legal ones."

Like many other early culture jammers, Rodriguez de Gerada soon extended his critiques beyond tobacco and alcohol ads to include rampant ad bombardment and commercialism in general, and, in many ways, he has the ambitiousness of branding itself to thank for this political evolution. As inner-city kids began stabbing each other for their Nike, Polo, Hilfiger and Nautica gear, it became clear that tobacco and alcohol companies are not the only marketers that prey on poor children's longing for escape. As we have seen, these fashion labels sold disadvantaged kids so successfully on their exaggerated representations of the good life – the country club, the yacht, the superstar celebrity – that logowear has become, in some parts of the Global City, both talisman and weapon. Meanwhile, the young feminists of Carly Stasko's generation whose sense of injustice had been awakened by Naomi Wolf's *Beauty Myth*, and Jean Kilbourne's documentary *Killing Us Softly*, also lived through the feeding frenzies around "alternative," Gen-X, hip-hop and rave culture. In the process, many became vividly aware that marketing affects communities not only by stereotyping them, but also – and equally powerfully – by hyping and chasing after them. This was a tangible shift from one generation of feminists to the next. When *Ms.* went ad-free in 1990, for instance, there was a belief that the corrosive advertising interference from which Gloria Steinem and Robin Morgan were determined to free their publication was a specifically female problem.[17] But as the politics of identity mesh with the burgeoning critique of corporate power, the demand has shifted from reforming problematic ad campaigns to questioning whether advertisers have any legitimate right to invade every nook and cranny of our mental and physical environment: it has become about the disappearance of space and the lack of meaningful choice. Ad culture has demonstrated its remarkable ability to absorb, accommodate and even profit from content critiques. In this context, it has become abundantly clear that the only attack that will actually shake this resilient industry is one leveled not at the pretty people in the pictures, but against the corporations that paid for them.

So for Carly Stasko, marketing has become more an environmental than a gender or self-esteem issue, and her environment is the streets, the university campus and the mass-media culture in which she, as an urbanite, lives her life. "I mean, this is my environment," she says, "and these ads are really directed at me. If these images can affect me, then I can affect them back."

The Washroom Ad as Political Catalyst

For many students coming of age in the late nineties, the turning point from focusing on the content of advertising to a preoccupation with the form itself occurred in the most private of places: in their university washrooms, staring at a car ad. The washroom ads first began appearing on North American campuses in 1997 and have been proliferating ever since. As we have already seen in Chapter 5, the administrators who allowed ads to creep onto their campuses told themselves that young people were already so bombarded with commercial messages that a few more wouldn't kill them, and the revenues would help fund valuable programs. But it seems there is such a thing as an ad that breaks the camel's back – and for many students, that was it.

The irony, of course, is that from the advertiser's perspective, niche nirvana had been attained. Short of eyelid implants, ads in college washrooms represent as captive a youth market as there is on earth. But from the students' perspective, there could have been no more literal metaphor for space closing in than an ad for Pizza Pizza or Chrysler Neon staring at them from over a urinal or from the door of a W.C. cubicle. Which is precisely why this misguided branding scheme created the opportunity for hundreds of North American students to take their first tentative steps toward direct anticorporate activism.

Looking back, school officials must see that there is something hilariously misguided about putting ads in private cubicles where students have been known to pull out their pens or eyeliners and scrawl desperate declarations of love, circulate unsubstantiated rumors, carry on the abortion debate and share deep philosophical insights. When the mini-billboards arrived, the bathroom became the first truly safe space in which to talk back to ads. In an instant, the direction of the scrutiny through the one-way glass of the focus group was reversed, and the target market took aim at the people behind the glass. The most creative response came from students at the University of Toronto. A handful of undergraduates landed part-time jobs with the washroom billboard company and kept conveniently losing the custom-made screwdrivers that opened the four hundred plastic frames. Pretty soon, a group calling themselves the Escher Appreciation Society were breaking into the "student-proof" frames and systematically replacing the bathroom ads with prints by Maurits Cornelis Escher. Rather than brushing up on the latest from Chrysler or Molson, students could learn to appreciate the Dutch graphic artist – chosen, the Escherites conceded, because his geometric work photocopies well.

The bathroom ads made it unmistakably clear to a generation of student activists that they don't need cooler, more progressive or more diverse ads – first and foremost, they need ads to shut up once in a while. Debate on campuses began to shift away from an evaluation of the content of ads to the fact that it was becoming impossible to escape from advertising's intrusive gaze.

Of course there are those among the culture jammers whose interest in advertising is less tapped into the new ethos of anti-branding rage and instead has much in common with the morality squads of the political correctness years. At times, *Adbusters* magazine feels like an only slightly hipper version of a Public Service Announcement about saying no to peer pressure or remembering to Reduce, Reuse and Recycle. The magazine is capable of lacerating wit, but its attacks on nicotine, alcohol and fast-food joints can be repetitive and obvious. Jams that change Absolut Vodka to "Absolut Hangover" or Ultra Kool cigarettes to "Utter Fool" cigarettes are enough to turn off would-be supporters who see the magazine crossing a fine line between information-age civil disobedience and puritanical finger-waving. Mark Dery, author of the original culture-jammers' manifesto and a former contributor to the magazine, says the anti-booze, -smoking and -fast-food emphasis reads as just plain patronizing – as if "the masses" cannot be trusted to "police their own desires."[18]

Listening to the Marketer Within

In a *New Yorker* article entitled "The Big Sellout," author John Seabrook discusses the phenomenon of "the marketer within." He argues persuasively that an emerging generation of artists will not concern themselves with old ethical dilemmas like "selling out" since they are a walking sales pitch for themselves already, intuitively understanding how to produce prepackaged art, to be their own brand. "The artists of the next generation will make their art with an internal marketing barometer already in place. The auteur as marketer, the artist in a suit of his own: the ultimate in vertical integration."[19]

Seabrook is right in his observation that the rhythm of the pitch is hardwired into the synapses of many young artists, but he is mistaken in assuming that the built-in marketing barometer will only be used to seek fame and fortune in the culture industries. As Carly Stasko points out, many people who grew up sold are so attuned to the tempo of marketing that as soon as they read or hear a new slogan, they begin to flip it and play with it in their minds, as she herself does. For Stasko, it is the adbuster that is within, and every ad campaign is a riddle just waiting for the right jam. So the skill Seabrook identifies, which allows artists to write the press bumpf for their own gallery openings and musicians to churn out metaphor-filled bios for their liner notes, is the same quality that makes for a deadly clever culture jammer. The culture jammer is the activist artist as *anti*marketer, using a childhood filled with Trix commercials, and an adolescence spent spotting the product placement on *Seinfeld*, to mess with a system that once saw itself as a specialized science. Jamie Batsy, a Toronto-area "hacktivist," puts it like this: "Advertisers and other opinion makers are now in a position where they are up against a generation of activists that were watching television before they could walk. This generation wants their brains back and mass media is their home turf."[20]

Culture jammers are drawn to the world of marketing like moths to a flame, and the high-gloss sheen on their work is achieved precisely because they still feel an affection – however deeply ambivalent – for media spectacle and the mechanics of persuasion. "I think a lot of people who are really interested in subverting advertising or studying advertising probably, at one time, wanted to be ad people themselves," says Carrie McLaren, editor of the

New York zine *Stay Free!*[21] You can see it in her own ad busts, which are painstakingly seamless in their design and savage in their content. In one issue, a full-page anti-ad shows a beat-up kid face down on the concrete with no shoes on. In the corner of the frame is a hand making away with his Nike sneakers. "Just do it," the slogan says.

Nowhere is the adbuster's ear for the pitch used to fuller effect than in the promotion of adbusting itself, a fact that might explain why culture jamming's truest believers often sound like an odd cross between used-car salesmen and tenured semiotics professors. Second only to Internet hucksters and rappers, adbusters are susceptible to a spiraling bravado and to a level of self-promotion that can be just plain silly. There is much fondness for claiming to be Marshall McLuhan's son, daughter, grandchild or bastard progeny. There is a strong tendency to exaggerate the power of wheatpaste and a damn good joke. And to overstate their own power: one culture-jamming manifesto, for instance, explains that "the billboard artist's goal is to throw a well aimed spanner into the media's gears, bringing the image factory to a shuddering halt."[22]

Adbusters has taken this hard-sell approach to such an extreme that it has raised hackles among rival culture jammers. Particularly galling to its critics is the magazine's line of anticonsumer products that they say has made the magazine less a culture-jamming clearinghouse than a home-shopping network for adbusting accessories. Culture-jammer "tool boxes" are listed for sale: posters, videos, stickers and postcards; most ironically, it used to sell calendars and T-shirts to coincide with Buy Nothing Day, though better sense eventually prevailed. "What comes out is no real alternative to our culture of consumption," Carrie McLaren writes. "Just a different brand." Fellow Vancouver jammers Guerrilla Media (GM) take a more vicious shot at *Adbusters* in the GM inaugural newsletter. "We promise there are no GM calendars, key chains or coffee mugs in the offing. We are, however, still working on those T-shirts that some of you ordered – we're just looking for that perfect sweatshop to produce them."[23]

Marketing the Antimarketers

The attacks are much the same as those lobbed at every punk band that signs a record deal and every zine that goes glossy: *Adbusters* has simply become too popular to have much cachet for the radicals who once dusted it off in their local secondhand bookstore like a precious stone. But beyond the standard-issue purism, the question of how best to "market" an antimarketing movement is a uniquely thorny dilemma. There is a sense among some adbusters that culture jamming, like punk itself, must remain something of a porcupine; that to defy its own inevitable commodification, it must keep its protective quills sharp. After the great Alternative and Girl Power™ cash-ins, the very process of naming a trend, or coining a catchphrase, is regarded by some with deep suspicion. "*Adbusters* jumped on it and were ready to claim this movement before it ever really existed," says McLaren, who complains bitterly in her own writing about the "USA Today/MTV-ization" of *Adbusters*. "It's become an advertisement for anti-advertising."[24]

There is another fear underlying this debate, one more confusing for its proponents than the prospect of culture jamming "selling out" to the dictates of marketing. What if, despite all the rhetorical flair its adherents can muster, culture jamming doesn't actually matter? What if there is no jujitsu, only semiotic shadowboxing? Kalle Lasn insists that his magazine has the power to "jolt postmodern society out of its media trance" and that his uncommercials threaten to shake network television to its core. "The television mindscape has been homogenized over the last 30 to 40 years. It's a space that is very safe for commercial messages. So, if you suddenly introduce a note of cognitive dissonance with a spot that says 'Don't buy a car,' or in the middle of a fashion show somebody suddenly says 'What about anorexia?' there's a powerful moment of truth."[25] But the real truth is that, as a culture, we seem to be capable of absorbing limitless amounts of cognitive dissonance on our TV sets. We culture jam manually every time we channel surf – catapulting from the desperate fundraising pleas of the Foster Parent Plan to infomercials for Buns of Steel; from Jerry Springer to Jerry Falwell; from New Country to Marilyn Manson. In these information-numb times, we are beyond being abruptly awakened by a startling image, a sharp juxtaposition or even a fabulously clever détournement.

Jaggi Singh is one activist who has become disillusioned with the jujitsu theory. "When you're jamming, you're sort of playing their game, and I think ultimately that playing field is stacked against us because they can saturate... we don't have the resources to do all those billboards, we don't have the resources to buy up all that time, and in a sense, it almost becomes pretty scientific – who can afford these feeds?"

Logo Overload

To add further evidence that culture jamming is more drop in the bucket than spanner in the works, marketers are increasingly deciding to join in the fun. When Kalle Lasn says culture jamming has the feeling of "a bit of a fad," he's not exaggerating.[26] It turns out that culture jamming – with its combination of hip-hop attitude, punk anti-authoritarianism and a well of visual gimmicks – has great sales potential.

Yahoo! already has an official culture-jamming site on the Internet, filed under "alternative." At Soho Down & Under on West Broadway in New York, Camden Market in London or any other high street where alterno gear is for sale, you can load up on logo-jammed T-shirts, stickers and badges. Recurring détournements – to use a word that seems suddenly misplaced – include Kraft changed to "Krap," Tide changed to "Jive," Ford changed to "Fucked" and Goodyear changed to "Goodbeer." It's not exactly trenchant social commentary, particularly since the jammed logos appear to be interchangeable with the corporate kitsch of unaltered Dubble Bubble and Tide T-shirts. In the rave scene, logo play is all the rage – in clothing, temporary tattoos, body paint and even ecstasy pills. Ecstasy dealers have taken to branding their tablets with famous logos: there is Big Mac E, Purple Nike Swirl E, X-Files E, and a mixture of uppers and downers called a "Happy Meal." Musician Jeff Renton explains the drug culture's appropriation of corporate logos as a revolt against invasive marketing. "I think it's a matter of: 'You come into our lives with your million-dollar advertising campaigns putting logos in places that make us feel uncomfortable, so we're going to take your logo back and use it in places that make you feel uncomfortable,'" he says.[27]

But after a while, what began as a way to talk back to the ads starts to feel more like evidence of our total colonization by them, and especially because the ad industry is proving that it is capable of cutting off the culture jammers at the pass. Examples of pre-jammed ads include a 1997 Nike campaign that used the slogan "I am not/A target market/I am an athlete" and Sprite's "Image Is Nothing" campaign, featuring a young black man saying that all his life he has been bombarded with media lies telling him that soft drinks will make him a better athlete or more attractive, until he realized that "image is nothing." Diesel jeans, however, has gone furthest in incorporating the political content of adbusting's anticorporate attacks. One of the most popular ways for artists and activists to highlight the inequalities of free-market globalization is by juxtaposing First World icons with Third World scenes: Marlboro Country in the war-torn rubble of Beirut (see image, page 10); an obviously malnourished Haitian girl wearing Mickey Mouse glasses; *Dynasty* playing on a TV set in an African hut; Indonesian students rioting in front of McDonald's arches. The power of these visual critiques of happy one-worldism is precisely what the Diesel clothing company's "Brand O" ad campaign attempts to co-opt. The campaign features ads within ads: a series of billboards flogging a fictional Brand O line of products in a nameless North Korean city. In one, a glamorous skinny blonde is pictured on the side of a bus that is overflowing with frail-looking workers. The ad is selling "Brand O Diet – There's no limit to how thin you can get." Another shows an Asian man huddled under a piece of cardboard. Above him towers a Ken and Barbie Brand O billboard.

Perhaps the point of no return came in 1997 when Mark Hosler of Negativland received a call from the ultra-hip ad agency Wieden & Kennedy asking if the band that coined the term "culture jamming" would do the soundtrack for a new Miller Genuine Draft commercial. The decision to turn down the request and the money was simple enough, but it still sent him spinning. "They utterly failed to grasp that our entire work is essentially in opposition to everything that they are connected to, and it made me really depressed because I had thought that our esthetic couldn't be absorbed into marketing," Hosler says.[28] Another rude awakening came when Hosler first saw Sprite's "Obey Your Thirst" campaign. "That commercial was a hair's breadth away from a song on our [*Dispepsi*] record. It was surreal. It's not just the fringe that's getting absorbed now – that's always happened. What's getting absorbed now is the idea that there's no opposition left, that any resistance is futile."[29]

I'm not so sure. Yes, some marketers have found a way to distill culture jamming into a particularly edgy kind of nonlinear advertising, and there is no doubt that Madison Avenue's embrace of the techniques of adbusting has succeeded in moving product off the superstore shelves. Since Diesel began its aggressively ironic "Reasons for Living" and "Brand O" campaigns in the U.S., sales have gone from $2 million to $23 million in four years,[30] and the Sprite "Image Is Nothing" campaign is credited with a 35 percent rise in sales in just three years.[31] That said, the success of these individual campaigns has done nothing to disarm the antimarketing rage that fueled adbusting in the first place. In fact, it may be having the opposite effect.

Ground Zero of the Cool Hunt

The prospect of young people turning against the hype of advertising and defining themselves against the big brands is a continuous threat coming from cool-hunting agencies like Sputnik, that infamous team of professional diary readers and generational snoops. "Intellectual crews," as Sputnik calls thinking young people, are aware and resentful of how useful they are to the marketers:

> They understand that mammoth corporations now seek their approval to continually deliver goods that will translate to megasales in the mainstream. Their stance of being intellectual says to each other, and to themselves, and most importantly to marketers—who spend innumerable dollars for in-your-face this-is-what-you-need advertisements—that they cannot be bought or fooled anymore by the hype. Being a head means that you won't sell out and be told what to wear, what to buy, what to eat or how to speak by anyone (or anything) other than yourself.[32]

But while the Sputnik writers inform their corporate readers about the radical ideas on the street, they appear to think that though these ideas will dramatically influence how young people will party, dress and talk, they will magically have no effect whatsoever on how young people will behave as political beings.

After they sound the alarm, the hunters always reassure their readers that all this anticorporate stuff is actually a meaningless pose that can be worked around with a hipper, edgier campaign. In other words, anticorporate rage is no more meaningful a street trend than a mild preference for the color orange. The happy underlying premise of the cool hunters' reports is that despite all the punk-rock talk, there is no belief that is a true belief and there are no rebels who cannot be tamed with an ad campaign or by a street promoter who *really speaks to them*. The unquestioned assumption is that there is no end point in this style cycle. There will always be new spaces to colonize – whether physical or mental – and there will always be an ad that will be able to penetrate the latest strain of consumer cynicism. Nothing new is taking place, the hunters tell each other: marketers have always extracted symbols and signs from the resistance movements of their day.

What they don't say is that previous waves of youth resistance were focused primarily on such foes as "the establishment," the government, the patriarchy and the military-industrial complex. Culture jamming is different – its rage encompasses the very type of marketing that the cool hunters and their clients are engaging in as they try to figure out how to use anti-marketing rage to sell products. The big brands' new ads must incorporate a youth cynicism not about products as status symbols, or about mass homogenization, but about multinational brands themselves as tireless culture vultures.

The admen and adwomen have met this new challenge without changing their course. They are busily hunting down and reselling the edge, just as they have always done, which is why Wieden & Kennedy thought there was nothing strange about asking Negativland to shill for Miller. After all, it was Wieden & Kennedy, a boutique ad agency based in Portland, Oregon, that made Nike a feminist sneaker. It was W&K who dreamed up the postindustrial alienation marketing plan for Coke's OK Cola; W&K who gave the world the immortal plaid-clad assertion that the Subaru Impreza was "like punk rock"; and it was W&K who brought Miller Beer into the age of irony. Masters at pitting the individual against various incarnations of mass-market bogeymen, Wieden & Kennedy sold cars to people who hated car ads, shoes to people who loathed image, soft drinks to the Prozac Nation and, most of all, ads to people who were "not a target market."

The agency was founded by two self-styled "beatnik artists," Dan Wieden and David Kennedy, whose technique, it seems, for quieting their own nagging fears that they were selling out has consistently been to drag the ideas and icons of the counterculture with them into the ad world. A quick tour through the agency's body of work is nothing short of a counterculture reunion – Woodstock meets the Beats meets Warhol's Factory. After putting Lou Reed in a Honda spot in the mid-eighties, W&K used the Beatles anthem "Revolution" in one Nike commercial, then carted out John Lennon's "Instant Karma" for another. They also paid proto-rock-and-roller Bo Diddley to do the "Bo Knows" Nike spots, and filmmaker Spike Lee to do an entire series of Air Jordan ads. W&K even got Jean-Luc Godard to direct a European Nike commercial. There were still more countercultural artifacts lying around: they stuck William Burroughs's face in a mini-TV-set in another Nike commercial and designed a campaign, nixed by Subaru before it made it to air, that used Jack Kerouac's *On the Road* as the voice-over text for an SVX commercial.

After making its name on the willingness of the avant-garde to set its price for the right mix of irony and dollars, W&K can hardly be blamed for thinking that culture jammers would also be thrilled to take part in the postmodern fun of a self-aware ad campaign. But the backlash against the brands, of which culture jamming is only one part, isn't about vague notions of alternativeness battling the mainstream. It has to do with the specific issues that have been the subject of this book so far: the loss of public space, corporate censorship and unethical labor practices, to name but three – issues less easily digested than tasty morsels like Girl Power and grunge.

Which is why Wieden & Kennedy hit a wall when they asked Negativland to mix for Miller, and why that was only the first in a string of defeats for the agency. The British political pop-band Chumbawamba turned down a $1.5 million contract that would have allowed Nike to use its hit song "Tubthumping" in a World Cup spot. Abstract notions about staying indie were not at issue (the band did allow the song to be used in the soundtrack for

Home Alone 3); at the center of their rejection was Nike's use of sweatshop labor. "It took everybody in the room under 30 seconds to say no," said band member Alice Nutter.[33] The political poet Martin Espada also got a call from one of Nike's smaller agencies, inviting him to take part in the "Nike Poetry Slam." If he accepted, he would be paid $2,500 and his poem would be read in a thirty-second commercial during the 1998 Winter Olympics in Nagano. Espada turned the agency down flat, offering up a host of reasons and ending with this one: "Ultimately, however, I am rejecting your offer as a protest against the brutal labor practices of the company. I will not associate myself with a company that engages in the well-documented exploitation of workers in sweatshops."[34] The rudest awakening came with Wieden & Kennedy's cleverest of schemes: in May 1999, with labor scandals still hanging over the swoosh, the agency approached Ralph Nader – the consumer-rights movement's most powerful leader and a folk hero for his attacks on multinational corporations – and asked him to do a Nike ad. The idea was simple: Nader would get $25,000 for holding up an Air 120 sneaker and saying, "Another shameless attempt by Nike to sell shoes." A letter sent to Nader's office from Nike headquarters explained that "what we are asking is for Ralph, as the country's most prominent consumer advocate, to take a light-hearted jab at us. This is a very Nike-like thing to do in our ads." Nader, never known for being light of heart, would only say, "Look at the gall of these guys."[35]

It was indeed a very Nike-like thing to do. Ads co-opt out of reflex – they do so because consuming is what consumer culture does. Madison Avenue is generally not too picky about what it will swallow, it doesn't avoid poison directed against itself but rather, as Wieden & Kennedy have shown, chomps down on whatever it finds along the path as it looks for the new "edge." The scenario that it appears unwilling to consider is that its admen and adwomen, the perennial teenage followers, may finally be following their target market off a cliff.

Adbusting in the Thirties: "Become a Toucher Upper!"

Of course the ad industry has disarmed backlashes before – from women complaining of sexism, gays claiming invisibility, ethnic minorities tired of gross caricatures. And that's not all. In the 1950s and again in the 1970s,

Western consumers became obsessed with the idea that they were being fooled by advertisers through the covert use of subliminal techniques. In 1957, Vance Packard published the runaway best-seller *The Hidden Persuaders*, which shocked Americans with allegations that social scientists were packing advertisements with messages invisible to the human eye. The issue re-emerged in 1973, when Wilson Bryan Key published *Subliminal Seduction*, a study of the lascivious messages tucked away in ice cubes. Key was so transported by his discovery that he made such bold claims as "the subliminal promise to anyone buying Gilbey's gin is simply a good old-fashioned sexual orgy."[36]

But all these antimarketing spasms had one thing in common: they focused exclusively on the content and techniques of advertising. These critics didn't want to be subliminally manipulated – and they *did* want African Americans in their cigarette ads and gays and lesbians selling jeans. Because the concerns were so specific, they were relatively easy for the ad world to address or absorb. For instance, the charge of hidden messages harbored in ice cubes, and other carefully cast shadows, spawned an irony-laden advertising subgenre that design historians Ellen Luton and J. Abbot Miller term "meta-subliminal" – ads that parody the charge that ads send secret messages. In 1990, Absolut Vodka launched the "Absolut Subliminal" campaign which showed a glass of vodka on the rocks with the word "absolut" clearly screened into the ice cubes. Seagram's and Tanqueray gin followed with their own subliminal in-jokes, as did the cast of *Saturday Night Live* with the recurring character Subliminal Man.

The critiques of advertising that have traditionally come out of academe have been equally unthreatening, though for different reasons. Most such criticism focuses not on the effects of marketing on public space, cultural freedom and democracy, but rather on ads' persuasive powers over seemingly clueless people. For the most part, marketing theory concentrates on the way ads implant false desires in the consuming public – making us buy things that are bad for us, pollute the planet or impoverish our souls. "Advertising," as George Orwell once said, "is the rattling of a stick inside a swill bucket." When such is the theorist's opinion of the public, it is no wonder that there is little potential for redemption in most media criticism: this sorry populace will never be in possession of the critical tools it needs to formulate a political response to marketing mania and media synergy.

The future is even bleaker for those academics who use advertising criticism for a thinly veiled attack on "consumer culture." As James Twitchell writes in *Adcult USA*, most advertising criticism reeks of contempt for the people who "want – ugh! – things."[37] Such a theory can never hope to form the intellectual foundation of an actual resistance movement against the branded life, since genuine political empowerment cannot be reconciled with a belief system that regards the public as a bunch of ad-fed cattle, held captive under commercial culture's hypnotic spell. What's the point of going through the trouble of trying to knock down the fence? Everyone knows the branded cows will just stand there looking dumb and chewing cud.

Interestingly, the last time that there was a successful attack on the practice of advertising – rather than a disagreement on its content or techniques – was during the Great Depression. In the 1930s the very idea of the happy, stable consumer society portrayed in advertising provoked a wave of resentment from the millions of Americans who found themselves on the outside of the dream of prosperity. An anti-advertising movement emerged that attacked ads not for faulty imagery but as the most public face of a deeply faulty economic system. People weren't incensed by the pictures in the ads, but rather by the cruelty of the obviously false promise that they represented – the lie of the American Dream that the happy consumer lifestyle was accessible to all. In the late twenties, and through the thirties, the frivolous promises of the ad world made for stomach-wrenching juxtapositions with the casualties of economic collapse, setting the stage for an unparalleled wave of consumer activism.

There was a short-lived magazine published in New York called *The Ballyhoo*, a sort of Depression-era *Adbusters*. In the wake of the 1929 stock-market crash, *The Ballyhoo* arrived as a cynical new voice, viciously mocking the "creative psychiatry" of cigarette and mouthwash ads, as well as the outright quackery used to sell all kinds of potions and lotions.[38] *The Ballyhoo* was an instant success, reaching a circulation of more than 1.5 million in 1931. James Rorty, a 1920s Mad Ave adman turned revolutionary socialist, explained the new magazine's appeal: "Whereas the stock in trade of the ordinary mass or class consumer magazine is reader-confidence in advertising, the stock in trade of *Ballyhoo* was reader-disgust with advertising, and with high-pressure salesmanship in general.... *Ballyhoo*, in turn, parasites on the grotesque, bloated body of advertising."[39]

Ballyhoo's culture jams include "Scramel" cigarettes ("they're so fresh they're insulting"), or the line of "69 different Zilch creams: What the well greased girls will wear. Absolutely indispensable (Ask any dispensary)." The editors encouraged readers to move beyond their snickers and go out and bust bothersome billboards themselves. A fake ad for the "Twitch Toucher Upper School" shows a drawing of a woman who has just painted a mustache on a glamorous cigarette model. The caption reads, "Become a Toucher Upper!" and goes on to say: "If you long to mess up advertisement: if your heart cries out to paint pipes in the mouths of beautiful ladies, try this 10-second test NOW! Our graduates make their marks all over the world! Good Toucher Uppers are always in demand" (see image on page 278). The magazine also created fake products to skewer the hypocrisy of the Hoover administration, like the "Lady Pipperal Bedsheet De Luxe" – made extra long to snugly fit on park benches when you become homeless. Or the "smilette" – two hooks that clamp on to either side of the mouth and force a happy expression. "Smile away the Depression! Smile us into Prosperity!"

The hard-core culture jammers of the era were not the *Ballyhoo* humorists, however, but photographers like Walker Evans, Dorothea Lange and Margaret Bourke-White. These political documentarians latched on to the hypocrisies of ad campaigns such as the National Association of Manufacturers' "There's No Way Like the American Way" by highlighting the harsh visual contrasts between the ads and the surrounding landscape. A popular technique was photographing billboards with slogans like "World's Highest Standard of Living" in their actual habitat: hanging surreally over breadlines and tenements. The manic grinning models piled into the family sedan were clearly blind to the tattered masses and squalid conditions below. The photographers of the era also scrupulously documented the fragility of the capitalist system by picturing fallen businessmen holding up "Will Work for Food" signs in the shadow of looming Coke billboards and peeling hoardings.

In 1934, advertisers began to use self-parody to deal with the mounting criticism they faced, a tactic that some saw as proof of the industry's state of disrepair. "It is contended by the broadcasters, and doubtless also by the movie producers, that this burlesque sales promotion takes the curse out of sales talk, and this is probably true to a degree," writes Rorty of the self-mockery. "But the prevalence of the trend gives rise to certain ominous suspicions...When the burlesque comedian mounts the pulpit of the Church of Advertising, it may be legitimately suspected that the edifice is doomed; that it will shortly be torn down or converted to secular uses."[40]

Of course the edifice survived, though not unscathed. New Deal politicians, under pressure from a wide range of populist movements, imposed lasting reforms on the industry. The adbusters and social documentary photographers were part of a massive grassroots public revolt against big business that included the farmers' uprising against the proliferation of supermarket chains, the establishing of consumer purchasing cooperatives, the rapid expansion of a network of trade unions and a crackdown on garment industry sweatshops (which had seen the ranks of the two U.S. garment workers' unions swell from 40,000 in 1931 to more than 300,000 in 1933). Most of all, the early ad critics were intimately linked to the burgeoning consumer movement that had been catalyzed by *One Hundred Million Guinea Pigs: Dangers in Everyday Foods, Drugs and Cosmetics* (1933), by F.J. Schlink and Arthur Kallet, and *Your Money's Worth: A Study in the Waste of the Consumer Dollar* (1927), written by Stuart Chase and F.J. Schlink. These books presented exhaustive catalogs of the way regular folks were getting lied to, cheated, poisoned and ripped off by America's captains of industry. The authors founded Consumer Research (later splintered off into the Consumers Union), which served both as an independent product-testing laboratory and a political group that lobbied the government for better grading and labeling of products. The CR believed objective testing and truthful labeling could make marketing so irrelevant it would become obsolete. According to Chase and Schlink's logic, if consumers had access to careful scientific research that compared the relative merits of the products on the market, everyone would simply make measured, rational decisions about what to buy. The advertisers, of course, were beside themselves, and terrified of the following F.J. Schlink had built up on the college campuses and among the New York intelligentsia. As adman C.B. Larrabee noted in 1934, "Some forty or fifty thousand persons won't so much as buy a box of dog biscuits unless F.J. gives his 'O.K.'... obviously they think most advertisers are dishonest, double-dealing shysters."[41]

Schlink and Chase's rationalist utopia of Spock-like consumerism never came to fruition, but their lobbying did force governments around the world to move to outlaw blatantly false claims in advertising, to establish quality standards for consumer goods, and to become actively involved in the grading and labeling of them. And the *Consumers Union Reports* is still the buyer's bible in America, though it long ago severed its ties to other social movements.

It is worth noting that the modern-day ad world's most extreme attempts to co-opt anticorporate rage have fed directly off images pioneered by the Depression-era documentary photographers. Diesel's Brand O is almost a direct replica of Margaret Bourke-White's "American Way" billboard series, both in style and composition. And when the Bank of Montreal ran an ad campaign in Canada in the late nineties, at the height of a popular backlash against soaring bank profits, it used images that recalled Walker Evans's photographs of 1930s businessmen holding up those "Will Work for Food" signs. The bank's campaign consisted of a series of grainy black-and-white photographs of ragged-looking people holding signs that asked, "Will I ever own my own home?" and "Are we going to be okay?" One sign simply read, "The little guy is on his own." The television spots blasted Depression-era gospel and ragtime over eerie industrial images of abandoned freight trains and dusty towns.

In other words, when the time came to fight fire with fire, the advertisers raced back to an era when they were never more loathed and only a world war could save them. It seems that this kind of psychic shock – a clothing company using the very images that have scarred the clothing industry; a bank trading on anti-bank rage – is the only technique left that will get the attention of us ad-resistant roaches. And this may well be true, from a marketing point of view, but there is also a larger context that reaches beyond imagery: Diesel produces many of its garments in Indonesia and other parts of the Far East, profiting from the very disparities illustrated in its clever Brand O ads. In fact, part of the edginess of the campaign is the clear sense that the company is flirting with a Nike-style public-relations meltdown. So far, the Diesel brand does not have a wide enough market reach to feel the full force of having its images slingshot back at its body corporate, but the bigger the company gets – and it is getting bigger every year – the more vulnerable it becomes.

That was the lesson in the responses to the Bank of Montreal's "Sign of the Times" campaign. The bank's use of powerful images of economic collapse at exactly the same time that it announced record profits of $986 million (up in 1998 to $1.3 billion) inspired a spontaneous wave of adbusting. The simple imagery of the campaign – people holding up angry signs – was easy for the bank's critics to replicate with parodies that skewered the bank's exorbitant service fees, its inaccessible loans officers and the closing of branches in low-income neighborhoods (after all, the bank's technique had been stolen from the activists in the first place). Everyone got in on the action: lone jammers, CBC television's satirical show *This Hour Has 22 Minutes*, *The Globe and Mail*'s *Report on Business Magazine*, and independent video collectives.

Clearly, these ad campaigns are tapping into powerful emotions. But by playing on sentiments that are already directed against them – for example, public resentment at profiteering banks or widening economic disparities – the process of co-optation runs the very real risk of amplifying the backlash, not disarming it. Above all, imagery appropriation appears to radicalize culture jammers and other anticorporate activists – a "co-opt this!" stance develops that becomes even harder to diffuse. For instance, when Chrysler ran a campaign of pre-jammed Neon ads (the one that added a faux aerosol "p," changing "Hi" to "Hip"), it inspired the Billboard Liberation Front to go on its biggest tear in years. The BLF defaced dozens of Bay Area Neon billboards by further altering "Hip" to "Hype," and adding, for good measure, a skull and crossbones. "We can't sit by while these companies co-opt our means of communication," Jack Napier said. "Besides... they're tacky."

Perhaps the gravest miscalculation on the part of both markets and media is the insistence on seeing culture jamming solely as harmless satire, a game that exists in isolation from a genuine political movement or ideology. Certainly for some jammers, parody is perceived, in rather grandiose fashion, as a powerful end in itself. But for many more, as we will see in the next chapters, it is simply a new tool for packaging anticorporate salvos, one that is more effective than most at breaking through the media barrage. And as we will also see, adbusters are currently at work on many different fronts: the people scaling billboards are frequently the same ones who are organizing against the Multilateral Agreement on Investment, staging protests on the streets of Geneva against the World Trade Organization and occupying banks to protest against the profits they are making from student debts. Adbusting is not an end in itself. It is simply a tool – one among many – that is being used, loaned and borrowed in a much broader political movement against the branded life.

GENTS HAIRDRESSER
INFORMATION

CHAPTER THIRTEEN

RECLAIM THE STREETS

I picture the reality in which we live in terms of military occupation. We are occupied the way the French and Norwegians were occupied by the Nazis during World War II, but this time by an army of marketeers. We have to reclaim our country from those who occupy it on behalf of their global masters.

– Ursula Franklin, Professor Emeritus, University of Toronto, 1998

This is not a protest. Repeat. This is not a protest. This is some kind of artistic expression. Over.

– A call that went out on Metro Toronto police radios on May 16, 1998, the date of the first Global Street Party

It is one of the ironies of our age that now, when the street has become the hottest commodity in advertising culture, street culture itself is under siege. From New York to Vancouver to London, police crackdowns on graffiti, postering, panhandling, sidewalk art, squeegee kids, community gardening and food vendors are rapidly criminalizing everything that is truly street-level in the life of a city.

This tension between the commodification and criminalization of street culture has unfolded in a particularly dramatic manner in England. In the early to mid-nineties, as the ad world leaped to harness the sounds and imagery of the rave scene to sell cars, airlines, soft drinks and newspapers, the lawmakers in Britain made raves all but illegal, through the 1994 Criminal Justice Act. The act gave police far-reaching powers to seize sound equipment and deal harshly with ravers in any public confrontations.

To fight the Criminal Justice Act, the club scene (previously preoccupied with searching out the next all-night dance site) forged new alliances with more politicized subcultures that were also alarmed by these new police powers. Ravers got together with squatters facing eviction, with the so-called New Age travelers facing crackdowns on their nomadic lifestyle, and with radical "eco-warriors" fighting the paving-over of Britain's woodland areas by building tree houses and digging tunnels in the bulldozers' paths. A common theme began to emerge among these struggling countercultures: the right to uncolonized space – for homes, for trees, for gathering, for dancing. What sprang out of these cultural collisions among deejays, anti-corporate activists, political and New Age artists and radical ecologists may well be the most vibrant and fastest-growing political movement since Paris '68: Reclaim the Streets (RTS).

Since 1995, RTS has been hijacking busy streets, major intersections and even stretches of highway for spontaneous gatherings. In an instant, a crowd of seemingly impromptu partyers transforms a traffic artery into a surrealist playpen. Here's how it works. Like the location of the original raves, the RTS party's venue is kept secret until the day. Thousands gather at the designated meeting place, from which they depart en masse to a destination known only to a handful of organizers. Before the crowds arrive, a van rigged up with a powerful sound system is surreptitiously parked on the soon-to-be-reclaimed street. Next, some theatrical means of blocking traffic is devised – for example, two old cars deliberately crash into each other and a mock fight is staged between the drivers. Another technique is to plant twenty-foot scaffolding tripods in the middle of the roadway with a brave lone activist suspended high up top – the tripod poles prevent cars from passing but people can weave between them freely; and since to knock the tripod over would send the person on top crashing to the ground, the police have no recourse but to stand by and watch the events unfold. With traffic safely blocked, the roadway is declared a "street now open." Signs go up that say "Breathe," "Car Free," and "Reclaim Space." The RTS flag – a bolt of lightning on different colored backdrops – goes up and the sound system begins to blast everything from the latest electronic offerings to Louis Armstrong's "What a Wonderful World."

Then seemingly out of nowhere comes the traveling carnival of RTSers: bikers, stilt walkers, ravers, drummers. At previous parties, jungle gyms have been set up in the middle of intersections, as well as giant sandboxes, swing sets, wading pools, couches, throw rugs and volleyball nets. Hundreds of Frisbees sail through the air, free food is circulated and the dancing begins – on cars, at bus stops, on roofs and near signposts. Organizers describe their road-napings as anything from the realization of "a collective daydream" to "a large-scale coincidence." Like adbusters, RTSers have transposed the language and tactics of radical ecology into the urban jungle, demanding uncommercialized space in the city as well as natural wilderness in the country or on the seas. In this spirit, the most theatrical RTS stunt occurred when 10,000 partyers took over London's M41, a six-lane highway. Two people dressed in elaborate carnival costumes sat thirty feet above the roadway, perched on scaffolding contraptions that were covered by huge hoop skirts (see image on page 310). The police standing by had no idea that underneath the skirts were guerrilla gardeners with jackhammers, drilling holes in the highway and planting saplings in the asphalt. The RTSers – die-hard Situationist fans – had made their point: "Beneath the tarmac... a forest," a reference to the Paris '68 slogan, "Beneath the cobblestones... a beach."

The crowd followed us and the road turned from a traffic jam to a road rave with hundreds of people shouting and demanding clean air, public transport and bicycle lanes.

– RTS E-mail report, Tel Aviv, Israel, May 16, 1998

The events take culture jamming's philosophy of reclaiming public space to another level. Rather than filling the space left by commerce with advertising parodies, the RTSers attempt to fill it with an alternative vision of what society might look like in the absence of commercial control.

The seeds of RTS's urban environmentalism were planted in 1993 on Claremont Road, a quiet London street slated to disappear under a new expressway. "The M11 Link Road," explains RTSer John Jordan, "will stretch from Wanstead to Hackney in East London. To build it, the Department of Transport had to knock down 350 houses, displace several thousand people, cut through one of London's last ancient woodlands and devastate a community with a six-lane-wide stretch of tarmac at the cost of 240 million pounds, apparently to save six minutes on a car journey."[1] When the city ignored fierce local opposition to the road, a group of activist artists took it upon themselves to try to block the bulldozers by turning Claremont Road into a living sculptural fortress. They pulled sofas into the streets, hung TVs from tree branches, painted a giant chessboard in the middle of the road and put up spoof suburban development billboards in front of the houses slated for demolition: "Welcome to Claremont Road – Ideal Homes." The activists moved into chestnut trees, occupied construction cranes, blasted music and blew kisses at the cops and demolition workers below. The now empty houses were transformed – connected to each other through underground tunnels and filled with art installations. Outside, old cars were painted with slogans and zebra stripes and turned into flower boxes. The cars were not only made beautiful, they also made effective barricades, as did a hundred-foot scaffolding tower built through the roof of one of the homes. The tactic, Jordan explains, was not the use of art to achieve political ends but the transformation of art into a pragmatic political tool "both beautiful and functional."[2]

When Claremont Road was leveled in November 1994, it had become the most creative, celebratory, vibrantly living street in London. It was "a kind of temporary microcosm of a truly liberated, ecological culture," according to Jordan.[3] By the time all the activists had been cherry-picked out of their tree houses and fortresses, the point of the action – that high-speed roads suck the life out of a city – could have had no more graphic or eloquent expression.

Though another group had used the same name some years earlier, the current incarnation of Reclaim the Streets was formed in May 1995, with the express purpose of turning what happened on Claremont Road into an airborne virus that could spread at any time, to any place in the city – a roving "temporary autonomous zone," to use a term coined by the American anarchist guru Hakim Bey. According to Jordan, the thinking was simple: "If we could no longer reclaim Claremont Road, we would reclaim the streets of London."[4]

Five hundred people showed up to the RTS party on Camden Street in May

1995 to dance to a bicycle-powered sound system, drums and whistles. With the Criminal Justice Act in full effect, the gathering caught the attention of the newly politicized rave scene and a key alliance was formed. At RTS's next event, three thousand people showed up to the party on Upper Street, Islington; this time they danced to electronic music blasting from two trucks equipped with club-quality sound systems.

The combination of rave and rage has proved contagious, spreading across Britain to Manchester, York, Oxford and Brighton, and in the largest single RTS event to date, drawing 20,000 people to Trafalgar Square in April 1997. By then, Reclaim the Street parties had gone international, popping up in cities as far away as Sydney, Helsinki and Tel Aviv. Each party is locally organized, but with the help of E-mail lists and linked Web sites, activists in different cities are able to read reports from events around the world, swap cop-dodging strategies, trade information on building effective roadblocks, and read each other's posters, press releases and flyers. Since video and digital cameras appear to be the accessories of choice at the street parties, RTSers also draw inspiration from watching footage of faraway parties, which is circulated through activist video networks, such as the Oxford-based Undercurrents, and uploaded onto several RTS Web sites.

Anarchists among the crowd took advantage of the opportunity to vent their fury on banks, jewelry shops and local branches of McDonald's. Windows were smashed, paint bombs hurled and anti-globalisation slogans graffitied.

– RTS E-mail report, Geneva, Switzerland May 16, 1998

In many cities, the street parties have dovetailed with another explosive new international movement – the Critical Mass bicycle rides. The idea started in San Francisco in 1992 and began spreading to cities across North America, Europe and Australia at roughly the same time as RTS. Critical Mass bicycle riders also favor the rhetoric of large-scale coincidence: in dozens of cities, on the last Friday of every month, anywhere from seventeen to seven thousand cyclists gather at a designated intersection and go for a ride together. By force of their numbers, the bikers form a critical mass and the cars must yield to them. "We're not blocking traffic," the Critical Mass riders say, "we are the traffic." Since there's a fair amount of overlap between RTS partyers and Critical Mass riders, it has become a popular tactic for the sites of street parties to be cleared of traffic by "spontaneous" Critical Mass rides that sweep through the area just moments before the blockades are set up and the partyers arrive.

Perhaps in light of these connections, the mainstream media almost invariably describe RTS events as "anti-car protests." Most RTSers, however, insist that this is a profound oversimplification of their goals.[5] The car is a symbol, they say – the most tangible manifestation of the loss of communal space, walkable streets and sites of free expression. Rather than simply opposing the use of automobiles, as Jordan says, "RTS has always tried to take the single issue of transportation and the car into a wider critique of society... to dream of reclaiming space for collective use, as commons."[6] To underline these wider connections, RTS organized one London street party in solidarity with striking London Underground workers. Another was a joint event with those darlings of British rock stars, soccer players and anarchists – the sacked Liverpool dock workers. Other actions have taken on the ecological and human rights records of Shell, BP and Mobil.

These coalitions make RTS extremely difficult to categorize. "Is a street party a political rally?" asks Jordan rhetorically. "A festival? A rave? Direct action? Or just a bloody good party?" In many ways, the parties have defied easy labeling: they camouflage identifiable leaders, and have no center or even a focal point. RTS parties "swirl," as Jordan says.

Playing Politics

Not only is the confusion deliberate, but it is precisely this absence of rigidity that has helped RTS to capture the imagination of thousands of young people around the world. Since the days when Abbie Hoffman and the Yippies infused self-conscious absurdity into their "happenings," political protest had lapsed into a ritualized affair, following a fairly unimaginative grid of repetitive chants and scripted police confrontation. Pop, in the meantime, had become equally formulaic in its refusal to let the perceived earnestness of political conviction enter its ironic play space. Which is where RTS comes in. The deliberate culture clashes of the street parties mix the earnest predictability of politics with the amused irony of pop. For many people in their teens and twenties, this presents the first opportunity to reconcile being creatures of their Saturday-morning-cartoon childhoods with a genuine political concern for their communities and environment. RTS is just playful and ironic enough to finally make earnestness possible.

In many ways, Reclaim the Streets is the urban centerpiece of England's thriving do-it-yourself subculture. Exiled to the economic margins by decades of Tory rule, and given little reason to return by the right-of-center policies of Tony Blair's New Labour Party, a largely self-reliant infrastructure of food co-ops, illegal squats, independent media and free music festivals has emerged across the country. Spontaneous street parties are an extension of the DIY lifestyle, asserting as they do that people can make their own fun without asking any state's permission or relying on any corporation's largesse. At a street party, just showing up makes you both a participant and part of the entertainment.

The street party is also at odds with the way our culture tends to imagine freedom. Whether it's hippies dropping out to live in rural communes, or yuppies escaping the urban jungle in sport utility vehicles, freedom is usually about abandoning the claustrophobia of the city. Freedom is Route 66, it's "On the Road." It's eco-travel. It's anywhere but here. RTS, on the other hand, doesn't write off the city or the present. It harnesses the urge for entertainment and raves (and its darker side – the desire to freak out and riot) and channels them into an act of civil disobedience that is also a festival. For a day, the longing for free space is not about escape but transformation of the here and now.

We visited the Virgin at the place of the cathedral, who certainly didn't expect us and therefore didn't join the dance. In spite of this we offered a very nice sunny show till later that night, past eleven o'clock, reclaiming the street for about five hours.

– RTS E-mail report, Valencia, Spain, May 16, 1998

Of course, if you want to be really cynical, RTS is also flowery eco-poetry about vandalism. It's high-minded talk about blocking traffic. It's wildly dressed and painted kids screeching at extremely confused and possibly well-meaning cops about the tyranny of "car culture." And when RTS events go wrong – because only a handful of people show up, or the antihierarchy anarchist organizers are unable or unwilling to communicate with the crowd – that's exactly what the party becomes: some jerk demanding the right to sit in the middle of the street for a loony reason known only to him. But at their best, RTS actions have been too joyful and humane to dismiss, cracking the cynicism of many onlookers, from the hip British music press, which declared the party at Trafalgar Square "the best illegal rave or dance music party in history,"[7] to one striking Liverpool docker who noted that "the others talk about doing something – this lot actually do it."[8]

And, as with all successful radical movements, some voice concern that the mass appeal of RTS has made it too fashionable, that the subtle theory of "applying radical poetry to radical politics" is getting drowned out by the pounding beat and the mob mentality. In October 1997, Jordan told me that RTS was going through a process of rigorous re-examination. He claimed that the 20,000-strong Trafalgar Square party was not the sort of climax RTS had been moving toward. When the police tried to impound the van containing the sound system, protestors didn't cheekily blow kisses as hoped, they hurled bottles and rocks and four people were charged with attempted murder (the charges were later dropped). Despite the organizers' best efforts, RTS was spiraling into soccer hooliganism and, as one RTS spokesperson told *The Daily Telegraph*, when the organizers tried to regain control, some rioters turned against them. "I saw some of our people actually trying to stop yobbos who had got tanked up on beer and were mindlessly throwing bottles and rocks. A few of our contingent actually put themselves into the firing line and one was beaten up...."[9] Such shades of gray, however, were lost on most in the British media who covered Trafalgar Square with headlines like "Riot Frenzy – Anarchist Thugs Bring Terror to London."[10]

"The Resistance Will Be as Transnational as Capital"

After Trafalgar Square, Jordan says, it became clear that "it was too easy for the street party to be seen as just fun, just a party with a hint of political action.... If people think that turning up to a street party once a year, getting out of your head and dancing your heart out on a recaptured piece of public land is enough, then we are failing to reach our potential." The next task, he said, is to imagine a takeover bigger than just one street. "The street party is only a beginning, a taster of future possibilities. To date there have been 30 street parties all over the country. Imagine that growing to 100, imagine each one of those happening on the same day, imagine each one lasting for days on end and growing.... Imagine the street party growing roots... *la fête permanente....*"[11]

I admit that at the time I spoke to Jordan I was skeptical that this movement could pull off that level of coordination. At the best of times, Reclaim the Streets walks a delicate line, flirting openly with the urge to riot but attempting to flip it into a more constructive protest. The London RTSers say that one of the goals of the parties is to "visualize industrial collapse" – the challenge, then, is for participants to inspire one another enough to dance and plant trees in the rubble, rather than to douse it with gasoline and drop a Zippo. But shortly after our interview, a notice went out on a couple of activist E-mail lists, floating the idea of a coordinated day of simultaneous street parties around the world. Seven months later, the first-ever Global Street Party was under way. To make absolutely sure that the political underpinning of the event didn't get lost, the date chosen for the Global Street Party was May 16, 1998 – the same day the G-8 leaders gathered for a summit in Birmingham, England, and two days before they would proceed to Geneva to celebrate the fiftieth anniversary of the World Trade Organization. With Indian farmers, landless Brazilian peasants, unemployed French, Italian and German workers and international human-rights groups planning simultaneous actions around the two summits, RTS took its place in a fledgling international grassroots movement against transnational corporations and their agenda of economic globalization. This was definitely not just about cars.

The police attack was so hard and brutal that even the Czech public was shocked... Sixty-four people were detained including 22 of those younger than 18 and 13 women. During the police action, innocent people (who were just walking around) were also beaten. All detainees were beaten, mistreated and humiliated until morning hours.

– RTS E-mail report, Prague, Czech Republic, May 16, 1998

Though rarely reported as more than isolated traffic snares, thirty RTS events were successfully mounted around the world, in twenty different countries. On May 16, more than eight hundred people blocked a six-lane highway in Utrecht, the Netherlands, dancing for five hours. In Turku, Finland, two thousand partyers peacefully occupied one of the main bridges in the city. Almost a thousand Berliners held a rave at a downtown intersection and in Berkeley, California, seven hundred people played Twister on Telegraph Avenue. By far the most successful of the Global Street Parties was in Sydney, Australia, where an illegal political rally cum music festival went off without a hitch; between three and four thousand people "kidnaped" a road, setting up three stages for live concerts with bands and half a dozen deejays. There were no Levi's, Borders, Pepsi or Revlon sponsorships (the sort of backing that supposedly makes high-priced festivals like Lilith Fair "possible") but, somehow, Sydney's RTS managed to offer "three chai stalls, a food fund-raiser, a skateboard skate rail, a five terminal sidewalk Internet station, two sandstone sculptors, poets, fire twirlers, street gardeners... and loads of mayhem and frivolity."[12]

Police reaction to the Global Street Party varied wildly from city to city. In Sydney, the officers stood back in awe, asking only for the sound to be turned down as the party stretched into the evening. In Utrecht, the police were so friendly that "at one point," reports a local organizer, "they mingled with the crowd, sat on the pavement waiting for the sound system to arrive. When it finally arrived, they really assisted in getting the generator going." Not surprisingly, these were the exceptions. In Toronto, at the party I attended, the police officers let the event go on for an hour, then went into the crowd of four hundred partyers with open knives and (absurdly) began stabbing brightly colored balloons and energetically slashing streamers. As a result, the party degenerated into a series of incoherent cops-are-pigs skirmishes that led the six o'clock news. But Toronto's crackdown was nothing compared with what happened in other cities. Five thousand people danced on the streets of Geneva, but by midnight the party "had turned into a full scale riot. One car was set alight and thousands of police charged the main

Next time it'll be bigger...

– RTS E-mail report, Berlin, Germany, May 16, 1998

encampment, firing tear-gas into the crowd. The demonstrators smashed hundreds of windows, mainly banks and corporate offices, until 5 a.m., causing over half a million pounds in damage." With protestors anticipating the arrival of world leaders and trade officials for the WTO anniversary, the rioting continued for several days.[13]

In Prague, three thousand people showed up for the Global Street Party in Wenceslaus Square, where four sound systems were rigged up and twenty deejays were ready to play. Before long, however, a police car drove into the crowd at full speed; the vehicle was surrounded and overturned and once again, the rave became a riot. After organizers officially dissolved the event, three hundred people, mostly teens, marched through the streets of Prague, some of them stopping to hurl rocks and bottles through the plate-glass windows of McDonald's and Kentucky Fried Chicken outlets. More bottle throwing took place at the Berkeley, California, RTS, as well as several other inane activities including throwing a foam mattress into a bonfire on Telegraph Avenue (creating toxic fumes at an environmental protest – brilliant!) and smashing the window of a local independent bookstore (way to get those corporate bad guys). The event had been billed as a celebration of "art, love and rebellion" but police called it "a riot" – "the biggest in eight years."[14] There were at least twenty-seven arrests in Cambridge, four in Toronto, four in Berkeley, three in Berlin, sixty-four in Prague, dozens in Brisbane and more than two hundred over the days of rioting in Geneva.

> *Sorry for the screw-up but since only about ten of us turned up, we decided, after a walk around town with placards and a drummer, to bugger off to the beach for the rest of the afternoon.*
>
> – RTS E-mail report, Darwin, Australia, May 16, 1998

In several key cities, the Global Street Party was most certainly not the "*fête permanente*" that John Jordan had envisioned. However, the immediate international response provoked by nothing more than a few E-mail notices proved that there is both the potential and the desire for a truly global protest against the loss of public space. If anything, the urge to reclaim that space from branded life speaks so directly to so many young people of different nationalities that its greatest liability is the very force of the emotions it inspires.

That emotion was in full sway on May 16 in Birmingham, headquarters of the Global Street Party. The eight most powerful politicians in the world were busy trading hockey jerseys, signing trade agreements and – one cringes – having their own global sing-along to "All You Need Is Love." Against that backdrop, eight thousand activists who had gathered from all over Britain gained control of a roundabout, hooked up a sound system, played street volleyball and recaptured the RTS spirit of celebration. As in other cities, there were confrontations with the police who surrounded the party with a line three officers deep. This time, however, creative absurdity won out, and instead of rocks and bottles, the weapon of choice was that increasingly popular piece of slapstick ammo: the custard pie. And a new banner – a huge red kite – was hoisted amid the tripods, signs and flags, bearing the names of all the cities where street parties were taking place simultaneously in twenty countries around the world.[15] "The resistance," one sign said, "will be as transnational as capital."

RTS AGITPROP

The privatization of public space in the form of the car continues the erosion of neighborhood and community that defines the metropolis. Road schemes, business "parks," shopping developments – all add up to the disintegration of community and the flattening of a locality. Everywhere becomes the same as everywhere else. Community becomes commodity – a shopping village, sedated and under constant surveillance. The desire for community is then fulfilled elsewhere, through spectacle, sold to us in simulated form. A TV soap "street" or "square" mimicking the area that concrete and capitalism are destroying. The real street, in this scenario, is sterile. A place to move through not to be in. It exists only as an aid to somewhere else – through a shop window, billboard or petrol tank.

– London RTS

What I've noticed is that all of these events and actions had one thing in common: RECLAIMING. Whether we were reclaiming the road from cars, reclaiming buildings for squatters, reclaiming surplus food for the homeless, reclaiming campuses as a place for protest and theatre, reclaiming our voice from the deep dark depths of corporate media, or reclaiming our visual environment from billboards, we were always reclaiming. Taking back what should have been ours all along. Not "ours" as in "our club" or "our group," but ours as in the people. All the people. "Ours" as in "not the governments" and "not the corporations."... We want power given back to the people as a collective. We want to Reclaim the Streets.

– Toronto RTS

WANTED:
KATHIE LEE:
PAY
YOUR WORKERS
NOW OR
GO TO
NIKETOWN
NIKE
SHOES
MAKE
WORKERS
LOSE
AMERICAN FRIENDS
SERVICE COMMITTEE
THAT
EMPOWER
PROFIT$
THAT DEVOUR

CHAPTER FOURTEEN

BAD MOOD RISING

The New Anticorporate Activism

The earth is not dying, it is being killed. And those that are killing it have names and addresses.

—Utah Phillips

How do we tell Steve that his dad owns a sweatshop?!?

— Tori Spelling, as the character Donna on *Beverly Hills 90210*, after discovering that her own line of designer clothing was being manufactured by immigrant women in an L.A. sweatshop, October 15, 1997

While the latter half of the 1990s has seen enormous growth in the brands' ubiquity, a parallel phenomenon has emerged on the margins: a network of environmental, labor and human-rights activists determined to expose the damage being done behind the slick veneer. Dozens of new organizations and publications have been founded for the sole purpose of "outing" corporations that are benefiting from repressive government policies around the globe. Older groups, previously focused on monitoring governments, have reconfigured their mandates so that their primary role is tracking violations committed by multinational corporations. As John Vidal, environmental editor of *The Guardian*, puts it, "A lot of activists are attaching themselves leech-like onto the sides of the bodies corporate."

This leech-like attachment takes many forms, from the socially respectable to the near-terrorist. Since 1994, the Massachusetts-based Program on Corporations, Law & Democracy, for instance, has been developing policy alternatives designed to "contest the authority of corporations to govern." The Oxford-based Corporate Watch, meanwhile, focuses on researching — and helping others to research – corporate crime. (Not to be confused with the San Francisco-based Corporate Watch, which sprang up at about the same time with a nearly identical mission for the U.S.) JUSTICE. DO IT NIKE! is a group of scrappy Oregon activists devoted to haranguing Nike about its labor practices in its own backyard. The Yellow Pages, on the other hand, is an underground international cabal of hackers who have declared war on the computer networks of those corporations that have successfully lobbied to delink human rights from trade with China. "In effect, businessmen started dictating foreign policy," says Blondie Wong, director of Hong Kong Blondes, a group of Chinese pro-democracy hackers now living in exile. "By taking the side of profit over conscience, business has set our struggle back so far that they have become our oppressors too."[1]

Taking a distinctly lower-tech (some might say primitive) approach is Belgian Noel Godin and his global band of political pie slingers. Although politicians and movie stars have faced flying pies, the corporate sector has been the primary target: Microsoft CEO Bill Gates, Monsanto CEO Robert Shapiro, Chevron CEO Ken Derr, World Trade Organization director Renato Ruggiero have all been hit, as well as that architect of global free trade, Milton Friedman. "To their lies, we respond with pies," says Agent Blueberry, of the Biotic Baking Brigade (see image, page 258).[2]

The fad got so out of hand that in May 1999, Tesco, one of the largest supermarket chains in England, conducted a series of tests on its pies to see which ones made for the best slinging. "We like to keep abreast of what the customers are doing, and that's why we have to do the testing," said company spokesperson Melodie Schuster. Her recommendation: "The custard tart gives total face coverage."[3] Oh, and rest assured that none of the Tesco tarts contain any ingredients that have been genetically modified. The chain banned those from its products a month earlier – a response to a groundswell of anticorporate sentiment directed at Monsanto and the other agribusiness giants.

As we will see in a later chapter, Tesco made its decision to disassociate itself from genetically modified foods after a series of protests against "Frankenfoods" were held on its doorstep – part of an increasingly popular strategy among activist groups. Political rallies, which once wound their predictable course in front of government buildings and consulates, are now just as likely to take place in front of the stores of the corporate giants: outside Nike Town (see image, page 324), Foot Locker, the Disney Store and Shell gas pumps; on the roof of the corporate headquarters of Monsanto or BP; through malls and around Gap outlets; and even at supermarkets.

In short, the triumph of economic globalization has inspired a wave of techno-savvy investigative activists who are as globally minded as the corporations they track. This powerful form of activism reaches well beyond traditional trade unions. Its members are young and old; they come from elementary schools and college campuses suffering from branding fatigue and from church groups with large investment portfolios worried that corporations are behaving "sinfully." They are parents worried about their children's slavish devotion to "logo tribes," and they are also the political intelligentsia and social marketers who are more concerned with the quality of community life than with increased sales. In fact, by October 1997 there were so many disparate anticorporate protests going on around the world – against Nike, Shell, Disney, McDonald's and Monsanto – that Earth First! printed up an impromptu calendar with all the key dates and declared it the first annual End Corporate Dominance Month. About a month later, *The Wall Street Journal* ran a story headlined "Hurry! There Are Only 27 More Protesting Days Until Christmas."

"The Year of the Sweatshop"

In North America, much of this activity can be traced back to 1995–96, the period that Andrew Ross, director of American Studies at New York University, has called "The Year of the Sweatshop." For a time that year, North Americans couldn't turn on their televisions without hearing shameful stories about the exploitative labor practices behind the most popular, mass-marketed labels on the brandscape. In August 1995, the Gap's freshly scrubbed façade was further exfoliated to reveal a lawless factory in El Salvador where the manager responded to a union drive by firing 150 people and vowing that "blood will flow" if organizing continued.[4] In May 1996, U.S. labor activists discovered that chat-show host Kathie Lee Gifford's eponymous line of sportswear (sold exclusively at Wal-Mart) was being stitched by a ghastly combination of child laborers in Honduras and illegal sweatshop workers in New York. At about the same time, Guess jeans, which had built its image with sultry black-and-white photographs of supermodel Claudia Schiffer, was in open warfare with the U.S. Department of Labor over a failure on the part of its California-based contractors to pay the minimum wage. Even Mickey Mouse was letting his sweatshops show after a Disney contractor in Haiti was caught making Pocahontas pajamas under such impoverished conditions that workers had to nourish their babies with sugar water.

More outrage flowed after NBC aired an investigation of Mattel and Disney just days before Christmas 1996. With the help of hidden cameras, the reporter showed that children in Indonesia and China were working in virtual slavery "so that children in America can put frilly dresses on America's favorite doll."[5] In June 1996, *Life* magazine created more waves with photographs of Pakistani kids – looking shockingly young and paid as little as six cents an hour – hunched over soccer balls that bore the unmistakable Nike swoosh. But it wasn't just Nike. Adidas, Reebok, Umbro, Mitre and Brine were all manufacturing balls in Pakistan where an estimated 10,000 children worked in the industry, many of them sold as indentured slave laborers to their employers and branded like livestock.[6] The *Life* images were so chilling that they galvanized parents, students and educators alike, many of whom made the photographs into placards and held them up in protest outside sporting-goods stores across the United States and Canada.

Running alongside all this was the story of Nike's sneakers. The Nike saga started before the Year of the Sweatshop began and has only grown stronger as other corporate controversies have slipped in and out of the public eye. Scandal has dogged Nike, with new revelations about factory conditions trailing the company's own global flight patterns. First came the reports of union crackdowns in South Korea; when the contractors fled and set up shop in Indonesia, the watchdogs followed, filing stories on starvation wages and military intimidation of workers. In March 1996, *The New York Times* reported that after a wildcat strike at one Javanese factory, twenty-two workers were fired and one man who had been singled out as an organizer was locked in a room inside the factory and interrogated by soldiers for seven days. When Nike began moving production to Vietnam, the accusations moved too, with videotaped testimony of wage cheating and workers being beaten over the head with shoe uppers. When production moved decisively to China, the controversies over wages and the factories' "boot camp" style of management were right behind.

It wasn't only the superbrands and their celebrity endorsers who felt the sting of the Year of the Sweatshop – clothing-store chains, big-box retailers and department stores also found themselves being held responsible for the conditions under which the toys and fashions on their racks were produced. The issue came home for America in August 1995, when an apartment complex in El Monte, California, was raided by the U.S. Department of Labor. Seventy-two Thai garment workers were being held in bonded slavery – some had been in the compound for as long as seven years. The factory owner was a minor player in the industry, but the clothes the women were sewing were sold by such retail giants as Target, Sears and Nordstrom.

It is Wal-Mart, however, that has taken it on the chin most frequently since sweatshops made their big nineties comeback. As the world's largest retailer Wal-Mart is the primary distributor of many of the branded goods attracting controversy: Kathie Lee Gifford's clothing line, Disney's Haitian-made pajamas, child-produced clothing from Bangladesh, sweatshop-produced toys and sports gear from Asia. Why, consumers demanded, if Wal-Mart had the power to lower prices, alter CD covers and influence magazine content, did it not also have the power to demand and enforce ethical labor standards from its suppliers?

Though the revelations came out in the press one at a time, the incidents coalesced to give us a rare look under the hood of branded America. Few liked what they saw. The unsettling combination of celebrated brand names and impoverished production conditions have turned Nike, Disney and Wal-Mart, among others, into powerful metaphors for a brutal new way of doing business. In a single image, the brand-name sweatshop tells the story of the obscene disparities of the global economy: corporate executives and celebrities raking in salaries so high they defy comprehension, billions of dollars spent on branding and advertising – all propped up by a system of shantytowns, squalid factories and the misery and trampled expectations of young women like the ones I met in Cavite, struggling to survive.

The Year of the Brand Attack

Gradually, the Year of the Sweatshop turned into the Year of the Brand Attack. Having been introduced to the laborers behind their toys and clothing, shoppers met the people who grew their coffee at the local Starbucks; according to the U.S. Guatemalan Labor Education Project, some of the coffee frothed at the chain was cultivated with the use of child labor, unsafe pesticides and sub-subsistence wages. But it was in a courtroom in London, England, that the branded world was most thoroughly turned inside out. The highly publicized McLibel Trial began with McDonald's 1990 attempt to suppress a leaflet that accused the company of a host of abuses – from busting unions to depleting rain forests and littering the city streets. McDonald's denied the allegations and sued two London-based environmental activists for libel. The activists defended themselves by subjecting McDonald's to the corporate equivalent of a colonoscopy: the case lasted for seven years, and no infraction committed by the company was considered too minor to bring up in court or to post on the Internet.

The McLibel defendants' allegations about food safety dovetailed with another anticorporate movement taking off across Europe at the same time: the campaign against Monsanto and its bio-engineered agricultural crops. At the center of this dispute was Monsanto's refusal to inform consumers which of the foods they bought at the supermarket were the product of genetic engineering, setting off a wave of direct action that included the uprooting of Monsanto test crops.

As if that weren't enough, multinationals also found themselves under the microscope for their involvement with some of the world's most violent and repressive regimes: Burma, Indonesia, Colombia, Nigeria and Chinese-occupied Tibet. The issue was by no means new, but like the McDonald's and Monsanto campaigns, it came to a new prominence in the mid- to late nineties, with much of the activity focusing on the host of familiar brand names operating in Burma (now officially known as Myanmar). The bloody coup that brought the current military regime to power in Burma took place in 1988, but international awareness about brutal conditions inside the Asian country skyrocketed in 1995 when opposition leader and Nobel laureate Aung San Suu Kyi was released from six years of house arrest. In a videotaped appeal smuggled out of the country, Suu Kyi condemned foreign investors for propping up the junta that had disregarded her party's overwhelming election victory in 1990. Companies operating in Burma, she stated, are directly or indirectly profiting from state-run slave-labor camps. "Foreign investors should realize there could be no economic growth and opportunities in Burma until there is agreement on the country's political future."[7]

The first response from human-rights activists was to lobby governments in North America, Europe and Scandinavia to impose trade sanctions on the Burmese government. When this failed to halt the flow of trade, they began targeting individual companies based in the activists' own home countries. In Denmark, the protests centered on the national brewer, Carlsberg, which had entered into a large contract to build a brewery in Burma. In Holland, the target was Heineken; in the U.S. and Canada, Liz Claiborne, Unocal, Disney, Pepsi and Ralph Lauren were in the crosshairs.

But the most significant landmark in the growth of anticorporate activism also came in 1995, when the world lost Ken Saro-Wiwa. The revered Nigerian writer and environmental leader was imprisoned by his country's oppressive regime for spearheading the Ogoni people's campaign against the devastating human and ecological effects of Royal Dutch/Shell's oil drilling in the Niger Delta. Human-rights groups rallied their governments to interfere, and some economic sanctions were imposed, but they had little effect. In November 1995, Saro-Wiwa and eight other Ogoni activists were executed by a military government who had enriched themselves with Shell's oil money and through their own people's repression.

The Year of the Brand Attack stretched into two years, then three and now shows no sign of receding. In February 1999, a new report revealed that workers sewing Disney clothes in several Chinese factories were earning as little as 13.5 cents an hour and were forced to put in hours of overtime.[8] In May 1999, ABC's *20/20* returned to the island of Saipan and brought back footage of young women locked inside sweatshop factories sewing for the Gap, Tommy Hilfiger and Polo Ralph Lauren. New revelations have also come out about violent clashes surrounding Chevron's drilling activities in the

Niger Delta, and about Talisman Energy's plans to drill on contested territory in war-torn Sudan.

The volume and the tenacity of public outrage directed against them has blindsided the corporations, in large part because the activities for which they were being condemned were not particularly new. McDonald's has never been a friend of the working poor; oil companies have a long and uninterrupted history of collaborating with repressive governments to extract valuable resources with little concern for the people who live near them; Nike has produced its sneakers in Asian sweatshops since the early seventies, and many of the clothing chains have been doing so for even longer. As *The Wall Street Journal*'s Bob Ortega writes, labor unions had been collecting evidence of child laborers in Bangladesh making clothing sold at Wal-Mart since 1991, "But even though the unions had photos of children on the assembly lines... the accusations didn't get much play, in print or on television."[9]

Obviously much of the current focus on corporate abuses has to do with the tenacity of activists organizing around these issues. But since so many of the abuses being highlighted have been going on for decades, the current groundswell of resistance raises the question, Why now? Why did 1995–96 become the Year of the Sweatshop, turning quickly into the Years of the Brand Attack? Why not 1976, 1984, 1988, or, perhaps most relevant of all, why not 1993? It was in May of that year that the Kader toy factory in Bangkok burned to the ground. The building was a textbook firetrap, and when the piles of plush fabric ignited, the flames raced through the locked factory, killing 188 workers and injuring 469 more. Kader was the worst fire in industrial history, taking more lives than the Triangle Shirtwaist Company fire that killed 146 young workers in New York City in 1911. The parallels between Triangle and Kader – separated from each other by half a world, and eighty-two years of so-called development – are chilling: it was as if time hadn't moved forward, but had simply shifted locations.

At Triangle, as at Kader, the workers were almost all young women – some as young as fourteen, but most about nineteen. A report issued after the Triangle fire found that most of the dead were Italian and Russian immigrants, and almost half had come to America ahead of their families, seeking employment to subsidize the journeys of parents and siblings – so similar to the situation of the migrant peasant girls who perished at Kader. Like the Kader factory, the Triangle building was an accident waiting to happen, complete with fake fire exits, mounds of flammable material and doors that stayed locked all day to keep out the union organizers. Like the young women at Kader, many of the girls at Triangle wrapped themselves in cloth and jumped out the factory windows to their deaths – that way, they reasoned, their families would at least be able to identify their bodies. A *New York World* reporter described the gruesome Triangle scene. "Suddenly something that looked like a bale of dark dress goods was hurled from an eighth-story window.... Then another seeming bundle of cloth came hurtling through the same window, but this time a breeze tossed open the cloth and from the crowd of five hundred persons came a cry of horror. The breeze disclosed the form of a girl shooting down to instant death."[10]

The Triangle Shirtwaist Company fire was the defining incident of the first anti-sweatshop movement in the United States. It catalyzed hundreds of thousands of workers into militancy and promoted a government response that eventually led to a fifty-four-hour weekly cap on overtime, no work past 9 p.m. and breakthroughs in health and fire regulations. Perhaps the most significant advance as a result of the fire was the introduction of what today would be called independent monitoring – the founding of the New York Factory Investigation Commission, which was authorized to stage surprise raids on suspected sweatshop operators.

So what did the 188 deaths in the Kader fire accomplish? Sadly, despite the fact that several international labor and development groups stepped in to denounce the unlawful factory operator, Kader didn't become a symbol of the desperate need for reform the way Triangle Shirtwaist had done. In *One World, Ready or Not*, William Greider describes visiting Thailand and meeting victims and activists who had been fighting hard for retribution. "Some of them were under the impression that a worldwide boycott of Kader products was underway, organized by conscience-stricken Americans and Europeans. I had to inform them that the civilized world had barely noticed their tragedy.... A fire in Bangkok was like a typhoon in Bangladesh, an earthquake in Turkey." Little wonder, then, that only six months after Kader, another devastating sweatshop fire – this one at the Zhili toy factory in Shenzhen, China – took the lives of another 87 young workers.

At the time, it didn't seem to register with the international community that the toys the Kader women had been sewing were destined for the joyful aisles of Toys 'R' Us, to be wrapped and placed under Christmas trees in Europe, the United States and Canada. Many news reports failed even to mention the names of the brands being stitched in the factory. As Greider writes, "The Kader fire might have been more meaningful for Americans if they could have seen the thousands of soot-stained dolls that spilled from the wreckage, macabre litter scattered among the dead. Bugs Bunny, Bart Simpson and the Muppets. Big Bird and other *Sesame Street* dolls. Playskool 'Water Pets.'"[11]

But in 1993 few people in the West – and certainly not in the Western media – were ready to make the connection between the burned-out building in Bangkok, buried on page six or ten of their newspapers, and the brand-name toys filling North American and European homes. That is no longer the case today. What happened in 1995 was a kind of collective "click" on the part of both the media and the public. The cumulative response to the horror stories of Chinese prison labor, the scenes of teenage girls being paid pennies in the Mexican maquiladoras, and burning in fires in Bangkok, has been a slow but noticeable shift in how people in the West see workers in the developing world. "They're getting our jobs" is giving way to a more humane reaction: "Our corporations are stealing their lives."

Much of this has to do with timing. Concerns expressed about child labor in India and Pakistan had remained at the level of a steady drone for more than a decade. But by 1995, the question of linking trade policies to human rights had been pushed so far off most governments' agendas that when thirteen-year-old Craig Kielburger deliberately disrupted Canadian prime minister Jean Chrétien's trade mission to India to talk about the children who were working there in bonded slavery, the issue seemed urgent and exotic. Moreover, in North America, the total usurpation of foreign policy by the free-trade agenda invited disruption – the world was ready to listen.

The same is true of corporate crime in general. It may be nothing new for consumer goods to be produced under oppressive conditions, but what clearly *is* new is the tremendously expanded role consumer-goods companies are playing in our culture. Anticorporate activism is on the rise because many of us feel the international brand-name connections that crisscross the globe more keenly than we ever have before – and we feel them precisely because we have never been as "branded" as we are today.

Branding, as we have seen, has taken a fairly straightforward relationship between buyer and seller and – through the quest to turn brands into media providers, arts producers, town squares and social philosophers – transformed it into something much more invasive and profound. For the past decade, multinationals like Nike, Microsoft and Starbucks have sought to become the chief communicators of all that is good and cherished in our culture: art, sports, community, connection, equality. But the more successful this project is, the more vulnerable these companies become: if brands are indeed intimately entangled with our culture and our identities, when they do wrong, their crimes are not dismissed as merely the misdemeanors of another corporation trying to make a buck. Instead, many of the people who inhabit their branded worlds feel complicit in their wrongs, both guilty and connected. But this connection is a volatile one: it is not the old-style loyalty between lifelong employee and corporate boss; rather, this is a connection more akin to the relationship of fan and celebrity: emotionally intense but shallow enough to turn on a dime.

This volatility is the unintended consequence of brand managers striving for unprecedented intimacy with the consumer while forging a more casual role with the workforce. In reaching brand-not-products nirvana, these companies have lost two things that may prove more precious in the long run: consumer detachment from their global activities and citizen investment in their economic success.

It has taken us a while, but if another Kader happened tomorrow, the first question journalists would ask would be, "What toys were being produced?" "Where were they being shipped?" and "Which companies hired the contractors?" Labor activists in Thailand would be in instant communication with solidarity groups in Hong Kong, Washington, Berlin, Amsterdam, Sydney, London and Toronto. E-mails would be fired off from Washington-based Campaign for Labor Rights, from the Clean Clothes Campaign out of Amsterdam, and forwarded through a network of Web sites, listserves and fax trees. The National Labor Committee, UNITE!, the Labour Behind the Label Coalition and the World Development Movement would be organizing protests outside Toys 'R' Us, shouting, "Our children don't need bloodstained toys!" University students would dress up as the cartoon characters of their childhood and hand out pamphlets comparing Bugs Bunny's payout for *Space Jam* to the cost of putting in a fire exit at Kader. Meetings would be scheduled with national associations of toy manufacturers; new and tougher codes of conduct would be highlighted for consideration. The public mind is not only able but eager to make the global connections that William Greider searched for but did not find after the Kader fire.

Though anticorporate activism is seeing a renewal unparalleled since the thirties, there have, of course, been some significant anticorporate campaigns scattered between the thirties and their present-day revival. The granddaddy of modern brand-based actions is the boycott against Nestlé, which peaked in the late seventies. The campaign targeted the Swiss company for its aggressive marketing of costly baby formula as a "safer" alternative to breast-feeding in the developing world. The Nestlé case has a strong parallel with the McLibel Trial (to be discussed in detail in Chapter 16), largely because the issue didn't really capture the world's attention until the food company made the mistake of suing a Swiss activist group for libel in 1976.[12] As with McLibel, the ensuing court case put Nestlé under intense scrutiny and led to an international boycott campaign, launched in 1977.

The eighties saw the largest industrial accident in human history: a massive toxic leak in 1984 at a Union Carbide pesticide factory in Bhopal, India, killed two thousand people immediately and has taken five thousand more lives in the years since. Today, graffiti on the wall of the dilapidated and abandoned factory reads "Bhopal = Hiroshima."[13] Despite this tragedy, widely recognized to be the result of weak safety precautions including a switched-off alarm system, the eighties were a dry spell for most political movements that questioned the beneficial power of capital. Although there was a broad recognition during the Central American wars that U.S. multinationals were propping up various dictatorships, solidarity work in North America focused primarily on the actions of governments, as opposed to multinational corporations. As one report on the subject notes, "attacking [corporations] tended to be seen as a hangover from the 'silly seventies.'"[14]

There was, however, one major exception to this rule: the anti-apartheid movement. Frustrated by the international community's refusal to impose meaningful trade sanctions on South Africa, anti-apartheid activists developed a series of alternative roadblocks designed, if not to prevent multinationals from profiting from the racist regime, at least to inconvenience them if they persisted in doing so. Students and faculty members at several universities set up tent cities demanding that schools divest themselves of their endowments from any company doing business with the African nation. Church groups disrupted corporate shareholder meetings with demands for immediate withdrawal, while more moderate investors pushed corporate boards to adopt the Sullivan principles – a set of rules for companies in South Africa that purported to minimize their complicity with the apartheid regime. Meanwhile, trade unions pulled their pensions and bank accounts from institutions issuing loans to the South African government, and dozens of municipal governments passed selective purchasing agreements canceling large contracts with companies invested in South Africa. The most creative blockades were erected by the international trade-union movement. Several times a year, the unions would call a day of action, during which dock workers refused to unload cargo that had come from South Africa, and airline ticket agents refused to book flights to and from Johannesburg. In the words of campaign organizer Ken Luckhardt, workers became "activists at the point of production."[15]

Though there are definite similarities, there is one key difference between the apartheid actions and the kind of anticorporate campaigning gaining momentum today. The South Africa boycott was an antiracist campaign that happened to use trade (whether the importing of wine or the exporting of General Motors dollars) as a tool to bring down the South African political system. Many of the current anticorporate campaigns are also rooted in a political attack – but what they are attacking is as much a global economic system as a national political one. During the years of apartheid, companies such as the Royal Bank of Canada, Barclays Bank in England and General

Motors were generally regarded as morally neutral forces that happened to be entangled with an aberrantly racist government. Today, more and more campaigners are treating multinationals, and the policies that give them free rein, as the root cause of political injustices around the globe. Sometimes the companies commit these violations in collusion with governments, sometimes they commit them despite a government's best efforts.

This systemic critique has been embraced, in recent years, by several established human-rights groups like Amnesty International, PEN and Human Rights Watch, as well as environmental rights organizations like the Sierra Club. For many of these organizations, this represents a significant shift in policy. Until the mid-eighties foreign corporate investment in the Third World was seen in the mainstream development community as a key to alleviating poverty and misery. By 1996, however, that concept was being openly questioned, and it was recognized that many governments in the developing world were protecting lucrative investments – mines, dams, oil fields, power plants and export processing zones – by deliberately turning a blind eye to egregious rights violations by foreign corporations against their people. And in the enthusiasm for increased trade, the Western nations where most of these offending corporations were based also chose to look the other way, unwilling to risk their own global competitiveness for some other country's problems. The bottom line was that in parts of Asia, Central and South America and Africa, the promise that investment would bring greater freedom and democracy was starting to look like a cruel hoax. And worse: in case after case, foreign corporations were found to be soliciting, even directly contracting, the local police and military to perform such unsavory tasks as evicting peasants and tribespeople from their land; cracking down on striking factory workers; and arresting and killing peaceful protestors – all in the name of safeguarding the smooth flow of trade. Corporations, in other words, were stunting human development, rather than contributing to it.

Arvind Ganesan, a researcher with Human Rights Watch, is blunt about what his organization refers to as "a shift in the terms of the debate over corporate responsibility for human rights."[16] Rather than improved human rights flowing from increased trade, "governments ignore human rights in favor of perceived trade advantages."[17] Ganesan points out that the severing of the connection between investment and human-rights improvements is today clearest in Nigeria, where the long-awaited transition to democracy has been coupled with a renewed wave of military brutality against Niger Delta communities protesting against the oil companies.

Amnesty International, in a departure from its focus on prisoners persecuted for either their religious or political beliefs, is also beginning to treat multinational corporations as major players in the denial of human rights worldwide. More and more, recent Amnesty reports have found that people such as the late Ken Saro-Wiwa have been persecuted for what a government sees as a destabilizing anticorporate stance. In a 1997 report, the group documents the fact that Indian villagers and tribal peoples were violently arrested, and some killed, for peacefully resisting the development of private power plants and luxury hotels on their lands. A democratic country, in other words, was becoming less democratic as a result of corporate intervention. "Development," Amnesty warned, is "being pursued at the expense of human rights. . . ."

> This pattern highlights the degree to which the central and state authorities in India are prepared to deploy state force and utilize provisions of the law in the interests of development projects, curtailing the right of freedom of association, expression and assembly. India's moves to liberalize its economy and develop new industries and infrastructure have in many areas marginalized and displaced communities and contributed to further violations of their human rights.[18]

India's situation, the report states, is not "the only or the worst" one, but is part of a trend toward the disregarding of human rights in favor of "development" in the global economy.

Where the Power Is

At the heart of this convergence of anticorporate activism and research is the recognition that corporations are much more than purveyors of the products we all want; they are also the most powerful political forces of our time. By now, we've all heard the statistics: how corporations like Shell and Wal-Mart bask in budgets bigger than the gross domestic product of most nations; how, of the top hundred economies, fifty-one are multinationals and only forty-nine are countries. We have read (or heard about) how a handful of powerful CEOs are writing the new rules for the global economy, engineering what Canadian writer John Ralston Saul has called "a *coup d'état* in slow motion." In his book about corporate power, *Silent Coup*, Tony Clark takes this theory one step further when he argues that citizens must go after corporations not because we don't like their products, but because corporations have become the ruling political bodies of our era, setting the agenda of globalization. We must confront them, in other words, because that is where the power is.

So although the media often describe campaigns like the one against Nike as "consumer boycotts," that tells only part of the story. It is more accurate to describe them as political campaigns that use consumer goods as readily accessible targets, as public-relations levers and as popular-education tools. In contrast to the consumer boycotts of the seventies, there is a more diffuse relationship between lifestyle choices (what to eat, what to smoke, what to wear) and the larger questions of how the global corporation – its size, political clout and lack of transparency – is reorganizing the world economy. Behind the protests outside Nike Town, behind the pie in Bill Gates's face and the bottle shattering the McDonald's window in Prague, there is something too visceral for most conventional measures to track – a kind of bad mood rising. And the corporate hijacking of political power is as responsible for this mood as the brands' cultural looting of public and mental space. I also like to think it has to do with the arrogance of branding itself: the seeds of discontent are part of its very DNA.

> *"Look, Mike, there's a real market for the truth about Nike.... Our debut product will be a proprietary database of Nike labor abuses! I see a Web presence and a CD-ROM of stats, worker affidavits, human rights reports and hidden camera video clips."*
>
> *"Kind of niche product, isn't it, babe?"*
>
> *"No. This will be huge!"*[19]

So goes an exchange in Gary Trudeau's Doonesbury cartoon strip – and it's a joke with a strong sting to it. The continuing attacks on brands like Nike, Shell and McDonald's not only reflect genuine indignation at sweatshops, oil spills and corporate censorship, they also reflect how large the antagonistic audience has become. The desire (and ability) to back up free-floating anti-corporate malaise with legitimate facts, figures and real-life anecdotes is so widespread that it even transcends old rivalries within the social and ecological movements. The United Food and Commercial Workers' union, which started targeting Wal-Mart because of its low wages and union-busting tactics, now collects and disseminates information on Wal-Mart stores being built on sacred Native burial grounds. Since when did a grocery-store workers' union weigh in on indigenous land claims? Since puncturing Wal-Mart became a cause in and of itself. Why did the London eco-anarchists behind the McLibel Trial – who don't believe in working for the Man in any form – take up the plight of teenage McDonald's workers? Because, for them, it's another angle from which to attack the golden beast.

The political backdrop to this phenomenon is well known. Many citizens' movements have tried to reverse conservative economic trends over the last decade by electing liberal, labor or democratic-socialist governments, only to find that economic policy remains unchanged or caters even more directly to the whims of global corporations. Centuries of democratic reforms that had won greater transparency in government suddenly appeared ineffective in the new climate of multinational power. What good was an open and accountable Parliament or Congress if opaque corporations were setting so much of the global political agenda in the back rooms?

Disillusionment with the political process has been even more pronounced on the international stage, where attempts to regulate multinationals through the United Nations and trade regulatory bodies have been blocked at every turn. A significant setback came in 1986 when the U.S. government effectively killed the little-known United Nations Commission on Transnational Corporations. Started in the mid-seventies, the commission set out to draft a universal code of conduct for multinational corporations. Its goals were preventing corporate abuses such as companies dumping, in the Third World, drugs that are illegal in the West; examining the environmental and labor impacts of export factories and resource extraction; and pushing the private sector toward greater transparency and accountability.

The merit of these goals seems self-evident today but the commission, in many ways, was a casualty of its time. American industry was opposed to its creation from the start and in the heat of Cold War mania managed to secure their government's withdrawal on the grounds that the commission was a Communist plot and that the Soviets were using it for espionage. Why, they demanded, were Soviet-bloc national enterprises not subject to the same probing as American companies? During this era, criticisms of the abuses of multinational corporations were so bound up in anti-Communist paranoia that when the Bhopal tragedy happened in 1984, the immediate response of a U.S. embassy official in New Delhi was not to express horror but to say, "This is a feast for the Communists. They'll go with it for weeks."[20]

More recently, attempts to force the World Trade Organization to include enforcement of basic labor laws as a condition of global trade have been dismissed by member nations who insist such enforcement is the job of the UN's International Labor Organization. The ILO "is the competent body to determine and deal with these standards, and we affirm our support for its work in promoting them," states the WTO's Singapore Ministerial Declaration of December 13, 1996. However, when the ILO embarked on an initiative to draft a meaningful corporate code of conduct, it too was blocked.

At first, these failures to regulate capital left many reform and opposition movements in a state of near-paralysis: citizens, it seemed, had lost their say. Slowly, however, a handful of nongovernmental organizations and groups of progressive intellectuals have been developing a political strategy that recognizes that multinational brands, because of their high profile, can be far more galvanizing targets than the politicians whom they bankroll. And once the corporations are feeling the heat, they have learned, it becomes much easier to get the attention of elected politicians. In explaining why he has chosen to focus his activism on the Nike corporation, Washington-based labor activist Jeff Ballinger says bluntly, "Because we have more influence on a brand name than we do with our own governments."[21] Besides, adds John Vidal, "Activists always target the people who have the power...so if the power moves from government to industry to transnational corporations, so the swivel will move onto these people."[22]

Already, a common imperative is emerging from the disparate movements taking on multinational corporations: the people's right to know. If multinationals have become larger and more powerful than governments, the argument goes, then why shouldn't they be subject to the same accountability controls and transparency that we demand of our public institutions? So anti-sweatshop activists have been demanding that Wal-Mart hand over lists of all the factories around the world that supply the chain with finished products. University students, as we will see in Chapter 17, are demanding the same information about factories that produce clothing with their school insignia. Environmentalists, meanwhile, have used the courts to X-ray the inner working of McDonald's. And all over the world, consumers are demanding that companies like Monsanto provide clear labeling of genetically modified food and open their research to outside scrutiny.

Placing demands like these on private companies, whose only legal duty is to their shareholders, has generated a surprising number of successes. The reason is that many multinationals have a rather sizable weak spot. As we will see in the next chapter, activists around the world are making liberal use of the very factor that has been the subject of this book so far: the brand. Brand image, the source of so much corporate wealth, is also, it turns out, the corporate Achilles' heel.

Top: Billboard Liberation Front jams an Apple campaign on the streets of San Francisco.
Bottom: The Gap falls victim to a "skulling" epidemic on Toronto outdoor ads.

CHAPTER FIFTEEN

THE BRAND BOOMERANG

The Tactics of Brand-Based Campaigns

It can take 100 years to build up a good brand and 30 days to knock it down.

– David D'Alessandro, president of John Hancock Mutual Life Insurance, January 6, 1999

Branding, as we have seen, is a balloon economy: it inflates with astonishing rapidity but it is full of hot air. It shouldn't be surprising that this formula has bred armies of pin-wielding critics, eager to pop the corporate balloon and watch the shreds fall to the ground. The more ambitious a company has been in branding the cultural landscape, and the more careless it has been in abandoning workers, the more likely it is to have generated a silent battalion of critics waiting to pounce. Moreover, the branding formula leaves corporations wide open to the most obvious tactic in the activist arsenal: bringing a brand's production secrets crashing into its marketing image. It's a tactic that has worked before.

Though marketing and production have not always been separated by so many bodies of water and layers of subcontractors, the two have never been exactly cozy. Ever since the first ad campaigns created folksy mascots to lend a homemade feel to mass-produced goods, it has been the very business of the advertising industry to distance products from the factories that make them. Helen Woodward, an influential copywriter in the 1920s, famously warned her co-workers that "if you are advertising any product, never see the factory in which it was made.... Don't watch the people at work... because, you see, when you know the truth about anything, the real inner truth – it is very hard to write the surface fluff which sells it."[1]

Back then, Dickensian images like those from the Triangle Shirtwaist Fire were still fresh in the minds of Western consumers. They didn't need to be reminded of the dark side of industrialization when they were buying soap, stockings, cars or any other product that promised happiness in the self and envy in everybody else. Besides, many of the consumers being targeted by advertising were themselves factory workers, and the last thing a fluff writer wanted to do was trigger a memory of the dreary monotony of the assembly line.

But as First World countries have shifted into "information economies," we have developed a certain nostalgia for the gritty authenticity of Woodward's era of industrialization. And so the factory, long marketing's greatest taboo, has recently found a place in advertising. The shop floor is featured in Saturn car ads, for example, where we meet empowered auto workers who can "stop the line" just because something looks a little dodgy. Interior shots of a factory also briefly appear in an early nineties Subaru ad – there to make the trademark Wieden & Kennedy point that cars really aren't about impressing your neighbors but driving "the best machine."

However, the factories featured in both the Saturn and Subaru campaigns aren't the sweat shop floor that Woodward warned her fellow ad writers to never lay eyes upon; these are New Age nostalgia factories – about as realistic as Intel's dancing techno-technicians. The role of these factories, like that of Aunt Jemima and the Quaker Oats mascot, is to associate Subaru and Saturn with a simpler time, a time when goods were made in the countries where they were consumed, when people still knew their neighbors and nobody had heard of an export processing zone. In the early nineties, at a time when car factories were closing in droves and the market was being flooded with cheap imports, the ads – though purporting to take us behind the glitz of advertising – were there not to illuminate the manufacturing process, but to obscure it.

In other words, Helen Woodward's rule holds truer now than ever: at no point has the double life of our branded goods been more conflicted. Despite the rhetoric of One Worldism, the planet remains sharply divided between producers and consumers, and the enormous profits raked in by the superbrands are premised upon these worlds remaining as separate from each other as possible. It is a tidy formula: because the contract factory owners in the free-trade zones don't sell a single Reebok sneaker or Mickey Mouse sweatshirt directly to the public, they have a limitless threshold for bad public relations. Building up a positive relationship with the shopping public, meanwhile, is left entirely in the hands of the brand-name multinationals. The only catch is that for the system to function smoothly, workers must know little of the marketed lives of the products they produce and consumers must remain sheltered from the production lives of the brands they buy.

The formula has worked for quite a while. For the first two decades of their existence, export processing zones were indeed globalization's dirty little secret – secured "labor warehouses" where the unsightly business of production was contained behind high walls and barbed wire. But the "brands, not products" mania that has gripped the business world since the early nineties is coming back to haunt the free-floating, incorporeal corporation. And no wonder. Severing brands so decisively from their sites of production and shuttling factories away into the industrial hellholes of the EPZs has created a potentially explosive situation. It's as if the global production chain is based on the belief that workers in the South and consumers in the North will never figure out a way to communicate with each other – that despite the info-tech hype, only corporations are capable of genuine global mobility. It is this supreme arrogance that has made brands like Nike and Disney so vulnerable to the two principal tactics employed by anticorporate campaigners: exposing the riches of the branded world to the tucked-away sites of production and bringing back the squalor of production to the doorstep of the blinkered consumer.

Designer Activism: The Logo Is the Star

I'm sitting in a crowded classroom in Berkeley, California, and somebody is turning up my collar to see the label. For a moment, I feel as if I am back in grade school with Romi the logo vigilante checking for impostors. Instead, it's 1997 and the person examining my collar is Lora Jo Foo, president of Sweatshop Watch. She is running a seminar called "Ending Sweatshops at Home and Abroad" as part of a conference on globalization.

Every time Foo runs a seminar on sweatshops, she pulls out a pair of scissors and asks everyone to cut the labels off their clothing. She then unfurls a map of the world made of white cloth. Our liberated brand names are sewn onto the map, which, over the course of many such gatherings in several countries, has become a crazy patchwork quilt of Liz Claiborne, Banana Republic, Victoria's Secret, Gap, Jones New York, Calvin Klein and Ralph Lauren logos. Most of the dense little rectangular patches are concentrated in Asia and Latin America. Foo then traces a company's global travel routes: she begins with when its products were still being produced in North America (only a few labels remain on that part of the map); then moves to Japan and South Korea; then to Indonesia and the Philippines; then to China and Vietnam. According to Foo, clothing logos make a great teaching aid; they take faraway, complex issues and plant them as close to home as the clothes on our backs.

It must be said that no one is more surprised by the power and appeal of brand-based activism than those who have spearheaded the campaigns. Many of the people leading the anti-sweatshop movement are longtime advocates on behalf of the Third World's poor and marginalized. In the eighties, they plugged away in near-total obscurity on behalf of Nicaragua's Sandinista rebels and El Salvador's FMLN opposition party. After the wars ended and the pace of globalization accelerated, they learned that the new war zone for Central America's poor was the sweatshop factory locked inside the military-guarded free-trade zone. But what they weren't prepared for was how sympathetic the public would be to this problem. "I think that what gives this issue such widespread appeal – makes it so much more real to people than the Central American wars were – is that people make a direct connection with their own lives; it's no longer something that's 'out there,'" says Trim Bissell of the Washington-based Campaign for Labor Rights. "If they eat at one of those chain outlets, they may well be putting into their bodies food that in one way or another depends on the oppression of someone else. If they buy toys for their children, those toys may have been made by children who have no childhood. It is so direct and so emotional and so human that people contact *us* and say 'How can I help?' In this work, we're not having to say 'There's a problem.' We're mostly saying, 'Here's a productive way you can direct your outrage.'"[2]

American author Lorraine Dusky described the dynamics of this personal connection in *USA Today*. Watching TV reports of the May 1998 riots in Indonesia, she found herself wondering whether her logos had anything to do with a young Indonesian girl shown wailing over the dead body of a fire victim. "Were my Nikes somehow to blame?" she writes. "That bereft young girl might still have a father if Nike had insisted that workers be better paid. Because if Nike had, other sweatshop employers might have followed suit." It may seem like a leap – blaming one's sneakers for a death in an Indonesian pro-democracy protest – but it did provide the connection necessary, as Dusky writes, to see that "globalization means more than the easy exchange of currency and goods; it means that we are all our sisters' and brothers' keepers."[3]

But while the effectiveness of brand-based campaigns may be in their immediate relevance to our own branded lives, there is another factor contributing to their appeal, particularly among young people. Anticorporate activism enjoys the priceless benefits of borrowed hipness and celebrity – borrowed, ironically enough, from the brands themselves. Logos that have been burned into our brains by the finest image campaigns money can buy, and lifted a little closer to the sun by their sponsorship of much-loved cultural events, are perpetually bathed in a glow – the "loglo," to borrow a term from science fiction writer Neal Stevenson.[4] As Alexis de Tocqueville predicted, it is fantastical creations like this that have the power to make us "regret the world of reality" – and no reality has come to seem more comparatively regrettable than that of people suffering under poverty and oppression in faraway places. So in the late seventies, as the loglo grew brighter, social-justice activism faded; its woefully unmarketable ways no longer held much appeal for energetic young people or for media obsessed with slick aesthetics.

But today, with so many anticorporate activists adopting the aesthetics and humor of culture jamming and the irreverent attitude of street reclaiming, that is beginning to change. From their new "leech-like" vantage point, the brands' detractors are benefiting from the loglo in an unanticipated way. The loglo is so bright that activists are able to enjoy its light, even as they are in the act of attacking a brand. This vicarious branding may seem to some like an erosion of their political purity but it also clearly helps to lure foot soldiers to the cause. Like a good ad bust, anticorporate campaigns draw energy from the power and mass appeal of marketing, at the same time as they hurl that energy right back at the brands that have so successfully colonized our everyday lives.

You can see this jujitsu strategy in action in what has become a staple of many anticorporate campaigns: inviting a worker from a Third World country to come visit a First World superstore – with plenty of cameras rolling. Few newscasts can resist the made-for-TV moment when an Indonesian Nike worker gasps as she learns that the sneakers she churned out for $2 a day sell for $120 at San Francisco Nike Town. Since 1994, there have been at least five separate tours of Indonesian Nike workers through North America and Europe – Cicih Sukaesih, who lost her job for trying to organize a union in a Nike factory, has been back three times, her trips sponsored by coalitions of labor, church and school groups. In August 1995, two Gap seamstresses – seventeen-year-old Claudia Leticia Molina from Honduras and eighteen-year-old Judith Yanira Viera from El Salvador – went on similar North American speaking tours, addressing crowds outside dozens of Gap outlets. Perhaps most memorably, shoppers were able to put a face to the issue of child labor when fifteen-year-old Wendy Diaz appeared before the U.S. Congress. She had been working in a Honduran factory sewing Kathie Lee Gifford pants since she was thirteen. Diaz testified to the presence of "about 100 minors like me – thirteen, fourteen, fifteen years old – some even twelve.... Sometimes they kept us all night long, working.... The supervisors scream at us and yell at us to work faster. Sometimes they throw the garment in your face, or grab and shove you.... Sometimes the managers touch the girls. Pretending it's a joke they touch our legs. Many of us would like to go to night school but we can't because they constantly force us to work overtime."[5]

No group has taken advantage of the branding economy's various leaks and cracks with more laser-like accuracy than the National Labor Committee, under its director, Charles Kernaghan. In the five years between 1994 and 1999, the NLC's three-person office in New York has used Greenpeace-style media antics to draw more public attention to the plight of sweatshop workers than the multimillion-dollar international trade union movement has achieved in almost a century. As the garment-industry bible *Women's Wear Daily* put it, "Charles Kernaghan and his anti-sweatshop battle have been shaking up the issue of labor abuses in the apparel industry like nothing since the Triangle Shirtwaist Fire."[6]

The NLC didn't achieve this rather remarkable feat by lobbying government or even by organizing workers. It did it by setting out to sully some of the most polished logos on the brandscape. Kernaghan's formula is simple enough. First, select America's most cartoonish icons, from literal ones like Mickey Mouse to virtual ones like Kathie Lee Gifford. Next, create head-on collisions between image and reality. "They live or die by their image," Kernaghan says of his corporate adversaries. "That gives you a certain power over them ... these companies are sitting ducks."[7]

Like the best culture jammers, Kernaghan has a natural feel for the pitch. He knew that he could "sell" overseas sweatshops to the U.S. media – notorious for its double-jeopardy bias against labor and problems in places where people don't speak English. But what he needed to do was steer clear of obscure labor laws and arcane trade agreements, and keep the focus squarely on the logos behind the violations. It's a formula that has brought the sweatshop story under serious scrutiny on *60 Minutes* and *20/20* and in *The New York Times* – and ultimately even to *Hard Copy*, which sent a crew to accompany Kernaghan on a tour of Nicaraguan sweatshops in fall 1997.

The tabloid news show and the gutsy labor group didn't make as strange bedfellows as one might think. We are a celebrity-obsessed culture, and such a culture is never in finer form than when one of its most loved icons is mired in scandal. What Kernaghan had seized upon is that the fanatical obsession with logos extends not only to building them up, but also to tearing them down. Though on a vastly different scale, Nike's sweatshops are to labor reporting what O.J. Simpson's trial was to the legal beat: designer dirt. And the NLC, for better or for worse (definitely for worse, say its critics), is indeed the *Hard Copy* of the labor movement, forever searching out that intersection between the dazzling celebrity stratosphere and real life on the mean streets.

So Kernaghan lays out the facts and figures of the global economy in Disney pajamas, Nike running shoes, Wal-Mart aisles and the personal riches of the individuals involved – and crunches the numbers into homemade statistical contraptions that he then wields like a mallet. For example: all 50,000 workers at the Yue Yen Nike Factory in China would have to work for nineteen years to earn what Nike spends on advertising in one year.[8] Wal-Mart's annual sales are worth 120 times more than Haiti's entire annual budget; Disney CEO Michael Eisner earns $9,783 an hour while a Haitian worker earns 28 cents an hour; it would take a Haitian worker 16.8 years to earn Eisner's hourly income; the $181 million in stock options Eisner exercised in 1996 is enough to take care of his 19,000 Haitian workers and their families for fourteen years.[9]

A typical Kernaghanism is to compare and contrast the plush living conditions of the dogs on the set of *101 Dalmatians* with the shacks in which the Haitian workers live who sewed Disney pajamas decorated with the movie's characters. The animals, he says, stayed in "doggie condos" fitted with cushy beds and heat lamps, were cared for by on-call vets and served beef and chicken. The Haitian workers live in malaria- and dysentery-infested hovels, sleep on cots and can rarely afford to buy meat or go to the doctor.[10] It is in this collision between the life of brand and the reality of production that Kernaghan works his own marketing magic.

The NLC's events – far from the usual gray labor rallies – take full advantage of the powers of the loglo. An October 1997 rally in New York City was a case in point: it began in Times Square across from Disney's flagship superstore, proceeded along Seventh Avenue, past Macy's Tommy Hilfiger window display, past Barnes & Noble, and Stern's department store. As the kickoff of "The Holiday Season of Conscience," the rally had as visual backdrop for the chants and speeches Manhattan's most enormous logos: a giant red swoosh on the skyline, the Maxell guy in his armchair getting "blown away" by digital sound and 3-D displays for *The Lion King* on Broadway. When Jay Mazur, president of UNITE, pronounced that "sweatshops are back and we know why," he did so with a towering, neon-lit Little Mermaid forming a halo over his head. At another NLC-sponsored protest, this one in March 1999, participants parked a giant rubber rat outside the Disney store. And because Kernaghan's tactics don't demand pop-cultural asceticism in exchange for participation, they have proven hugely appealing to students, many of whom show up for these rallies as walking culture jams. Echoing the cartoonish aesthetics of rave culture, high-school kids and college students dress in fuzzy animal costumes: a six-foot pink pig holding up a sign that reads "Pigs Against Greed," the Cookie Monster sporting a "No Justice, No Cookie" placard.

For the NLC, logos are both targets and props. Which is why, when Kernaghan speaks to a crowd – at college campuses, labor rallies or international conferences – he is never without his signature shopping bag brimming with Disney clothes, Kathie Lee Gifford pants and other logo gear. During his presentations, he holds up the pay slips and price tags to illustrate the vast discrepancies between what workers are paid to make the items and what we pay to buy them. He also takes his shopping bag with him when he visits the export processing zones in Haiti and El Salvador, pulling out items from his bag of tricks to show workers the actual price tags of the goods they sew. In a letter to Michael Eisner, he describes a typical reaction:

> Prior to leaving for Haiti, I went to a Wal-Mart store on Long Island and purchased several Disney garments which had been made in Haiti. I showed these to the crowd of workers, who immediately recognized the clothing they had made... I held up a size four Pocahontas T-shirt. I showed them the Wal-Mart price tag indicating $10.97. But it was only when I translated the $10.97 into the local currency – 172.26 gourdes – that, all at once, in unison, the workers screamed with shock, disbelief, anger, and a mixture of pain and sadness, as their eyes remained fixed on the Pocahontas shirt.... In a single day, they worked on hundreds of Disney shirts. Yet the sales price of just one shirt in the U.S. amounted to nearly five days of their wages![11]

The moment when the Haitian Disney workers cried out in disbelief was captured by one of Kernaghan's colleagues on video and included in the NLC-produced documentary *Mickey Mouse Goes to Haiti*. Since then, the documentary has been shown in hundreds of schools and community centers in North America and Europe, and many young activists say that scene played a critical role in persuading them to join the global struggle against sweatshops.

Another Kind of Logo Traffic

Information about the disparity between wages and retail prices can also prove radicalizing to the workers in the factories who, as I learned in Cavite, know little about the value of the goods they produce. At the All Asia factory in the Cavite Export Processing Zone, for instance, the boss used to leave the price tags for the Sassoon skirts in plain view – $52, they said. "Those price tags were put beside the buttons, and we were able to see the prices when we passed through the packing section," one seamstress told me. "So we computed the amount in pesos and the workers were saying, 'So the company is having this kind of sales? Then why are we getting this small pay?'" After management got wind of these covert discussions there were no more price tags left lying around at All Asia.

In fact, I discovered that even finding out which brand names are being produced behind the locked gates of the Cavite zone requires a fair bit of detective work, work that has been embraced by the Workers' Assistance Center outside the zone. One of the center's walls is covered by a bulletin board that looks remarkably like Lora Jo Foo's logo quilt. Clothing labels are pinned all over the board: Liz Claiborne, Eddie Bauer, Izod, Guess, Gap, Ellen Tracy, Sassoon, Old Navy. Beside each label on the board is the name of the factory it came from: V.T. Fashion, All Asia, Du Young. The organizers at WAC believe that this information connecting brands to work is crucial in their attempt to empower zone workers to stand up for their legal rights, particularly since the factory bosses are forever crying poor. When workers learn, for instance, that the Old Navy jeans they are sewing for pennies apiece are sold by a famous company called the Gap and will sell for $50 in America, they are more likely to demand overtime pay, or even long-promised health coverage. Many workers are eager for this information too, which is why they have taken the great risk of smuggling these clothing tags out of their factories; they slip them into their pockets at work, hope that the guards don't find the scraps when they get searched at the gate, and then bring them over to the center. The next task for WAC is to find out something about the company that owns these names – not always easy since many of the brands aren't even available for sale in the Philippines, and those that are can only be found in the high-priced malls of Manila's tourist district.

In the last few years, however, gathering this information has become a little easier, in part as a result of a marked increase in activist traffic around the world. With the aid of travel subsidies from well-funded nongovernmental organizations and unions, representatives from the tiny Workers' Assistance Center in Rosario have gone to conferences all over Asia as well as in Germany and Belgium. And only two months after I first met her in Cavite, I saw WAC organizer Cecille Tuico again in Vancouver at the November 1997 People's Summit on APEC. The conference was attended by several thousand activists from forty countries and was timed to coincide with a meeting of the leaders of the eighteen Asia-Pacific economies – from Bill Clinton to Jiang Zemin – who were gathering in Vancouver that week.

On the last day of the summit, Cecille and I skipped out of the seminars and spent an afternoon on busy Robson Street, popping in and out of chain stores that sell many of the brand names produced in the Cavite zone. We scoured the racks of fleece Baby Gap sleepers and booties, Banana Republic jackets, Liz Claiborne blouses and Izod Lacoste shirts, and when we came across a "Made in the Philippines" tag, we scribbled down the style numbers and prices. When she returned to Cavite, Cecille converted the prices into pesos (taking into account her country's plummeting currency rate) and carefully pinned them next to the labels on the bulletin board in the WAC office. She and her colleagues point to these figures when workers drop by the center distressed about an illegal firing, back wages owed or an endless string of overnight shifts. Together, they calculate how many weeks a zone seamstress would have to work to be able to afford one Baby Gap sleeper for her child, and workers whisper this shocking figure to each other when they return to their cramped dormitory rooms, or break for lunch at their sweltering factories. The news spreads through the zone like wildfire.

I remembered our "sweatshopping" trip (as *The Nation* writer Eyal Press calls these odd excursions) when I received an E-mail from Cecille some months later telling me that WAC has finally succeeded in unionizing two garment factories inside the zone. The logos on the labels? Gap, Arizona Jeans, Izod, J.C. Penney and Liz Claiborne.

Act Globally

Ever since the politics of representation first captured the imagination of feminists in the early seventies, there have been women urging their movement sisters to look beyond how the fashion and beauty industries oppress Westerners as consumers, and to consider the plight of the women around the world who sweated to keep them in style. During the twenties and thirties, Emma Goldman and the International Ladies' Garment Workers' Union rallied the women's movement behind sweatshop workers, but in recent decades, these connections have seemed somewhat out of step with the times. Though there has always been a component of second-wave feminism that sought to forge political connections with women in developing countries, the struggle for internationalism never quite took hold of the movement in the way that pay equity, media representation or abortion rights did. Somehow, the seventies rallying cry that "the personal is political" seemed more related to the issue of how fashion made women feel about themselves than to the global mechanisms of how the garment industry made other women work.

In 1983, the American academic Cynthia Enloe was one of the voices in the wilderness. She insisted that the "Made in Hong Kong" and "Made in Indonesia" labels that were appearing with greater frequency inside her clothing provided a non-abstract starting point for women who wanted to understand the complexities of global economics. "We can become more able to talk about, and to make sense of such alleged 'abstractions' as 'international capital' and the 'international sexual division of labour.' Both of these concepts, so long the presumed intellectual preserve of male theoreticians (most of whom never ask who weaves and who sews) are in reality only as 'abstract' as the jeans in our closets and the underwear in our dresser drawers," she wrote.[12]

At the time, thanks to a combination of too little awareness, cultural barriers and First World parochialism, few were ready to listen. But many are listening today. Once again, this shift may be an unexpected by-product of branding ubiquity. Now that the corporations have spun their own global rainbow of logos and labels, the infrastructure for genuine international solidarity is there for everyone to see and use. The logo network may have been designed to maximize consumption and minimize production costs, but regular people can now turn themselves into "spiders" (as the members of the Free Burma Coalition call themselves) and travel across its web as easily as the corporations that spun it. Which is where Lora Jo Foo's logo map comes in – and Cecille Tuico's bulletin board and Charles Kernaghan's shopping bag and Lorraine Dusky's sneaker epiphany. It's like the Internet in general: it may have been built by the Pentagon, but it quickly became the playground of activists and hackers.

So while cultural homogenization – the idea of everyone eating at Burger King, wearing Nike shoes and watching Backstreet Boys videos – may inspire global claustrophobia, it has also provided a basis for meaningful global communication. Thanks to the branded web, McDonald's workers around the world are able to swap stories on the Internet about working under the arches; club kids in London, Berlin and Tel Aviv can commiserate about the corporate co-optation of the rave scene; and North American journalists can talk with poor rural factory workers in Indonesia about how much Michael Jordan gets paid to do Nike commercials. This logo web has the unprecedented power to connect students who face ad bombardment in their university washrooms with sweatshop workers who make the goods in the ads and frustrated McWorkers who sell them. They may not all speak the same language, but they now have enough common ground to begin a discussion. Playing on the Benetton slogan, one Reclaim the Streets organizer described these new global networks as the "United Colors of Resistance."

A world united by Benetton slogans, Nike sweatshops and McDonald's jobs might not be anyone's utopian global village, but its fiber-optic cables and shared cultural references are nonetheless laying the foundations for the first truly international people's movement. That may mean fighting Wal-Mart when it comes to town, but it also means using the Net to network with the other fifty-odd communities in North America that have fought the same battle; it means bringing resolutions about global labor offenses to the local city council meeting, and joining the international fight against the Multilateral Agreement on Investment. It also means making sure that the cries from a toy factory fire in Bangkok can be heard loud and clear outside the Toys 'R' Us at the mall.

Following the Logo Trail

As global brand-based connections gain popularity, that trail from the mall to the sweatshop becomes better traveled. I certainly wasn't the first foreign journalist to pick through the laundry of the Cavite Export Processing Zone. In the few months before I arrived, there had been, among others, a German television crew and a couple of Italian documentary filmmakers who hoped to dig up some scandal on their homegrown brand, Benetton. In Indonesia, so many journalists have wanted to visit Nike's infamous factories that by the time I arrived in Jakarta in August 1997 the staff at the labor-rights group Yakoma were starting to feel like professional tour guides. Every week another journalist – or "human-rights tourist," as Gary Trudeau calls them in his cartoons – descended upon the area. The situation was the same at a factory I tried to visit outside Medan, where child laborers were stitching Barbie's itsy-bitsy party outfits. I met with local activists at the Indonesian Institute for Children's Advocacy and they pulled out a photo album filled with pictures of the NBC crew that had been there. "It won awards," program director Muhammad Joni proudly informs me of the *Dateline* documentary. "They dressed up as importers. Hidden cameras – very professional."[13] Joni glances down at my little tape recorder and at the batik sundress I bought the week before on the beach, unimpressed.

After four years of research, what I find most shocking is that so many supposed "dirty little secrets" are crammed into the global broom closet with such a casual attitude. In the EPZs, labor violations are a dime a dozen – they come tumbling out as soon as you open the door even a crack. As *The Wall Street Journal*'s Bob Ortega writes, "in truth, the entire apparel industry was one continuing and underreported scandal."[14]

With such corporate carelessness at play, no public-relations budget has proved rich enough to clearly dissociate the brand from the factory. And the wider the disparity between the image and the reality, the harder the company seems to get hit. Family-oriented brands like Disney, Wal-Mart and

Kathie Lee Gifford have been forced to confront the conditions under which real families produced their wares. And when the McLibel crew released many of their most gruesome tidbits about McDonald's-tortured chickens, and hamburgers infested with E. coli bacteria, they displayed these facts over an image of the manic plastic face of Ronald McDonald. The logo adopted by the McLibel defendants was a cigar-chomping fat cat hiding behind a clown mask because, as the McLibelers put it, "Children love a secret, and Ronald's is especially disgusting."[15]

When the brand being targeted is anchored by a well-known personality, as is increasingly the case in the era of the superbrand, these collisions between image and reality are potentially even more explosive. For instance, when Kathie Lee Gifford was exposed for using sweatshops, she didn't have the option of reacting like a corporate CEO – whom we expect to be motivated exclusively by shareholder returns. The bubbly talk-show host is the human Aspartame of daytime TV. She could hardly start talking like a callous capitalist cowboy when fifteen-year-old Wendy Diaz publicly pleaded, "If I could talk to Kathie Lee I would ask her to help us, to end all the maltreatment, so that they would stop yelling at us and hitting us, and so they would let us go to night school and let us organize to protect our rights."[16] After all, five minutes before, Gifford might have been confessing to the free world that a child's illness had moved her to such copious tears that she was forced to reduce the swelling under her eyes with Preparation H. She is, as Andrew Ross writes, "a perfect foil for revelations about child labor." Confronted with Diaz's words, Gifford had two options: throw away her entire multimillion-dollar TV-Mom persona, or become the fairy godmother of the maquiladoras. The choice was simple enough. "It took Gifford only two weeks to ascend to the saintly rank of labor crusader," Ross recounts.[17]

In an odd twist of marketing fate, corporate sponsorship itself has become an important lever for activists. And why shouldn't it? When the International Olympic Committee (IOC) became mired in bribery and doping scandals in late 1998, the media immediately focused on how the controversy would affect the games' corporate sponsors – companies that claimed to be aghast at the IOC's innocence lost. "It goes to the heart of why we're involved in the Olympics. Anything that affects the positive image of the Olympics affects us," said a spokesperson for Coca-Cola.[18]

But surely that theory cuts both ways: if sponsors can be tarnished by corruption in the events they sponsor, those events can also be tarnished by the dubious activities of their sponsors. This is a connection that is being made with increasing frequency as the sponsorship industry balloons. In August 1998, Celine Dion's concert tour was picketed by human-rights activists in Boston, Philadelphia and Washington, D.C. Although she was unaware of it, her tour sponsor – Ericsson cellular – was among Burma's most intransigent foreign investors, refusing to cease its dealings with the junta despite the campaign for an international boycott. But when the brand bashing moved beyond Ericsson proper and began to spill onto Dion's diva image, it took only a week of protests to induce Ericsson to announce its immediate withdrawal from Burma. Meanwhile, sponsors that fail to shield performers from anticorporate campaigns being waged against them have also found themselves under attack from all sides. For instance, at Suzuki's Rock 'n' Roll Marathon in San Diego, California, in May 1999, the bands mutinied against their corporate sponsor. Hootie and the Blowfish – hardly known for their radical political views – decided to join forces with the campaigners who were targeting the event because of Suzuki's business dealings with the Burmese junta. Band members insisted that a Suzuki banner be taken down before they got on stage and then performed wearing "Suzuki out of Burma" T-shirts and stickers.[19]

In addition to aggressive sponsorship, another marketing trend that has begun to backfire is the commercial co-optation of identity politics, discussed in Chapter 5. Rather than softening its image, Nike's feminist-themed ads and antiracist slogans have only served to enrage women's groups and civil-rights leaders, who insist that a company that got rich off the backs of young women in the Third World has no business using the ideals of feminism and racial equality to sell more shoes. "I think people feel uneasy about the repackaging of social justice images as commercials from the start," U.S. media critic Makani Themba explains, "but they're not sure why. Then you hear these charges and you're ready to pounce on Nike as hypocritical."[20]

Which might explain why the first company to feel the heat of the sweatshop police was one that had seemed to be a paragon of ethical corporatism, Levi Strauss. In 1992 Levi's became the first company to adopt a corporate code of conduct after some of its contractors overseas were found to be treating their workers as indentured slaves. This was not the image the company had presented back home, with its commitment to nonhierarchical collective decision making and, later, its high-profile sponsorship of such girl-power events as the Lilith Fair festival. Similarly, the Body Shop – though it may well be the most progressive multinational on the planet – still has a tendency to display its good deeds in its store windows before getting its corporate house in order. Anita Roddick's company has been the subject of numerous damning investigations in the press, which have challenged the company's use of chemicals, its stand on unions and even its claim that its products have not been tested on animals.

We have heard the same refrain over and over again from Nike, Reebok, the Body Shop, Starbucks, Levi's and the Gap: "Why are you picking on us? We're the good ones!" The answer is simple. They are singled out because the politics they have associated themselves with, which have made them rich – feminism, ecology, inner-city empowerment – were not just random pieces of effective ad copy that their brand managers found lying around. They are complex, essential social ideas, for which many people have spent lifetimes fighting. That's what lends righteousness to the rage of activists campaigning against what they see as cynical distortions of those ideas. Al Dunlap, the notorious job-slasher-for-hire who built his reputation on ruthless layoffs, may be able to respond to calls for corporate accountability with a rev of his chainsaw, but companies such as Levi's and the Body Shop can't shrug them off, because they publicly presented social accountability as the foundation of their corporate philosophy from the first. Over and over again, it is when the advertising teams creatively overreach themselves that – like Icarus – they fall.

Injustice – in Synergy

By some accident of fate, on February 25, 1997, the multiple layers of anti-corporate rage converged over the Mighty Ducks hockey arena in Anaheim,

California. It was Disney's annual meeting and about 10,000 shareholders crowded into the arena to rake Michael Eisner over the coals. They were upset that he had paid more than $100 million in a severance package to Hollywood superagent Michael Ovitz, who'd lasted only fourteen scandal-racked months at Disney as second in command. Eisner was further attacked for his own $400 million multiyear pay package, as well as for stacking the board with friends and paid Disney consultants. As if shareholders weren't angry enough, the obscene amounts of money lavished on Ovitz and Eisner were thrown into harsh relief by an unrelated shareholders' resolution chiding Disney for paying starvation wages to workers in its overseas factories, and calling for independent monitoring of these practices. Outside the arena, dozens of National Labor Committee supporters were shouting and waving placards about the plight of Disney's Haitian workforce. Of course the monitoring resolution was trounced, but the way the issues of sweatshop labor and executive compensation played off one another must have been music to Charles Kernaghan's ears.

Eisner, who apparently expected the gathering to be little more than a pep rally, was clearly caught off guard by this confluence of events. Wasn't he simply playing by the rules – making his shareholders rich and himself richer? Weren't profits up a healthy 16 percent from the year before? Wasn't the entertainment industry, as Eisner himself reminded the restless gathering, "extremely competitive"? Ever the expert at speaking to children, Eisner ventured, "I don't think people understand executive compensation."[21]

Or maybe they understood it all too well. As one shareholder commented – to much applause – "Nobody argues that Mr. Eisner hasn't done a fantastic job. But that's more in one year than someone like me will get in a lifetime. It's more than the President of the United States makes – and look what he runs!"[22] Still, Eisner's confusion is understandable. He is by no means the only CEO paid truly goofy amounts of money – compared to some executives, he's positively roughing it, with an annual base salary of only $750,000 (with bonuses and stock options on top of that, of course). And Disney certainly isn't alone in its sweatshop woes. According to the U.S. Investor Responsibility Research Center, there were seventy-nine anti-sweatshop shareholder resolutions on the books for major American multinationals between 1996 and 1998, including Dayton Hudson, Nike, the Gap, Land's End, J.C. Penney and Toys 'R' Us.[23] It's clear that what was being put on trial at that rowdy meeting in Anaheim was far more than the excesses of a single company – at issue was the central question of global economic disparity: disparity between executive and worker, between North and South, between consumer and producer, and even between individual shareholders and the boss. The Mouse's family values provided a helpful glass structure at which to lob stones, but the truth is that virtually any Fortune 500 company could have been in the hot seat that day.

Self Serve
SHELL
KILLS
OGONI

CHAPTER SIXTEEN

A TALE OF THREE LOGOS

The Swoosh, the Shell and the Arches

Dozens of brand-based campaigns have succeeded in rattling their corporate targets, in several cases pushing them to substantially alter their policies. But three campaigns stand out for having reached well beyond activist circles and deep into public consciousness. The tactics they have developed – among them the use of the courts to force transparency on corporations, and the Internet to bypass traditional media – are revolutionizing the future of political engagement. By now it should come as no surprise that the targets of these influential campaigns are three of the most familiar and best-tended logos on the brandscape: the Swoosh, the Shell and the Arches.

The Swoosh: The Fight for Good Jobs

Nike CEO Phil Knight has long been a hero of the business schools. Prestigious academic publications such as *The Harvard Business Review* have lauded his pioneering marketing techniques, his understanding of branding and his early use of outsourcing. Countless MBA candidates and other students of marketing and communications have studied the Nike formula of "brands, not products." So when Phil Knight was invited to be a guest speaker at the Stanford University Business School – Knight's own alma mater – in May 1997, the visit was expected to be one in a long line of Nike love-ins. Instead, Knight was greeted by a crowd of picketing students, and when he approached the microphone he was taunted with chants of "Hey Phil, off the stage. Pay your workers a living wage." The Nike honeymoon had come to a grinding halt.

No story illustrates the growing distrust of the culture of corporate branding more than the international anti-Nike movement – the most publicized and tenacious of the brand-based campaigns. Nike's sweatshop scandals have been the subject of over 1,500 news articles and opinion columns. Its Asian factories have been probed by cameras from nearly every major media organization, from CBS to Disney's sports station, ESPN. On top of all that, it has been the subject of a series of Doonesbury cartoon strips and the butt of Michael Moore's documentary *The Big One*. As a result, several people in Nike's PR department work full time dealing with the sweatshop controversy – fielding complaints, meeting with local groups and developing Nike's response – and the company has created a new executive position: vice president for corporate responsibility. Nike has received hundreds and thousands of letters of protest, faced hundreds of both small and large groups of demonstrators, and is the target of a dozen critical Web sites.

For the last two years, anti-Nike forces in North America and Europe have attempted to focus all the scattered swoosh bashing on a single day. Every six months they have declared an International Nike Day of Action, and brought their demands for fair wages and independent monitoring directly to Nike's customers, shoppers at flagship Nike Towns in urban centers or the less glamorous Foot Locker outlets in suburban malls. According to Campaign for Labor Rights, the largest anti-Nike event so far took place on October 18, 1997: eighty-five cities in thirteen countries participated. Not all the protests have attracted large crowds, but since the movement is so decentralized, the sheer number of individual anti-Nike events has left the company's public-relations department scrambling to get its spin onto dozens of local newscasts. Though you'd never know it from its branding ubiquity, even Nike can't be everywhere at once.

Since so many of the stores that sell Nike products are located in malls, protests often end with a security guard escorting participants into the parking lot. Jeff Smith, an activist from Grand Rapids, Michigan, reported that "when we asked if private property rights ruled over free speech rights, the [security] officer hesitated and then emphatically said YES!" (Though in the economically depressed city of St. John's, Newfoundland, anti-Nike campaigners reported that after being thrown out of a mall, "they were approached by a security guard who asked to sign their petition."[1]) But there's plenty that can be done on the sidewalk or in the mall parking lot.

SLOGANS FROM THE INTERNATIONAL ANTI-NIKE MOVEMENT:

Just Don't Do It *Just Don't* *Nike, Do It Just* *Justice. Do it, Nike*

Campaigners have dramatized Nike's labor practices through what they call "sweatshop fashion shows," and "The Transnational Capital Auction: A Game of Survival" (the lowest bidder wins), and a global economy treadmill (run fast, stay in the same place). In Australia, anti-Nike protestors have been known to parade around in calico bags painted with the slogan "Rather wear a bag than Nike." Students at the University of Colorado in Boulder dramatized the difference between the legal minimum wage and a living wage by holding a fundraising run in which "participants pay an entrance fee of $1.60 (daily wages for a Nike worker in Vietnam) and the winner will receive $2.10 (the price of three square meals in Vietnam)."[2] Meanwhile, activists in Austin, Texas, made a giant papier-mâché Nike sneaker piñata, and a protest outside a Regina, Saskatchewan, shopping center featured a deface-the-swoosh booth. The last stunt is something of a running theme in all the anti-Nike actions: Nike's logo and slogan have been jammed so many times – on T-shirts, stickers, placards, banners and pins – that the semiotic bruises have turned them black and blue (see list below).

Tellingly, the anti-Nike movement is at its strongest inside the company's home state of Oregon, even though the area has reaped substantial economic benefits from Nike's success (Nike is the largest employer in Portland and a significant local philanthropist). Phil Knight's neighbors, nonetheless, have not all rushed to his defense in his hour of need. In fact, since the *Life* magazine soccer-ball story broke, many Oregonians have been out for blood. The demonstrations outside the Portland Nike Town are among the largest and most militant in the country, sometimes sporting a menacing giant Phil Knight puppet with dollar signs for eyes or a twelve-foot Nike swoosh dragged by small children (to dramatize child labor). And in contravention of the principles of nonviolence that govern the anti-Nike movement, one protest in Eugene, Oregon, led to acts of vandalism including the tearing-down of a fence surrounding the construction of a new Nike Town, gear pulled off shelves at an existing Nike store and, according to one eyewitness, "an entire rack of clothes ... dumped off a balcony into a fountain below."[3]

Local papers in Oregon have aggressively (sometimes gleefully) followed Knight's sweatshop scandals, and the daily paper *The Oregonian* sent a reporter to Southeast Asia to do its own lengthy investigation of the factories.

The Swooshtika   *Just Boycott It*   *Ban the Swoosh*   *Nike – Fair Play?*

Mark Zusman, editor of the Oregon newspaper *The Willamette Week*, publicly admonished Knight in a 1996 "memo": "Frankly, Phil, it's time to get a little more sophisticated about this media orgy... Oregonians already have suffered through the shame of Tonya Harding, Bob Packwood and Wes Cooley. Spare us the added humiliation of being known as the home of the most exploitative capitalist in the free world."[4]

Even Nike's charitable donations have become controversial. In the midst of a critical fundraising drive to try to address a $15 million shortfall, the Portland School Board was torn apart by debate about whether to accept Nike's gift of $500,000 in cash and swooshed athletic gear. The board ended up accepting the donation, but not before looking their gift horse publicly in the mouth. "I asked myself," school board trustee Joseph Tam told *The Oregonian*, "Nike contributed this money so my children can have a better education, but at whose expense? At the expense of children who work for six cents an hour?... As an immigrant and as an Asian I have to face this moral and ethical dilemma."[5]

Nike's sponsorship scandals have reached far beyond the company's home state. In Edmonton, Alberta, teachers, parents and some students tried to block Nike from sponsoring a children's street hockey program because "a company which profits from child labor in Pakistan ought not to be held up as a hero to Edmonton children."[6] At least one school involved in the citywide program sent back its swooshed equipment to Nike headquarters. And when Nike approached the City of Ottawa Council in March 1998 to suggest building one of its swooshed gymnasium floors in a local community center, it faced questions about "blood money." Nike withdrew its offer and gave the court to a more grateful center, run by the Boys and Girls Clubs. The dilemma of accepting Nike sponsorship money has also exploded on university campuses, as we will see in the next chapter.

At first, much of the outrage stemmed from the fact that when the sweatshop scandal hit the papers, Nike wasn't really acting all that sorry about it. While Kathie Lee Gifford and the Gap had at least displayed contrition when they got caught with their sweatshops showing, Phil Knight had practically stonewalled: denying responsibility, attacking journalists, blaming rogue

Nike, Nein, Ich Kaufe Es Nicht! *(Nike – No, I Don't Buy It!)*

contractors and sending out flacks to speak for the company. While Kathie Lee was crying on TV, Michael Jordan was shrugging his shoulders and saying that his job was to shoot hoop, not play politics. And while the Gap agreed to allow a particularly controversial factory in El Salvador to be monitored by local human-rights groups, Nike was paying lip service to a code of conduct that its Asian workers, when interviewed, had never even heard of.

But there was a critical difference between Nike and the Gap at this stage. Nike didn't panic when its scandals hit the middle-American mall, because the mall, while it is indeed where most Nike products are sold, is not where Nike's image was made. Unlike the Gap, Nike has drawn on the inner cities, merging, as we've seen, with the styles of poor black and Latino youth to load up on imagery and attitude. Nike's branding power is thoroughly intertwined with the African-American heroes who have endorsed its products since the mid-eighties: Michael Jordan, Charles Barkley, Scottie Pippen, Michael Johnson, Spike Lee, Tiger Woods, Bo Jackson – not to mention the rappers who wear Nike gear on stage. While hip-hop style was the major influence at the mall, Phil Knight must have known that as long as Nike was King Brand with Jordan fans in Compton and the Bronx, he could be stirred but not shaken. Sure, their parents, teachers and church leaders might be tut-tutting over sweatshops, but as far as Nike's core demographic of thirteen- to seventeen-year-old kids was concerned, the swoosh was still made of Teflon.

By 1997, it had become clear to Nike's critics that if they were serious about taking on the swoosh in an image war, they would have to get at the source of the brand's cachet – and as Nick Alexander of the multicultural *Third Force* magazine wrote in the summer of that year, they weren't even close. "Nobody has figured out how to make Nike break down and cry. The reason is that nobody has engaged African Americans in the fight.... To gain significant support from communities of color, corporate campaigns need to make connections between Nike's overseas operations and conditions here at home."[7]

The connections were there to be made. It is the cruelest irony of Nike's "brands, not products" formula that the people who have done the most to infuse the swoosh with cutting-edge meaning are the very people most hurt by the company's pumped-up prices and nonexistent manufacturing base. It is inner-city youth who have most directly felt the impact of Nike's decision

Nike Soyez Sport! (Nike Be a Sport) *Just Duit (It's just money)*

to manufacture its products outside the U.S., both in high unemployment rates and in the erosion of the community tax base (which sets the stage for the deterioration of local public schools).

Instead of jobs for their parents, what the inner-city kids get from Nike is the occasional visit from its marketers and designers on "bro-ing" pilgrimages. "Hey, bro, what do you think of these new Jordans – are they fresh or what?" The effect of high-priced cool hunters whipping up brand frenzy on the cracked asphalt basketball courts of Harlem, the Bronx and Compton has already been discussed: kids incorporate the brands into gang-wear uniforms; some want the gear so badly they are willing to sell drugs, steal, mug, even kill for it. Jessie Collins, executive director of the Edenwald–Gun Hill Neighborhood Center in the northeast Bronx, tells me that it's sometimes drug or gang money, but more often it's the mothers' minimum-wage salaries or welfare checks that are spent on disposable status wear. When I asked her about the media reports of kids stabbing each other for their $150 Air Jordans she said dryly, "It's enough to beat up on your mother for... $150 is a hell of a lot of money."[8]

Shoe-store owners like Steven Roth of Essex House of Fashion are often uncomfortable with the way so-called street fashions play out for real on the postindustrial streets of Newark, New Jersey, where his store is located.

> I do get weary and worn down from it all. I'm always forced to face the fact that I make my money from poor people. A lot of them are on welfare. Sometimes a mother will come in here with a kid, and the kid is dirty and poorly dressed. But the kid wants a hundred-twenty-buck pair of shoes and that stupid mother buys them for him. I can feel that kid's inner need – this desire to own these things and have the feelings that go with them – but it hurts me that this is the way things are.[9]

It's easy to blame the parents for giving in, but that "deep inner need" for designer gear has grown so intense that it has confounded everyone from community leaders to the police. Everyone pretty much agrees that brands like Nike are playing a powerful surrogate role in the ghetto, subbing for everything from self-esteem to African-American cultural history to political power. What they are far less sure about is how to fill that need with empowerment and a sense of self-worth that does not necessarily come with a logo attached. Even broaching the subject of brand fetishism to these kids is risky. With so much emotion invested in celebrity consumer goods, many kids take criticism of Nike or Tommy as a personal attack, as grave a transgression as insulting someone's mother to his face.

Not surprisingly, Nike sees its appeal among disadvantaged kids differently. By supporting sports programs in Boys and Girls Clubs, by paying to repave urban basketball courts and by turning high-performance sports gear into street fashions, the company claims it is sending out the inspirational message that even poor kids can "Just Do It." In its press material and ads, there is an almost messianic quality to Nike's portrayal of its role in the inner cities: troubled kids will have higher self-esteem, fewer unwanted pregnancies and more ambition – all because at Nike "We see them as athletes." For Nike, its $150 Air Jordans are not a shoe but a kind of talisman with which poor kids can run out of the ghetto and better their lives. Nike's magic slippers will help them fly – just as they made Michael Jordan fly.

A remarkable, subversive accomplishment? Maybe. But one can't help thinking that one of the main reasons black urban youth can get out of the ghetto only by rapping or shooting hoops is that Nike and the other multinationals are reinforcing stereotypical images of black youth and simultaneously taking all the jobs away. As U.S. Congressman Bernie Sanders and Congresswoman Marcy Kaptur stated in a letter to the company, Nike has played a pivotal part in the industrial exodus from urban centers. "Nike has led the way in abandoning the manufacturing workers of the United States and their families.... Apparently, Nike believes that workers in the United States are good enough to purchase your shoe products, but are no longer worthy enough to manufacture them."[10]

And when the company's urban branding strategy is taken in conjunction with this employment record, Nike ceases to be the savior of the inner city and turns into the guy who steals your job, then sells you a pair of overpriced sneakers and yells, "Run like hell!" Hey, it's the only way out of the ghetto, kid. Just do it.

That's what Mike Gitelson thought, anyway. A social worker at the Bronx's Edenwald–Gun Hill Neighborhood Center, he was unimpressed with the swoosh's powers as a self-help guru in the projects and "sick of seeing kids wearing sneakers they couldn't afford and which their parents couldn't afford."[11] Nike's critics on college campuses and in the labor movement may be fueled largely by moral outrage, but Mike Gitelson and his colleagues simply feel ripped off. So rather than lecturing the kids on the virtues of frugality, they began telling them about how Nike made the shoes that they wanted so badly. Gitelson told them about the workers in Indonesia who earned $2 a day, he told them that it cost Nike only $5 to make the shoes they bought for between $100 and $180, and he told them about how Nike didn't make any of its shoes in the U.S. – which was part of the reason their parents had such a tough time finding work. "We got really angry," says Gitelson, "because they were taking so much money from us here and then going to other countries and exploiting people even worse.... We want our kids to see how it affects them here on the streets, but also how here on the streets affects people in Southeast Asia." His colleague at the center, youth worker Leo Johnson, lays out the issue using the kids' own lingo. "Yo, dude," he tells his preteen audiences, "you're being suckered if you pay $100 for a sneaker that costs $5 to make. If somebody did that to you on the block, you know where it's going."[12]

The kids at the center were upset to learn about the sweatshops but they were clearly most pissed off that Phil Knight and Michael Jordan were playing them for chumps. They sent Phil Knight a hundred letters about how much money they had spent on Nike gear over the years – and how, the way they figured it, Nike owed them big time. "I just bought a pair of Nikes for $100," one kid wrote. "It's not right what you're doing. A fair price would have been $30. Could you please send me back $70?" When the company answered the kids with a form letter, "That's when we got really angry and started putting together the protest," Gitelson says.

They decided the protest would take the form of a "shoe-in" at the Nike Town at Fifth Avenue and Fifty-seventh Street. Since most of the kids at the center are full-fledged swooshaholics, their closets are jam-packed with old Air Jordans and Air Carnivores that they would no longer even consider wearing. To put the obsolete shoes to practical use, they decided to gather them together in garbage bags and dump them on the doorstep of Nike Town.

When Nike executives got wind that a bunch of black and Latino kids from the Bronx were planning to publicly diss their company, the form letters came to an abrupt halt. Up to that point, Nike had met most criticism by attacking its critics as members of "fringe groups," but this was different: if a backlash took root in the inner cities, it could sink the brand at the mall. As Gitelson puts it, "Our kids are exactly who Nike depends upon to set the trends for them so that the rest of the country buys their sneakers. White middle-class adults who are fighting them, well, it's almost okay. But when youth of color start speaking out against Nike, they start getting scared."[13]

The executives in Oregon also knew, no doubt, that Edenwald was only the tip of the iceberg. For the past couple of years, debates have been raging in hip-hop scenes about rappers "label whoring for Nike and Tommy" instead of supporting black-owned clothing companies like FUBU (For Us By Us). And rapper KRS-One planned to launch the Temple of Hip Hop, a project that promised to wrest the culture of African-American youth away from white record and clothing labels and return it to the communities that built it. It was against this backdrop that, on September 10, 1997 – two weeks before the shoe-in protest was scheduled to take place – Nike's chief of public relations, Vada Manager, made the unprecedented move of flying in from Oregon with a colleague to try to convince the center that the swoosh was a friend of the projects.

"He was working overtime to put the spins on us," says Gitelson. It didn't work. At the meeting, the center laid out three very concrete demands:

1. Those who work for Nike overseas should be paid a living wage, with independent monitoring to ensure that it is occurring.
2. Nike sneakers should be sold less expensively here in America with no concessions to American workforce (i.e. no downsizing, or loss of benefits)
3. Nike should seriously re-invest in the inner city in America, especially New York City since we have been the subject of much of their advertising.[14]

Gitelson may have recognized that Nike was scared – but not *that* scared. Once it became clear that the two parties were at an impasse, the meeting turned into a scolding session as the two Nike executives were required to listen to Edenwald director Jessie Collins comparing the company's Asian sweatshops with her experience as a young girl picking cotton in the sharecropping South. Back in Alabama, she told Manager, she earned $2 a day, just like the Indonesians. "And maybe a lot of Americans can't identify with those workers' situation, but I certainly can."[15]

Vada Manager returned to Oregon defeated and the protest went off as planned, with two hundred participants from eleven community centers around New York. The kids – most of whom were between eleven and thirteen years old – hooted and hollered and dumped several clear garbage bags of smelly old Nikes at the feet of a line of security guards who had been brought in on special assignment to protect the sacred Nike premises. Vada Manager again flew to New York to run damage control, but there was little he could do. Local TV crews covered the event, as did an ABC news team and *The New York Times*.

In a harsh bit of bad timing for the company, the *Times* piece ran on a page facing another story about Nike. Graphically underlining the urgency of the protest, this story reported that a fourteen-year-old boy from Crown Heights had just been murdered by a fifteen-year-old boy who beat him and left him on the subway tracks with a train approaching. "Police Say Teenager Died for His Sneakers and Beeper," the headline read. And the brand of his sneakers? Air Jordans. The article quoted the killer's mother saying that her son had got mixed up with gangs because he wanted to "have nice things." A friend of the victim explained that wearing designer clothes and carrying a beeper had become a way for poor kids to "feel important."

The African-American and Latino kids outside Nike Town on Fifth Avenue – the ones swarmed by cameras and surrounded by curious onlookers – were feeling pretty important, too. Taking on Nike "toe to toe," as they said, turned out to be even more fun than wearing Nikes. With the Fox News camera pointed in his face, one of the young activists – a thirteen-year-old boy from the Bronx – stared into the lens and delivered a message to Phil Knight: "Nike, we made you. We can break you."

What is perhaps most remarkable about the Nike backlash is its durability. After four solid years in the public eye, the Nike story still has legs (so too, of course, does the Nike brand). Still, most corporate scandals are successfully faced down with a statement of "regret" and a few glossy ads of children playing happily under the offending logo. Not with Nike. The news reports, labor studies and academic research documenting the sweat behind the swoosh have yet to slow down, and Nike critics remain tireless at dissecting the steady stream of materials churned out by Nike's PR machine. They were unmoved by Phil Knight's presence on the White House Task Force on Sweatshops – despite his priceless photo op standing beside President Clinton at the Rose Garden press conference. They sliced and diced the report Nike commissioned from civil-rights leader Andrew Young, pointing out that Young completely dodged the question of whether Nike's factory wages are inhumanely exploitative, and attacking him for relying on translators provided by Nike itself when he visited the factories in Indonesia and Vietnam. As for Nike's other study-for-hire – this one by a group of Dartmouth business students who concluded that workers in Vietnam were living the good life on less than $2 a day – well, everyone pretty much ignored that one altogether.

Finally, in May 1998, Phil Knight stepped out from behind the curtain of spin doctors and called a press conference in Washington to address his critics directly. Knight began by saying that he had been painted as a "corporate crook, the perfect corporate villain for these times." He acknowledged that his shoes "have become synonymous with slave wages, forced overtime and arbitrary abuse." Then, to much fanfare, he unveiled a plan to improve working conditions in Asia. It contained some tough new regulations on factory air quality and the use of petroleum-based chemicals. It promised to provide classes inside some Indonesian factories and promised not to hire anyone under eighteen years old in the shoe factories. But there was still nothing substantial in the plan about allowing independent outside monitors to inspect the factories, and there were no wage raises for the workers. Knight did promise, however, that Nike's contractors would no longer be permitted to appeal to the Indonesian government for a waiver on the minimum wage.

It wasn't enough. That September the San Francisco human-rights group

Global Exchange, one of the company's harshest critics, released an alarming report on the status of Nike's Indonesian workers in the midst of the country's economic and political crisis. "While workers producing Nike shoes were low paid before their currency, the rupiah, began plummeting in late 1997, the dollar value of their wages has dropped from $2.47/day in 1997 to 80 cents/day in 1998." Meanwhile, the report noted that with soaring commodity prices, workers "estimated that their cost of living had gone up anywhere from 100 to 300 per cent."[16] Global Exchange called on Nike to double the wages of its Indonesian workforce, an exercise that would cost it $20 million a year – exactly what Michael Jordan is paid annually to endorse the company.

Not surprisingly, Nike did not double the wages, but it did, three weeks later, give 30 percent of the Indonesian workforce a 25 percent raise.[17] That, too, failed to silence the crowds outside the superstores, and five months later Nike came forward again, this time with what vice president of corporate responsibility Maria Eitel called "an aggressive corporate responsibility agenda at Nike."[18] As of April 1, 1999, workers would get another 6 percent raise. The company had also opened up a Vietnamese factory near Ho Chi Minh City to outside health and safety monitors, who found conditions much improved. Dara O'Rourke of the University of California at Berkeley reported that the factory had "implemented important changes over the past 18 months which appear to have significantly reduced worker exposures to toxic solvents, adhesives and other chemicals." What made the report all the more remarkable was that O'Rourke's inspection was a genuinely independent one: in fact, less than two years earlier, he had enraged the company by leaking a report conducted by Ernst & Young that showed that Nike was ignoring widespread violations at that same factory.

O'Rourke's findings weren't all glowing. There were still persistent problems with air quality, factory overheating and safety gear – and he had visited only the one factory.[19] As well, Nike's much-heralded 6 percent pay raise for Indonesian workers still left much to be desired; it amounted to an increase of one cent an hour and, with inflation and currency fluctuation, only brought wages to about half of what Nike paychecks were worth before the economic crisis. Even so, these were significant gestures coming from a company that two years earlier was playing the role of the powerless global shopper, claiming that contractors alone had the authority to set wages and make the rules.

The resilience of the Nike campaign in the face of the public-relations onslaught is persuasive evidence that invasive marketing, coupled with worker abandonment, strikes a wide range of people from different walks of life as grossly unfair and unsustainable. Moreover, many of those people are not interested in letting Nike off the hook simply because this formula has become the standard one for capitalism-as-usual. On the contrary, there seems to be a part of the public psyche that likes kicking the most macho and extreme of all the sporting-goods companies in the shins—I mean *really* likes it. Nike's critics have shown that they don't want this story to be brushed under the rug with a reassuring bit of corporate PR; they want it out in the open, where they can keep a close eye on it.

In large part, this is because Nike's critics know that the company's sweatshop scandals are not the result of a series of freak accidents: they know that the criticisms leveled at Nike apply to all the brand-based shoe companies contracting out to a global maze of firms. But rather than this serving as a justification, Nike—as the market leader—has become a lightning rod for this broader resentment. It has been latched on to as the essential story of the extremes of the current global economy: the disparities between those who profit from Nike's success and those who are exploited by it are so gaping that a child could understand what is wrong with this picture and indeed (as we will see in the next chapter) it is children and teenagers who most readily do.

So, when does the total boycott of Nike products begin? Not soon, apparently. A cursory glance around any city in the world shows that the swoosh is still ubiquitous; some athletes still tattoo it on their navels, and plenty of high-school students still deck themselves out in the coveted gear. But at the same time, there can be little doubt that the millions of dollars that Nike has saved in labor costs over the years are beginning to bite back, and take a toll on its bottom line. "We didn't think that the Nike situation would be as bad as it seems to be," said Nikko stock analyst Tim Finucane in *The Wall Street Journal* in March 1998.[20] Wall Street really had no choice but to turn on the company that had been its darling for so many years. Despite the fact that Asia's plummeting currencies meant that Nike's labor costs in Indonesia, for instance, were a quarter of what they were before the crash, the company was still suffering. Nike's profits were down, orders were down, stock prices were *way* down, and after an average annual growth of 34 percent since 1995, quarterly earnings were suddenly down 70 percent. By the third quarter, which ended in February 1999, Nike's profits were once again up 70 percent – but by the company's own account, the recovery was not the result of rebounding sales but rather of Nike's decision to cut jobs and contracts. In fact, Nike's revenues and future orders were down in 1999 for the second year in a row.[21]

Nike has blamed its financial problems on everything *but* the human-rights campaign. The Asian currency crisis was the reason Nikes weren't selling well in Japan and South Korea; or it was because Americans were buying "brown shoes" (walking shoes and hiking boots) as opposed to big white sneakers. But the brown-shoe excuse rang hollow. Nike makes plenty of brown shoes – it has a line of hiking boots, and it owns Cole Haan (and recently saved millions by closing down the Cole Haan factory in Portand, Maine, and moving production to Mexico and Brazil).[22] More to the point, Adidas staged a massive comeback during the very year that Nike was free-falling. In the quarter when Nike nose-dived, Adidas sales were up 42 percent, its net income was up 48 percent, to $255 million, and its stock price had tripled in two years. The German company, as we have seen, turned its fortunes around by copying Nike's production structure and all but Xeroxing its approach to marketing and sponsorships (the political implications of that will be dealt with in Chapter 18). In 1997–98, Adidas even redesigned its basketball shoes so they looked just like Nikes: big, white and ultra high tech. But unlike Nikes, they sold briskly. So much for the brown-shoe theory.

Over the years Nike has tried dozens of tactics to silence the cries of its critics, but the most ironic by far has been the company's desperate attempt to hide behind its product. "We're not political activists. We are a footwear manufacturer," said Nike spokeswoman Donna Gibbs, when the sweatshop scandal first began to erupt.[23] A footwear manufacturer? This from the company that made a concerted decision in the mid-eighties not to be about boring corporeal stuff like footwear – and certainly nothing as crass as manufacturing. Nike wanted to be about sports, Knight told us, it wanted to be about the idea of sports, then the idea of transcendence through sports; then it wanted to be about self-empowerment, women's rights, racial equality. It wanted its stores to be temples, its ads a religion, its customers a nation, its workers a tribe. After taking us all on such a branded ride, to turn around and say "Don't look at us, we just make shoes" rings laughably hollow.

Nike was the most inflated of all the balloon brands, and the bigger it grew, the louder it popped.

The Shell: The Fight for Open Space

In North America, Nike has been at the forefront of the burgeoning political movement taking aim at the power of multinationals, but in Britain, Germany and the Netherlands, that dubious honor has belonged to Royal Dutch/Shell.

It began in February 1995 when Shell finalized its plans to dispose of a rusted and obsolete oil-storage platform, known as Brent Spar, by sinking it in the Atlantic Ocean, 150 miles off the coast of Scotland. The environmental group Greenpeace was against the plan, claiming the 14,500-ton rig should be towed to land, where the oil sludge could be contained and the rig's parts recycled. Shell countered that land disposal was unsafe, not to mention impossible. Then, on April 30, just as Shell began towing the platform to its watery grave, a group of Greenpeace activists showed up in a helicopter and tried to land on the Brent Spar. Shell fended off the aircraft with water cannons, but the entire episode was captured on videotape, and the images were sent via satellite to TV stations around the world.

It was vintage Greenpeace, ever the made-for-TV activists. But the impact those images from the Brent Spar had on the European public took even Greenpeace by surprise. Before the Brent Spar incident, the group was teetering on the brink of obsolescence – the eco movement had been under attack, and appeared to be sputtering out in the wake of recession, and

Greenpeace itself had lost credibility because of internal divisions and questionable financial and tactical policies. When Greenpeace decided to launch a campaign against the sinking of the Brent Spar, it had no idea that this rather arcane issue would become a cause célèbre. As Robin Grove-White, chairman of Greenpeace U.K., readily admits, "No one, and certainly not people within Greenpeace, anticipated the profound and continuing reverberations."[24]

Unlike the environmentally disastrous *Exxon Valdez* oil spill four years earlier (a clear-cut case of negligence involving a drunken captain), it wasn't as if Shell was doing anything illegal. The plan had received full approval from John Major's governing Conservatives, and sinking had become a standard way of disposing of old platforms. Besides, it was even debatable whether Greenpeace's land-disposal alternative was more ecologically sound than Shell's proposed deep-sea dunk. But the image that Greenpeace generated – of an ugly, giant, rusted pollution generator fending off the good green activists that were buzzing it like dogged mosquitoes – caught people's attention, and gave them a timely and rare opportunity to stop and think about what was being proposed. And much of the public decided that Shell wanted to sink its hunk of metal and sludge because the most profitable corporation in the world was too cheap to come up with a better plan to dispose of its garbage. This view was reinforced by a damning study that found that land disposal of the Brent Spar would cost Shell US$70 million, while sinking it would cost a mere US$16 million. Coming from a $128 billion company, this apparent penny-pinching did not impress the gasoline-buying public at all.

That Shell's actions were legal and Greenpeace's were not seemed to be entirely beside the point. In the eyes of many Europeans, Shell was morally wrong. Thousands of people protested outside its gas stations, and in Germany the Shell office reported a sales drop of between 20 and 50 percent after the scandal began – "the worst we have experienced," said the oil multinational's German head, Peter Duncan.[25] A firebomb exploded at a Shell station in Hamburg ("Don't sink the Brent Spar Oil Platform" was the message left behind), and there was a drive-by shooting at a Frankfurt outlet. (No one was injured.) The unofficial boycott also spread through Britain to Denmark, Austria and the Netherlands.

Four months after the protests began, on June 20, 1995, something unprecedented happened: Shell backed down. It would spend the extra millions to tow the platform to Norway, where it would be dismantled on land. According to *The Wall Street Journal*, it was "a humiliating and painful U-turn."[26] Grove-White articulates the extent of the Brent Spar victory: "For the first time, an environmental group had catalyzed international opinion to bring about the kind of change of policy that unsettled the very basis of executive authority. However briefly, the world turned upside down – the rule book had been rewritten."[27]

Before the Brent Spar campaign was launched, there had been internal battles within Greenpeace about whether the group could "sell" the disposal of an old industrial hunk of junk as a galvanizing, media-friendly issue. Dutch Greenpeace campaigner Giys Thieme recalls the concerns inside the organization: "It wasn't an oil campaign, it wasn't an atmosphere campaign, it wasn't a chlorine campaign."[28] Neither was it a fight for fish, or whales, or even cute baby seals. Brent Spar, it turns out, was about the idea of preserving untouched space, just as the anti-logging protests in British Columbia's Clayoquot Sound a year earlier had been about protecting one of the last remaining stands of ancient, virgin forest. Clayoquot was about biodiversity, but it was also about preserving the idea of wilderness, and Brent Spar was much the same. Although Greenpeace presented scientific studies on the ecological impact the oil platform would have on the ocean floor (getting some of its facts wrong along the way), the fight was not so much about environmental protection in the traditional sense as it was about the need to keep the Atlantic Ocean floor from being used as a junkyard. Shell's plans to bury the monstrosity in the depths of the sea resonated in the public psyche worldwide: here was proof that if multinationals were left to their own devices, there would be no open space left on earth – even the depths of the ocean, the last great wilderness, would be colonized.

Shell, the British government and much of the business press pointed out that this reaction was entirely irrational. "Science Loses to Joe Six-Pack" a headline in *The Wall Street Journal* announced, while *The Economist* declared "A Defeat for Rational Decision Making." They were right, in a way. This concept of protecting the unknowable – for no empirical reason in the short term except that it comforts us that it is there – was indeed amorphous, but it was also powerful. As *Guardian* columnist Suzanne Moore wrote, Brent Spar had at least as much to do with mysticism as with science: "In the depths strange species lurk, and though we may never ever see them, we feel in our hearts that they should be left alone. Why must they share the great dark deep with bits and bobs from a dismembered oil platform?"[29]

The lesson Greenpeace took away from its Brent Spar victory, writes Grove-White, was about the sanctity of "the global commons" – places not named on any map, not owned by any private interest and thus belonging to everybody. The group also learned another lesson, something the anti-Nike campaigners had also discovered: targeting a big, rich, ubiquitous multinational corporation is to the late nineties what saving the whales was to the late eighties. It is populist and it is popular, and it was enough to bring Greenpeace back from the brink of death. After Brent Spar, the group was showered with members and money and, as *The Guardian* reported, it was even bequeathed estates. "One woman had phoned to say she had changed her will. 'Left all estate to Greenpeace,' says the note. Wants us to 'buy an inflatable with it and bash Shell.'"[30] In its Brent Spar postmortem *The Wall Street Journal* noted gravely that in the current climate, "economic warfare may be the best way to wage eco-warfare."[31]

Shell's capitulation also provided activists with another lesson. After going to the wall defending the appropriateness and inescapability of Shell's original plan, Prime Minister John Major was left looking like a corporate lap dog – and an unloved one at that. When Shell reversed its position, Major could only mutter that the executives were "wimps" for caving in to public pressure. His position was so compromised that it may well have played a role in his decision, only two days after Shell's U-turn, to step down as head of the Conservative Party and force a vote on his leadership. In this way, Brent Spar proved that corporations – even a notoriously cagey and cloistered company like Royal Dutch/Shell – are sometimes as vulnerable to public pressure as democratically elected governments (occasionally more so).

The lesson proved particularly relevant in the next Shell challenge – the need to focus world attention on the multinational's role in the despoliation of Nigeria, under the protection of the corrupt government of the late

General Sani Abacha. If the general wasn't vulnerable to pressure, Shell certainly was.

From the Ocean as Trash Pit to the Land as Oil Slick

Since the 1950s, Shell Nigeria has extracted $30 billion worth of oil from the land of the Ogoni people, in the Niger Delta. Oil revenue makes up 80 percent of the Nigerian economy – $10 billion annually – and, of that, more than half comes from Shell. But not only have the Ogoni people been deprived of the profits from their rich natural resource, many still live without running water or electricity, and their land and water have been poisoned by open pipelines, oil spills and gas fires.

Under the leadership of the writer and Nobel Peace Prize nominee Ken Saro-Wiwa, the Movement for the Survival of the Ogoni People (MOSOP) campaigned for reform, and demanded compensation from Shell. In response, and in order to keep the oil profits flowing into the government's coffers, General Sani Abacha directed the Nigerian military to take aim at the Ogoni. They killed and tortured thousands. The Ogoni not only blamed Abacha for the attacks, they also accused Shell of treating the Nigerian military as a private police force, paying it to quash peaceful protest on Ogoni land, in addition to giving financial support and legitimacy to the Abacha regime.

Facing mounting protests within Nigeria, Shell withdrew from Ogoni land in 1993 – a move that only put further pressure on the military to remove the Ogoni threat. A leaked memo from the head of the Rivers State Internal Security Force of the Nigerian Army was quite explicit: "Shell operations still impossible unless ruthless military operations are undertaken for smooth economic activities to commence.... Recommendations: Wasting operations during MOSOP and other gatherings making constant military presence justifiable. Wasting targets cutting across communities and leadership cadres especially vocal individuals of various groups."[32]

On May 10, 1994 – five days after the memo was written – Ken Saro-Wiwa said, "This is it. They [the Nigerian military] are going to arrest us all and execute us. All for Shell."[33] Twelve days later, he was arrested and tried for murder. Before receiving his sentence, Saro-Wiwa told the tribunal, "I and my colleagues are not the only ones on trial. Shell is here on trial.... The company has, indeed, ducked this particular trial, but its day will surely come." Then, on November 10, 1995 – despite pressure from the international community, including the Canadian and Australian governments, and to a lesser extent the governments of Germany and France – the Nigerian military government executed Saro-Wiwa along with eight other Ogoni leaders who had protested against Shell. It became an international incident and, once again, people took their protests to their Shell stations, widely boycotting the company. In San Francisco Greenpeaceniks staged a re-enactment of Saro-Wiwa's murder, with the noose fastened around the towering Shell sign (see image, page 364).

As Reclaim the Streets' John Jordan said of multinationals: "Inadvertently, they have helped us see the whole problem as one system." And here was that interconnected system in action: Shell, intent on sinking a monstrous oil platform off the coast of Britain, was simultaneously entangled in a human-rights debacle in Nigeria, in the same year that it laid off workers (despite earning huge profits), all so that it could pump gas into the cars of London – the very issue that had launched Reclaim the Streets. Because Ken Saro-Wiwa was a poet and playwright, his case was also claimed by the international freedom-of-expression group, PEN. Writers, including the English playwright Harold Pinter and the Nobel Prize–winning novelist Nadine Gordimer, took up the cause of Saro-Wiwa's right to express his views against Shell, and turned his persecution into the highest-profile free-expression case since the government of Iran declared a *fatwa* against Salman Rushdie, offering a bounty on his head. In an article for *The New York Times*, Gordimer wrote that "to buy Nigeria's oil under the conditions that prevail is to buy oil in exchange for blood. Other people's blood; the exaction of the death penalty on Nigerians."[34]

The convergence of social-justice, labor and environmental issues in the two Shell campaigns was not a fluke – it goes to the very heart of the emerging spirit of global activism. Ken Saro-Wiwa was killed for fighting to protect his environment, but an environment that encompassed more than the physical landscape that was being ravaged and despoiled by Shell's invasion of the delta. Shell's mistreatment of Ogoni land is both an environmental and a social issue, because natural-resource companies are notorious for lowering their standards when they drill and mine in the Third World. Shell's opponents readily draw parallels between the company's actions in Nigeria, its history of collaborating with the former apartheid government in South Africa, its ongoing presence in the Timor Gap in Indonesian-occupied East Timor and its violent clashes with the Nahua people in the Peruvian Amazon – as well as its standoff with the U'wa people of the Colombian Andes, who threatened in January 1997 to commit mass suicide if Shell went ahead with its drilling plans.

In Saro-Wiwa, civil liberties came together with antiracism; anticapitalism with environmentalism; ecology with labor rights. The bright yellow bulbous logo of Shell – Saro-Wiwa's Goliath of an opponent – became a common enemy for all concerned citizens, to the extent that their governments around the world were required to put the matter on the international agenda. PEN protested against Shell, as did the campaign department at the Body Shop, the activist shareholders who placed the Ogoni plight on the agenda of three consecutive Shell annual meetings and thousands upon thousands of others. In June 1998, Owens Wiwa, Ken's brother, wrote this of the company's situation:

> For centuries, corporations have declared huge profits from evil practices like the trans-Atlantic slave trade, colonialism, Apartheid and [from] dictatorships whose actions are genocidal. They have often gotten away with their loot, leaving governments to apologize. In this case, at the twilight of the 20th century, Shell has been caught in the triangle of ecosystem destruction, human rights abuse and health impairment of the Ogoni people. An apology will not be enough. We anticipate a clean-up of our soil, water and air; adequate, fair and equitable compensation for (a) the environmental damages, (b) human rights abuses due directly and indirectly to Shell activities and (c) the negative health impacts their services have on the people.[35]

To hear Shell tell it, these reparations are already well under way. "Shell continues to invest in community and environmental projects in Nigeria," R.B. Blakely, a spokesperson for Shell Canada, informed me. "Last year, Shell spent $20 million establishing hospitals, schools, educational programs and scholarships" (MOSOP put the figure at closer to $9 million, and says only a fraction of this amount was spent on Ogoni land). The company has also, according to Blakely, revised its "statement of business principles. These principles, which include the company's environmental performance as well as its responsibilities to the communities where we operate, apply to all companies in the Shell Group in all parts of the world."[36]

To arrive at these principles, Shell has looked deep into its corporate psyche and has focus grouped and deconstructed itself into a pulp. It has put its employees through a kind of New Age–consultancy boot camp, resulting in some awfully silly displays from such a grand old firm. In the interest of reinvention, Shell executives, according to *Fortune* magazine, have "helped each other climb walls in the freezing Dutch rain. They've dug dirt at low-income housing projects and made videotapes of themselves walking around blindfolded. They've tracked their time to figure out whether they're adding any value. They've even taken Myers-Briggs personality tests to see who fits in at the new Shell and who doesn't."[37]

Part of Shell's image overhaul has involved reaching out to black communities in Europe and North America, a strategy that has created bitter divisions in poor neighborhoods that are desperate for funding but suspicious of Shell's motives. For instance, in August 1997, the Oakland School Board in California hotly debated the ethics of accepting a donation from Shell worth $2 million – $100,000 for scholarships and the rest for the creation of a Shell Youth Training Academy. Since Oakland has a large African-American population that includes exiled Nigerians, the debate was wrenching. "Children in Nigeria don't have an opportunity to get a scholarship from Shell," said Tunde Okorodudu, an Oakland parent and a Nigerian pro-democracy activist. "We really need money for the children but we don't want blood money."[38] After months of stalemate, the board (like the Portland School Board that debated whether or not to accept Nike's donation) eventually voted to accept the money.

But even as the new Shell goes Zen, tossing around trendy management terms like the "new ethical paradigm," "change agents," the "third bottom line," and the "stakeholder economy," and even as Shell Nigeria speaks of "healing the wounds," the old Shell remains.[39] Although it has not yet succeeded in returning to Ogoni land, Shell continues to operate in other parts of the Niger Delta, and in the fall of 1998 tensions in the area once again erupted. The issues were all too familiar: communities complained of polluted lands, devastated fisheries, gas fires and flaring, and of seeing enormous profits pumped out of their oil-rich land while they continued to live in poverty. "You go to the flow stations, you see they are very well equipped, with all modern facilities. You go to the neighboring village, there is no water to drink, no food to eat. That is bringing about the protests," explained Paul Orieware, a local politician.[40] Only this time, Shell was up against foes far less committed to nonviolence than the Ogoni. In October, Nigerian protestors seized two Shell helicopters, nine Shell relay stations and a drilling rig, halting, according to Associated Press, "the transfer of some 250,000 barrels of crude a day."[41] More Shell stations were stormed and occupied in March 1999. Shell denied any wrongdoing and blamed the violence on ethnic conflicts.

The Arches: The Fight for Choice

At the same time as the anti-Shell campaigns were breaking out, the McLibel Trial, which had been in the docket for a few years already, was turning into an international situation. In June 1995, the trial was coming up to its first anniversary in court, when the two defendants, Helen Steel and Dave Morris, held a press conference outside the London courthouse. They announced that McDonald's (which had sued them for libel) had made a settlement offer. The company offered to donate money to a cause of Steel and Morris's choice if the two outspoken environmentalists on trial would stop criticizing McDonald's; then everyone would leave the whole messy nightmare behind them.

Steel and Morris defiantly refused the offer. They saw no reason to give in now. The trial, which had been designed to stem the flow of negative publicity – and to gag and bankrupt Steel and Morris – had been an epic public-relations disaster for McDonald's. It had done almost as much as mad cow disease to promote vegetarianism, had certainly done more to raise the issue of labor conditions in the McJob sector than any union drive and had sparked a more profound debate about corporate censorship than any other free-speech case in recent memory.

The pamphlet at the center of the suit was first published in 1986 by London Greenpeace, a splinter group of Greenpeace International (which the hard-core Londoners deemed too centralized and mainstream for their tastes). It was an early case study in using a single brand name to connect all the dots on the social agenda: issues of rain-forest depletion (to raise the cattle), Third World poverty (forcing peasants off their farms to make way for export crops and McDonald's livestock needs), animal cruelty (in treatment of the livestock), waste production (disposable packaging and litter), health (fried fatty foods), poor labor conditions (low wages and union busting in the McJob sector) and exploitative advertising (in McDonald's target marketing to children).

But the truth is, McLibel was never really about the contents of the pamphlet. In many ways, the case against McDonald's is less compelling than the ones against Nike and Shell, both of which are supported by hard evidence of large-scale human suffering. With McDonald's the evidence was less direct and, in some ways, the issues more dated. The concern about litter-producing fast-food restaurants reached its peak in the late eighties and London Greenpeace's campaign against the company clearly came from the standpoint of meat-is-murder vegetarianism: a valid perspective, but one for which there is a limited political constituency. What made McLibel take off as a campaign on a par with the ones targeting Nike and Shell was not what the fast-food chain did to cows, forests or even its own workers. The McLibel movement took off because of what McDonald's did to Helen Steel and Dave Morris.

Franny Armstrong, who produced a documentary about the trial, points out that Britain's libel law was changed in 1993 "so that governmental bodies such as local councils are no longer able to sue for libel. This was to protect people's right to criticize public bodies. Multinationals are fast becoming more powerful than governments – and even less accountable – so shouldn't the same rules apply? With advertising budgets in the billions, it's not as though they need to turn to the law to ensure their point of view is heard."[42] In other words, for many of its supporters, Steel and Morris's case was less about the merits of fast food than about the need to protect freedom of speech in a climate of mounting corporate control. If Brent Spar was about loss of space, and Nike was about the loss of good jobs, McLibel was about loss of voice – it was about corporate censorship.

When McDonald's issued libel writs against five Greenpeace activists in 1990 over the contents of the now-notorious leaflet, three members of the group did what most people would do when faced with the prospect of going up against an $11 billion corporation: they apologized. The company had a long and successful history with this strategy. According to *The Guardian*: "Over the past 15 years, McDonald's has threatened legal action against more than 90 organizations in the U.K., including the BBC, Channel 4, the *Guardian*, the *Sun*, the Scottish TUC, the New Leaf Tea Shop, student newspapers and a children's theatre group. Even Prince Philip received a stiff letter. All of them backed down and many formally apologised in court."[43]

But Helen Steel and Dave Morris made another choice. They used the trial to launch a seven-year experiment in riding the golden arches around the global economy. For 313 days in court – the longest trial in English history – an unemployed postal worker (Morris) and a community gardener (Steel) went to war with chief executives from the largest food empire in the world.

Over the course of the trial, Steel and Morris meticulously elaborated every one of the pamphlet's claims, with the assistance of nutritional and environmental experts and scientific studies. With 180 witnesses called to the stand, the company endured humiliation after humiliation as the court heard stories of food poisoning, failure to pay legal overtime, bogus recycling claims and corporate spies sent to infiltrate the ranks of London Greenpeace. In one particularly telling incident, McDonald's executives were challenged on the company's claim that it serves "nutritious food": David Green, senior vice president of marketing, expressed his opinion that Coca-Cola is nutritious because it is "providing water, and I think that is part of a balanced diet."[44] In another embarrassing exchange, McDonald's executive Ed Oakley explained to Steel that the McDonald's garbage stuffed into landfills is "a benefit, otherwise you will end up with lots of vast empty gravel pits all over the country."[45]

On June 19, 1997, the judge finally handed down the verdict. The courtroom was packed with an odd assortment of corporate executives, pink-haired vegan anarchists and rows of journalists. It felt like an eternity to most of us sitting there, as Judge Rodger Bell read out his forty-five-page ruling – a summary of the actual verdict, which was over a thousand pages long. Although the judge deemed most of the pamphlet's claims too hyperbolic to be acceptable (he was particularly unconvinced by its direct linking of McDonald's to "hunger in the 'Third World'"), he deemed others to be based on pure fact. Among the decisions that went in Steel and Morris's favor were that McDonald's "exploit(s) children" by "using them, as more susceptible subjects of advertising"; that its treatment of some animals has been "cruel"; that it is anti-union and pays "low wages"; that its management can be "autocratic" and "most unfair"; and that a consistent diet of McDonald's food contributes to the risk of heart disease. Steel and Morris were ordered to pay damages to McDonald's in the amount of US$95,490. But in March 1999 an appeals court judge found that Judge Bell had been overly harsh and sided more forcefully with Steel and Morris on the claims "concerning nutrition and health risks and on the allegations about pay and conditions for McDonald's employees." Still finding that their claims about food poisoning, cancer and world poverty were unproven, the court nonetheless lowered the amount of damages to $61,300.[46] McDonald's has never tried to collect its settlement and has decided not to apply for an injunction to halt the further dissemination of the leaflet.

After the first verdict, McDonald's was quick to declare victory, but few were convinced. "Not since Pyrrhus has a victor emerged so bedraggled," read *The Guardian*'s editorial the next day. "As P.R. fiascos go, this action takes the prize for ill-judged and disproportionate response to public criticism."[47] In fact, while all this was going on, the original pamphlet had gathered the cachet of a collector's item, with three million copies distributed in the U.K. alone. John Vidal had published his critically acclaimed book *McLibel: Burger Culture on Trial*; *60 Minutes* had produced a lengthy segment about the trial; England's Channel 4 had run a three-hour dramatization of it; and Franny Armstrong's documentary *McLibel: Two Worlds Collide* had made the rounds of the independent film circuit (having been turned down by every major broadcaster because of – ironically – libel concerns).

For Helen Steel, Dave Morris and their supporters, McLibel was never solely about winning in court – it was about using the courts to win over the public. And judging by the crowds outside the McDonald's outlets two days after the verdict came down, they had every right to be declaring victory. Standing outside their neighborhood McDonald's in North London on a Saturday afternoon, Steel and Morris could barely keep up with the demand for "What's wrong with McDonald's?" the leaflet that started it all. Passersby requested copies, drivers pulled over to get their McLibel mementos and mothers with toddlers stopped to talk to Helen Steel about how difficult it can be for a busy parent when her child demands unhealthy food – what can a mother do?

Across the United Kingdom, a similar scene was playing itself out at more than five hundred McDonald's outlets, all of which were simultaneously picketed on June 21, 1997, along with thirty in North America. As with the Nike protests, every event was different. At one British franchise, the community put on a street performance featuring an ax-wielding Ronald McDonald, a cow and lots of ketchup. At another, people passed out free vegetarian food. At all of them, supporters handed out the famous leaflet: 400,000 copies that weekend alone. "They were flying out of their hands," said Dan Mills of the McLibel Support Campaign, amused at the irony: before McDonald's decided to sue, London Greenpeace's campaign was winding down, and only a few hundred copies of the contentious leaflet had ever been distributed. It has now been translated into twenty-six languages and is one of the hottest properties in cyberspace.

Lessons of the Big Three: Use the Courts as a Tool

It's a good bet that many brand-name giants besides McDonald's have paid close attention to the goings-on in that British courtroom. In 1996, Guess dropped a libel suit it had launched against the L.A. women's group Common Threads, in response to a poetry reading about the plight of garment workers sewing Guess jeans.[48] Similarly, though Nike consistently accuses its critics of fabrication, it has stayed away from trying to clear its name in court. And no wonder: the courtroom is the only place where private corporations are forced to open shuttered windows and let the public look in.

As Helen Steel and Dave Morris write,

> If companies do choose to use oppressive laws against their critics then court cases do not have to only be about legal procedures and verdicts. They can be turned into a public forum and focus for protest, and for the wider dissemination of the truth. This is what happened with McLibel... Maybe for the first time in history, a powerful institution (it just happened to be a fast-food chain, but in some ways could've been any financial organization or state department) was subject to lengthy, detailed and critical public scrutiny. That can only be a good thing![49]

The message has not been lost on Steel and Morris's fellow activists around the world; everyone who followed McLibel saw how effective a long, dramatic trial could be at building up a body of evidence and stoking sentiment against a corporate opponent. Some campaigners, not waiting to be sued themselves, are taking their corporate opponents to court instead. For instance, in January 1999, when U.S. labor activists decided they wanted to draw attention to the ongoing sweatshop violations in the U.S. territory of Saipan, they launched an unconventional lawsuit in California court against seventeen American retailers, including the Gap and Tommy Hilfiger. The suit, filed on behalf of thousands of Saipan garment workers, accuses the brand-name retailers and manufacturers of participating in a "racketeering conspiracy" in which young women from Southeast Asia are lured to Saipan with promises of well-paid jobs in the United States. What they get instead is wage cheating and "America's worst sweatshop," in the words of Al Meyerhoff, lead attorney on the case. A companion lawsuit further alleges that by labeling goods from Saipan "Made in the U.S.A." or "Made in the Northern Marianas, U.S.A.," the companies are engaging in false advertising, leaving customers with the impression that the manufacturers were subject to U.S. labor laws, when they were not.[50]

Meanwhile, the Center for Constitutional Rights has taken a similar tack with Royal Dutch/Shell, filing a federal lawsuit against the company in a New York court on the first anniversary of Ken Saro-Wiwa's death. According to the Center's David A. Love, "The suit – filed on behalf of Ken Saro-Wiwa and the other Ogoni activists who were executed by Nigeria's military regime in November 1995 – alleges that the executions were carried out with 'the knowledge, consent, and/or support' of Shell Oil." It further alleges that the hangings were part of a conspiracy "to violently and ruthlessly suppress any opposition to Royal Dutch/Shell's conduct in its exploitation of oil and natural gas resources in Ogoni and in the Niger Delta." Shell denies the charges and is challenging the legitimacy of the suit. At the time of writing, neither the Saipan case nor the Shell case had been settled.[51]

Lessons of the Big Three: Use the Net to Shine a McSpotlight

If the courts are becoming a popular tool to pry open closed corporations, it is the Internet that has rapidly become the tool of choice for spreading information about multinationals around the world. All three of the campaigns described in this chapter have distinguished themselves by a pioneering use of information technologies, an approach that continues to unnerve their corporate targets.

Each day, information about Nike flows freely via E-mail between the U.S. National Labor Committee and Campaign for Labor Rights; the Dutch-based Clean Clothes Campaign; the Australian Fairwear Campaign; the Hong Kong–based Asian Monitoring and Resource Centre; the British Labour Behind the Label Coalition and Christian Aid; the French Agir Ici and Artisans du Monde; the German Werkstatt Okonomie; the Belgian Les Magasins de Monde; and the Canadian Maquila Solidarity Network – to name but a few of the players. In a September 1997 press release, Nike attacked its critics as "fringe groups, which are again using the Internet and fax modems to promote mistruths and distortions for their own purposes." But by March 1998, Nike was ready to treat its on-line critics with a little more respect. In explaining why it had just introduced yet another package of labor reforms, company spokesman Vada Manager said, "You make changes because it's the right thing to do. But obviously our actions have clearly been accelerated because of the World Wide Web."[52]

Shell was similarly humbled by the mobility of both the Brent Spar campaign and the Ogoni support movement. Natural-resource companies had grown accustomed to dealing with activists who could not escape the confines of their nationhood: a pipeline or mine could spark a peasants' revolt in the Philippines or the Congo, but it would remain contained, reported only by the local media and known only to people in the area. But today, every time Shell sneezes, a report goes out on the hyperactive "shell-nigeria-action" listserve, bouncing into the in-boxes of all the far-flung organizers involved in the campaign, from Nigerian leaders living in exile to student activists around the world. And when a group of activists occupied part of Shell's U.K. headquarters in January 1999, they made sure to bring a digital camera with a cellular linkup, allowing them to broadcast their sit-in on the Web, even after Shell officials turned off the electricity and phones.

Shell has responded to the rise of Net activism with an aggressive Internet strategy of its own: in 1996, it hired Simon May, a twenty-nine-year-old "Internet manager." According to May, "There has been a shift in the balance of power, activists are no longer entirely dependent on the existing media. Shell learned it the hard way with the Brent Spar, when a lot of information was disseminated outside the regular channels."[53] But if the power balance has shifted, it is May's job to shift it back in Shell's favor: he oversees the monitoring of all on-line mentions of the company, responds to E-mail queries about social issues and has helped to establish Shell's on-line "social concerns" discussion forum on the company Web site.

Big companies are big targets... Thousands of companies are or have been the targets of anti-corporate activism online. With WeberWorks/Monitor powered by eWatch, not only will WeberWorks/Monitor clients be alerted when they become a target, they will also receive critical insights for how to effectively handle the situation.

–James M. Alexander, president of eWatch an Internet monitoring company, May 1998

The Internet played a similar role during the McLibel Trial, catapulting London's grassroots anti-McDonald's movement into an arena as global as the one in which its multinational opponent operates. "We had so much information about McDonald's, we thought we should start a library," Dave Morris explains, and with this in mind, a group of Internet activists launched the McSpotlight Web site. The site not only has the controversial pamphlet on-line, it contains the complete 20,000-page transcript of the trial, and offers a debating room where McDonald's workers can exchange horror stories about

McWork under the Golden Arches. The site, one of the most popular destinations on the Web, has been accessed approximately sixty-five million times.[54]

Ben, one of the studiously low-profile programmers for McSpotlight told me that "this is a medium that doesn't require campaigners to jump through hoops doing publicity stunts, or depend on the good will of an editor to get their message across."[55] It's also less vulnerable to libel suits than more traditional media. Ben explains that while McSpotlight's server is located in the Netherlands, it has "mirror sites" in Finland, the U.S., New Zealand and Australia. That means that if a server in one country is targeted by McDonald's lawyers, the site will still be available around the world from the other mirrors. In the meantime, everyone visiting the site is invited to give their opinion on whether or not McSpotlight will get sued. "Is McSpotlight next in court? Click on yes or no."

Once again, the broader corporate world is scrambling to learn the lessons of these campaigns. Speaking in Brussels at a June 1998 conference on the growing power of anticorporate groups, Peter Verhille of the PR firm Entente International noted that "one of the major strengths of pressure groups – in fact the leveling factor in their confrontation with powerful companies – is their ability to exploit the instruments of the telecommunication revolution. Their agile use of global tools such as the Internet reduces the advantage that corporate budgets once provided."[56] Indeed, the beauty of the Net for activists is that it allows coordinated international actions with minimal resources and bureaucracy. For instance, for the International Nike Days of Action, local activists simply download information pamphlets from the Campaign for Labor Rights Web site to hand out at their protests, then file detailed E-mail reports from Sweden, Australia, the U.S. and Canada, which are then forwarded to all participating groups.

A similar electronic clearinghouse model was used to coordinate both the Reclaim the Streets global street parties and the picketing outside McDonald's outlets after the McLibel verdict. The McSpotlight programmers posted a list of all 793 McDonald's franchises in Britain and in the weeks before the verdict came down, local activists signed up to "adopt a store (and teach it some respect)" on the day of protest. More than half were adopted. I had been following all of this closely from Canada, but when I finally got a chance to see the London headquarters of the McLibel Support Campaign – the hub from which hundreds of political actions had been launched around the world, linking up thousands of protestors and becoming a living archive for all things anti-McDonald's – I was shocked. In my mind, I had pictured an office crammed with people tapping away on high-tech equipment. I should have known better: McLibel's head office is nothing more than a tiny room at the back of a London flat with graffiti in the stairwells. The office walls are papered in subvertisements and anarchist agitprop. Helen Steel, Dave Morris, Dan Mills and a few dozen volunteers had gone head to head with McDonald's for seven years with a rickety PC, an old modem, one telephone and a fax machine. Dan Mills apologized to me for the absence of an extra chair.

Tony Juniper of Britain's environmental group Friends of the Earth calls the Internet "the most potent weapon in the toolbox of resistance."[57] That may well be so, but the Net is more than an organizing tool – it has become an organizing model, a blueprint for decentralized but cooperative decision making. It facilitates the process of information sharing to such a degree that many groups can work in concert with one another without the need to achieve monolithic consensus (which is often impossible, anyway, given the nature of activist organizations). And because it is so decentralized, these movements are still in the process of forging links with their various wings around the world, continually surprising themselves with how far unreported little victories have traveled, how thoroughly bits of research have been recycled and absorbed. These movements are only now starting to feel their own reach and, as the students and local communities profiled in the next chapter will show, their own power.

CHAPTER SEVENTEEN

LOCAL FOREIGN POLICY

Students and Communities Join the Fray

Pretty soon we'll have to do our own offshore drilling.

— Berkeley, California, city councilor Polly Armstrong on her council's decision to outlaw municipal gasoline purchases from all the major oil companies

"Okay. I need people on each door. Let's go!" shouted Sean Hayes in the distinctive clipped baritone of a high-school basketball coach, which, as it happens, he is. "Let's go!" Coach Hayes bellowed again, clapping his meaty hands loud enough for the sound to bounce off the walls of the huge gymnasium of St. Mary's Secondary School in Pickering, Ontario (a town best known for its proximity to a nuclear power plant of questionable quality).

Hayes had invited me to participate in the school's first "Sweatshop Fashion Show," an event he began planning when he discovered that the basketball team's made-in-Indonesia Nike sneakers had likely been manufactured under sweatshop conditions. He's an unapologetic jock with a conscience and, together with a handful of do-gooder students, had organized today's event to get the other two thousand kids at St. Mary's to think about the clothes they wear in terms beyond "cool" or "lame."

The plan was simple: as student models decked out in logowear strutted down a makeshift runway, another student off to the side would read a prepared narration about the lives of the Third World workers who made the gear. The students would quickly follow that with scenes from *Mickey Mouse Goes to Haiti* and a skit about how teenagers often feel "unloved, unwanted, unacceptable and unpopular if you do not have the right clothes." My part would come at the end, when I was to give a short speech about my research in export processing zones, and then facilitate a question-and-answer period. It sounded straightforward enough.

While we were waiting for the bell to ring and the students to stream in, Hayes turned to me and said, with a forced smile: "I hope the kids actually hear the message and don't think it's just a regular fashion show." Having read the students' prepared narration I couldn't help thinking that his concern sounded, frankly, paranoid. True, fashion shows have become such a high-school stalwart that they now rival car washes as the prom fundraiser of choice. But did Hayes actually think his students were so heartless that they could listen to testimony about starvation wages and physical abuse and expect that the clothing in question would be on sale at a discount after the assembly? Just then, a couple of teenage boys poked their heads in the door and checked out the frantic preparations. "Yo, guys," one of them said. "I'm guessing fashion show – this should be a joke." Coach Hayes looked nervous.

As two thousand students piled onto the bleachers, the room came alive with the giddiness that accompanies all mass reprieves from class, whether for school plays, AIDS education lectures, teachers' strikes or fire alarms. A quick scan of the room turned up no logos on these kids, but that was definitely not by choice. St. Mary's is a Catholic school and the students wear uniforms – bland affairs that they were nonetheless working for all they were worth. It's hard to make gray flannel slacks and acrylic navy sweaters look like gangsta gear but the guys were doing their best, wearing their pants pulled down halfway to their knees with patterned boxer shorts bunched over their belts. The girls were pushing the envelope too, pairing their drab tunics with platform loafers and black lipstick.

As it turned out, Coach Hayes's concerns were well founded. As the hip-hop started playing and the first kids bounded down the runway in Nike shoes and workout wear, the assembly broke into cheers and applause. The moment the young woman saddled with reading the earnest voice-over began, "Welcome to the world of Nike…" she was drowned out by hoots and whistles. It didn't take much to figure out that they weren't cheering for her but rather at the mere mention of the word Nike – everyone's favorite celebrity brand.

Waiting for my cue, I was ready to flee the modern teenage world forever, but after some booming threats from Coach Hayes, the crowd finally quieted down. My speech was at least not booed and the discussion that followed was among the liveliest I've ever witnessed. The first question (as at all Sweatshop 101 events) was "What brands are sweatshop-free?" – Adidas? they asked. Reebok? The Gap? I told the St. Mary's students that shopping for an exploitation-free wardrobe at the mall is next to impossible, given the way all the large brands produce. The best way to make a difference, I told them, is to stay informed by surfing the Net, and by letting companies know what you think by writing letters and asking lots of questions at the store. The St. Mary's kids were deeply skeptical of this non-answer. "Look, I don't have time to be some kind of major political activist every time I go to the mall," one girl said, right hand planted firmly on right hip. "Just tell me what kind of shoes are okay to buy, okay?"

Another girl, who looked about sixteen, sashayed to the microphone. "I'd just like to say that this is capitalism, okay, and people are allowed to make money and if you don't like it maybe you're just jealous."

The hands shot up in response. "No, *I'd* just like to say that you are totally screwed up and just because everyone is doing something doesn't mean it's right – you've got to stand up for what you believe in instead of just standing in front of the mirror trying to look good!"

After watching thousands of Ricki and Oprah episodes, these kids take to the talk-show format as naturally as Elizabeth Dole. Just as they had cheered for Nike moments before, the students now cheered for each other – dog-pound style, with lots of "you-go-girls." Moments before the bell for next period, Coach Hayes made time for one last question. A boy in saggy slacks sauntered across the gym holding his standard-issue navy blue sweater away from his lanky body with two fingers, as if he detected a foul odor. Then, he slouched down to the mike and said, in an impeccable teenage monotone, "Umm, Coach Hayes, if working conditions are so bad in Indonesia, then why do we have to wear these uniforms? We buy thousands of these things and it says right here that they are 'Made in Indonesia.' I'd just like to know, how do you know they weren't made in sweatshops?"

The auditorium exploded. It was a serious burn. Another student rushed to the mike and suggested that the students should try to find out who makes their uniforms, a project for which there was no shortage of volunteers. When I left St. Mary's that day, the school had its work cut out for it.

There's no denying that the motivation behind the St. Mary's students' new-found concern over Indonesian labor conditions was that they had just discovered a high-minded excuse to refuse to wear their lame-ass uniforms – not an entirely selfless concern. But even if it was inadvertent, they had also stumbled across one of the most powerful levers being used to pry reform out of seemingly amoral multinational corporations.

When high schools, universities, places of worship, unions, city councils and other levels of government apply ethical standards to their bulk purchasing decisions, it takes anticorporate campaigning a significant step beyond the mostly symbolic warfare of adbusting and superstore protesting. Such community institutions are not only collections of individual consumers, they are also consumers themselves – and powerful ones at that. Thousands of schools like St. Mary's ordering thousands of uniforms each – it adds up to a lot of uniforms. They also buy sports equipment for their teams, food for their cafeterias and drinks for their vending machines. Municipal governments buy uniforms for their police forces, gas for their garbage trucks and computers for their offices; and they also invest their pension funds on the stock market. Universities, for their part, select telecommunications companies for their Internet portals, use banks to hold their money and invest endowments that can represent billions of dollars. And, of course, they are also increasingly involved in direct sponsorship arrangements with corporations. Most important, bulk institutional purchases and sponsorship deals are among the most sought after contracts in the marketplace, and corporations are forever trying to outbid one another to land them.

What all these business arrangements have in common is that they exist at a distinctive intersection between civic life (ostensibly governed by principles of "public good") and the corporate profit-making motive. When corporations sponsor an event on a university campus or sign a deal with a municipal government, they cross an important line between private and public space – a line that is not part of a consumer's interaction with a corporation as an individual shopper. We don't expect morality at the mall but, to some extent, we do still expect it in our public spaces – in our schools, national parks and municipal playgrounds.

So while it may be cold comfort to some, there is a positive side effect of the fact that, increasingly, private corporations are staking a claim to these public spaces. Over the past four years, there has been a collective realization among many public, civic and religious institutions that having a multinational corporation as a guest in your house – whether as a supplier or a sponsor – presents an important political opportunity. With their huge buying power, public and non-profit institutions can exert real public-interest pressure on otherwise freewheeling private corporations. This is nowhere more true than in the schools and universities.

Students Teach the Brands a Lesson

As we have already seen, soft-drink, sneaker and fast-food companies have been forging a flurry of exclusive logo allegiances with high schools, colleges and universities. Like the Olympic games, many universities have "official" airlines, banks, long-distance carriers and computer suppliers. For the sponsoring companies, these exclusive arrangements offer opportunities to foster warm and fuzzy logo loyalties during those formative college years – not to mention a chance to pick up some quasi-academic legitimacy. (Being the official supplier of a top-flight university sounds almost as if a panel of tenured professors got together and scientifically determined that Coke Is It! or Our Fries Are Crispier! For some lucky corporations, it can be like getting an honorary degree.)

However, these same corporations have at times discovered that there can be an unanticipated downside to these "partnerships": that the sense of ownership that goes along with sponsoring is not always the kind of passive consumer allegiance that the companies had bargained for. In a climate of mounting concern about corporate ethics, students are finding that a great way to grab the attention of aloof multinationals is to kick up a fuss about the extracurricular activities of their university's official brand – whether Coke, Pepsi, Nike, McDonald's, Starbucks or Northern Telecom. Rather than simply complaining about amorphous "corporatization," young activists have begun to use their status as sought-after sponsorees to retaliate against forces they considered invasive on their campuses to begin with. In this volatile context, a particularly aggressive sponsorship deal can act as a political catalyst, instigating wide-ranging debate on everything from unfair labor conditions to trading with dictators. Just ask Pepsi.

Pepsi (as we saw in Chapter 4) has been at the forefront of the drive to purchase students as a captive market. Its exclusive vending arrangements have paved the way for copycat deals, and fast-food outlets owned by PepsiCo were among the first to establish a presence in high schools and on university campuses in North America. One of Pepsi's first campus vending deals was with Ottawa's Carleton University in 1993. Since marketing on campus was still somewhat jarring back then, many students were immediately resentful at being forced into this tacit product endorsement, and were determined not to give their official drink a warm welcome. Members of the university's chapter of the Public Interest and Research Group – a network of campus social-justice organizations stretching across North America known as PIRGs – discovered that PepsiCo was producing and selling its soft drinks in Burma, the brutal dictatorship now called Myanmar. The Carleton students weren't sure how to deal with the information, so they posted a notice about Pepsi's involvement in Burma on a few on-line bulletin boards that covered student issues. Gradually, other universities where Pepsi was the official drink started requesting more information. Pretty soon, the Ottawa group had developed and distributed hundreds of "campus action kits," with pamphlets, petitions, and "Gotta Boycott" and "Pepsi, Stuff It" stickers. "How can you help free Burma?" one pamphlet asks. "Pressure schools to terminate food or beverage contracts selling PepsiCo products until it leaves Burma."

Many students did just that. As a result, in April 1996 Harvard rejected a proposed $1 million vending deal with Pepsi, citing the company's Burma holdings. Stanford University cost Pepsi an estimated $800,000 when a petition signed by two thousand students blocked the construction of a PepsiCo-owned Taco Bell restaurant. The stakes were even higher in Britain where campus soft-drink contracts are coordinated centrally through the National Union of Students' services wing. "Pepsi had just beat out Coke for the contract," recalls Guy Hughes, a campaigner with the London-based group Third World First. "Pepsi was being sold in eight hundred student unions across the U.K., so we used the consortium as a lever to pressure Pepsi. When [the student union] met with the company, one factor for Pepsi was that the boycott had become international."[1]

Aung San Suu Kyi, the leader of Burma's opposition party that was elected to power in 1990, only to be prevented from taking office by the military, has offered encouragement to this nascent movement. In 1997, in a speech read by her husband (who has since died) at the American University in Washington, D.C., she singled out students in the call to put pressure on multinational corporations that are invested in Burma. "Please use your liberty to promote ours," she said. "Take a principled stand against companies which are doing business with the military regime of Burma."[2]

After the campus boycotts made it into *The New York Times*, Pepsi sold its shares in a controversial Burmese bottling plant whose owner, Thien Tun, had publicly called for Suu Kyi's democracy movement to be "ostracized and crushed." Student activists, however, dismissed the move as a "paper shuffle" because Pepsi products were still being sold and produced in Burma. Finally, facing continued pressure, Pepsi announced its "total disengagement" from Burma on January 24, 1997. When Zar Ni, the coordinator of the American student movement, heard the news, he sent an E-mail out on the Free Burma Coalition listserve: "We finally tied the Pepsi Animal down! We did it!! We all did it!!!... We now KNOW we have the grassroots power to yank one of the most powerful corporations in the world."

If there is a moral to this story, it is that Pepsi's drive to capture the campus market landed the company at the center of a debate in which it had no desire to participate. It wanted university students to be its poster children – its real live Generation Next – but instead, the students turned the tables and made Pepsi the poster corporation for their campus Free Burma movement. Sein Win, a leader in exile of Burma's elected National League for Democracy, observed that "PepsiCo very much takes care of its image. It wanted to press the drink's image as 'the taste of a young generation,' so when the young generation participates in boycotts, it hurts the effort."[3] Simon Billenness, an ethical investment specialist who spearheaded the

Burma campaign, is more blunt: "Pepsi," he says, "was under siege from its own target market."[4] And Reid Cooper, coordinator of the Carleton University campaign, notes that without Pepsi's thirst for campus branding, Burma's plight might never have become an issue on campuses. "Pepsi tried to go into the schools," he tells me in an interview, "and from there it was spontaneous combustion."

Not surprisingly, the Pepsi victory has emboldened the Free Burma campaign on the campuses. The students have adopted the slogan "Burma: South Africa of the Nineties" and claim to be "The largest human rights campaign in cyberspace."[5] Today, more than one hundred colleges and twenty high schools around the world are part of the Free Burma Coalition. The extent to which the country's liberty has become a student cause célèbre became apparent when, in August 1998, eighteen foreign activists – most of them university students – were arrested in Rangoon for handing out leaflets expressing support for Burma's democracy movement. Not surprisingly, the event caught the attention of the international media. The court sentenced the activists to five years of hard labor, but at the last minute deported them instead of imprisoning them.

Other student campaigns have focused on different corporations and different dictators. With Pepsi out of Burma, attention began to shift on campuses to Coca-Cola's investments in Nigeria. At Kent State University and other schools where Coke won the campus cola war, students argued that Coke's high-profile presence in Nigeria offered an air of legitimacy to the country's illegitimate military regime (which, at the time, was still in power). Once again, the issue of Nigerian human rights might never have reached much beyond KSU's Amnesty International Club, but because Coke and the school had entered into a sponsorship-style arrangement, the campaign took off and students began shouting that their university had blood on its hands.

There have also been a number of food fights, most of them related to McDonald's expanding presence on college campuses. In 1997, the British National Union of Students entered into an agreement with McDonald's to distribute "privilege cards" to all undergraduates in the U.K. When students showed the card, they got a free cheeseburger every time they ordered a Big

Mac, fries and drink. But campus environmentalists opposed the deal, forcing the student association to bow out of the marketing alliance in March 1998. In providing its reasons for the change of heart, the association cited the company's "anti-union practices, exploitation of employees, its contribution to the destruction of the environment, animal cruelty and the promotion of unhealthy food products" – all carefully worded references to the McLibel judge's findings.[6]

As the brand backlash spreads, students are beginning to question not only sponsorship arrangements with the likes of McDonald's and Pepsi, but also the less flashy partnerships that their universities have with the private sector. Whether it's bankers on the board of governors, corporate-endowed professorships or the naming of campus buildings after benefactors, all are facing scrutiny from a more economically politicized student body. British students have stepped up a campaign to pressure their universities to stop accepting grant money from the oil industry, and in British Columbia, the University of Victoria Senate voted in November 1998 to refuse scholarship money from Shell. This agenda of corporate resistance is gradually becoming more structured, as students from across North America come together at annual conferences such as the 1997 "Democracy Teach-In: Campus Democracy vs. Corporate Control" at the University of Chicago, where they attend seminars like "Research: For People or Profit?" "Investigating Your Campus" and "What Is a Corporation and Why Is There a Problem?" In June 1999, student activists again came together, this time in Toledo, Ohio, in the newly formed Student Alliance to Reform Corporations. The purpose of the gathering was to launch a national campaign to force universities to invest their money only with companies that respect human rights and do not degrade the environment.

It should come as no surprise that by far the most controversial campus-corporate partnerships have been ones involving that most controversial of companies: Nike. Since the shoe industry's use of sweatshop labor became common knowledge, the deals that Nike had signed with hundreds of athletic departments in universities have become among the most contentious issues on campuses today, with "Ban the Swoosh" buttons rivaling women's symbols as the undergraduate accessory of choice. And in what Nike must see as the ultimate slap in the face, college campuses where the company has paid out millions of dollars to sponsor sports teams (University of North Carolina, Duke University, Stanford, Penn State and Arizona State, to name just a few) have become the hottest spots of the international anti-Nike campaign. According to the Campaign for Labor Rights, "These contracts, which are a centerpiece of Nike marketing, have now turned into a public relations nightmare for the company. Nike's aggressive campus marketing has now been forced into a defensive posture."[7]

At the University of Arizona, students attempted to get their university president to reconsider the school's endorsement of Nike products by delivering a pile of old Nike shoes to his office (followed by cameras from two local television stations). According to a student organizer, James Tracy, "each pair of shoes had a tale of Nike's abuse attached to them for the president to consider."[8] At Stanford University, similar protests greeted the athletic department's decision to sign a four-year, $5 million contract with Nike. In fact, bashing Nike has become such a popular sport on campus that at Florida State University – a major jock college – a group of students built an anti-Nike float for the 1997 homecoming parade.

Most of these universities are locked into multiyear sponsorship deals with Nike, but at the University of California at Irvine, students went after the company when its contract with the women's basketball team was up for renewal. Faced with mounting pressure from the student body, the school's athletic department decided to switch to Converse. On another campus, soccer coach Kim Keady was unable to persuade his employer, St. John's University, to stop forcing its team to use Nike gear. So, in the summer of 1998, he quit his job as assistant coach in protest.[9]

University of North Carolina student Marion Traub-Werner explains the appeal of this new movement: "Obviously there's the labor issue. But we're also concerned about Nike's intrusion into our campus culture. The swoosh is everywhere – in addition to all the uniforms, it's on the game schedules, it's on all the posters and it dominates the clothing section in the campus store."[10] Like no other company, Nike has branded this generation, and so if students now have the chance to brand Nike as an exploiter – well, the chance is too good to pass up.

The Real Brand U

While many campuses are busily taking on the brand-name interlopers, others are realizing that their universities are themselves brand names. Ivy League universities, and colleges with all-star sports teams, have extensive clothing lines, several of which rival the market share of many commercial designers. They also share many of the same labor problems. In 1998, the UNITE garment workers union published a report on the BJ&B factory in an export processing zone in the Dominican Republic. Workers at BJ&B, one of the world's largest manufacturers of baseball hats, embroider the school logos and crests of at least nine large American universities, including Cornell, Duke, Georgetown, Harvard and University of Michigan. The conditions at BJ&B were signature free-trade-zone ones: long hours of forced overtime, fierce union busting (including layoffs of organizers), short-term contracts, paychecks insufficient to feed a family, pregnancy tests, sexual harassment, abusive management, unsafe drinking water and huge markups (while the hats sold, on average, for $19.95, workers saw only 8 cents of that).[11] And of course, most of the workers were young women, a fact that was brought home when the union sponsored a trip to the U.S. for two former employees of the factory: nineteen-year-old Kenia Rodriguez and twenty-year-old Roselio Reyes. The two workers visited many of the universities whose logos they used to stitch on caps, speaking to gatherings of students who were exactly their age. "In the name of the 2,050 workers in this factory, and the people in this town, we ask for your support," Reyes said to an audience of students at the University of Illinois.[12]

These revelations about factory conditions were hardly surprising. College licensing is big business, and the players – Fruit of the Loom, Champion, Russell – have all shifted to contract factories with the rest of the garment industry, and make liberal use of free-trade zones around the world. In the U.S., the licensing of college names is a $2.5 billion annual industry, much of it brokered through the Collegiate Licensing Company. Duke University alone sells around $25 million worth of clothing associated with its winning basketball team every year. To meet the demand, it has seven hundred licensees who contract to hundreds of plants in the U.S. and in ten other countries.[13]

Because of Duke's leading role as a campus apparel manufacturer, a group of activists decided to turn the school into a model of ethical manufacturing – not only for other schools, but for the scandal-racked garment industry as a whole. In March 1998, Duke University unveiled a landmark policy requiring that all companies making T-shirts, baseball hats and sweatshirts bearing the "Duke" name agree to a set of clear labor standards. The code required that contractors pay the legal minimum wage, maintain safe working conditions and allow workers to form unions, no matter where the factories were located. What makes the policy more substantial than most other codes in the garment sector is that it requires factories to undergo inspections from independent monitors – a provision that sent Nike and Shell screaming from the negotiating table, despite overwhelming evidence that their stated standards are being disregarded on the ground. Brown University followed two months later with a tough code of its own.

Tico Almeida, a senior at Duke University, explains that many students have a powerful reaction when they learn about the workers who produce their team clothing in free-trade zones. "You have two groups of people, roughly the same age, who are getting such different experiences out of the same institutions," he says. And once again, says David Tannenbaum, an undergraduate at Princeton, the logo (this time a school logo) provides the global link. "While the workers are making our clothes thousands of miles away, in other ways we're close to it – we're wearing these clothes every day."[14]

The summer after the Duke and Brown codes were passed was filled with activity. In July, anti-sweatshop organizers from campuses across the country gathered in New York and organized themselves into a coalition, United Students Against Sweatshops. In August, a delegation of eight students, including Tico Almeida, went on a fact-finding mission to free-trade zones in Nicaragua, El Salvador and Honduras. Almeida told me he was hoping to find Duke sweatshirts because he had seen the "Made in Honduras" tag on clothing sold on his campus. But he soon discovered what most people do when they visit free-trade zones: that a potent combination of secrecy, deferred responsibility and militarism forms a protective barricade around much of the global garment industry. "It was like taking random stabs in the dark," he recalls.

When classes resumed in September 1998 and the student travelers were back on campus, the issue of sweatshop labor exploded into what *The New York Times* described as "the biggest surge in campus activism in nearly two decades."[15] At Duke, Georgetown, Wisconsin, North Carolina, Arizona, Michigan, Princeton, Stanford, Harvard, Brown, Cornell and University of California at Berkeley there were conferences, teach-ins, protests and sit-ins – some lasting three and four days. At Yale University, students held a "knit-in." All the demonstrations led to agreements from school administrators to demand higher labor standards from the companies that manufacture their wares.

This fast-growing movement has a somewhat unlikely rallying cry: "Corporate disclosure." The central demand is for the companies that produce college-affiliated clothing to hand over the names and addresses of all their factories around the world and open themselves up to monitoring. Who makes your school clothing, the students say, should not be a mystery. They argue that with the garment industry being the global, contracted-out maze that it is, the onus must be on companies to prove their goods *aren't* made in sweatshops – not on investigative activists to prove that they are. The students are also pushing for their schools to demand that contractors pay a "living wage," as opposed to the legal minimum wage. By May 1999, at least four administrations had agreed in principle to push their suppliers on the living-wage issue. As we will see in the next chapter, there is no agreement about how to turn those well-meaning commitments into real changes in the export factories. Everyone involved in the anti-sweatshop movement does agree, however, that even getting issues like disclosure and a living wage on the negotiating table with manufacturers represents a major victory, one that has eluded campaigners for many years.

In a smaller but equally precedent-setting initiative, Archbishop Theodore McCarrick announced in October 1997 that his Newark, New Jersey, archdiocese would become a "no sweat" zone. The initiative includes introducing an anti-sweatshop curriculum into all 185 Catholic schools in the area, identifying the manufacturers of all their school uniforms and monitoring them to make sure the clothes are being produced under fair labor conditions—just as the students at St. Mary's in Pickering, Ontario, decided to do.

All in all, students have picked up the gauntlet on the sweatshop issue with an enthusiasm that has taken the aging labor movement by storm. United Students Against Sweatshops, after only one year in existence, claimed chapters on a hundred U.S. campuses and a sister network in Canada. Free the Children, young Craig Kielburger's Toronto-based anti-child-labor organization (he was the thirteen-year-old who challenged the Canadian prime minister to review child-labor practices in India) has meanwhile gained strength in high schools and grade schools around the world. Charles Kernaghan, with his "outing" of Kathie Lee Gifford and Mickey Mouse, may have started this wave of labor organizing, but by the end of the 1998–99 academic year, he knew he was no longer driving it. In a letter to the United Students Against Sweatshops, he wrote: "Right now it is your student movement which is leading the way and carrying the heaviest weight in the struggle to end sweatshop abuses and child labor. Your effectiveness is forcing the companies to respond."[16]

Times have changed. As William Cahn writes in his history of the Lawrence Mill sweatshop strike of 1912, "Nearby Harvard University allowed students credit for their midterm examinations if they agreed to serve in the militia against the strikers. 'Insolent, well-fed Harvard men,' the *New York Call* reported, 'parade up and down, their rifles loaded... their bayonets glittering.'"[17] Today, students are squarely on the other side of sweatshop labor disputes: as the target market for everything from Guess jeans to Nike soccer balls and Duke-embossed baseball hats, young people are taking the sweatshop issue personally.

Community Action: Pulling the Selective Purchasing Lever

Since federal governments in North America and Europe have been largely unwilling to impose meaningful sanctions on such documented human-rights violators as Burma, Nigeria, Indonesia and China, preferring instead to "constructively engage" these nations with trade, there has been a move for more local levels of government to step in where the feds have stepped aside. In the U.S., town councils, city councils, school boards and even some individual state governments have been quietly doing an end run around the trade-mission cheerleading that now passes for foreign policy, and drafting their own, local, foreign policy.

Local legislators know that they can't keep multinationals from channeling funds to dictatorships in Nigeria and Burma, and they cannot prevent imports from companies that use child and prison labor in Pakistan and China, but they can do something else. They can collectively refuse to buy goods and services from these companies when they select their business partners for everything from cellular services to Little League soccer balls. The goal of "selective purchasing agreements," as these ethical trading policies are called, is twofold. First, the agreements may lead individual companies to decide that it is not cost-effective to continue to do business under unethical conditions abroad – for instance, if it is going to cost them contracts at home. Second, actions by local governments can put pressure on federal governments to take more principled positions in their foreign policy agendas.

Modeled after similar initiatives during the anti-apartheid years, the current local foreign policy "fad" (as one Republican commentator snidely called it) began, like so many other U.S. social-justice movements, in Berkeley, California.[18] In February 1995, Berkeley's city council passed a resolution barring the purchase of goods or services from companies invested in Burma. Of course such companies could still sell their wares inside Berkeley – just not to municipal agencies, such as the police force or sanitation services. The move set off a domino effect across the country – at last count, twenty-two cities, one county and two states had selective purchasing agreements relating to companies in Burma, and a handful of cities had disqualified purchases from companies with investments in Nigeria.

Though each law has slight variations in wording, the gist is summed up in this one, passed unanimously by the City of Cambridge, Massachusetts, on June 8, 1998:

> WHEREAS
> The city of Cambridge declares the right to measure the moral character of its business partners in determining with whom it seeks to have business relationships; now therefore be it

> RESOLVED
> That as a matter of public policy the City of Cambridge declares that it will not purchase goods, services or commodities from any company or corporation that does business in the nation of Burma...

The most significant move came in June 1996 when the Massachusetts State Legislature passed the Massachusetts Burma Law, making it far more difficult for companies doing business in the dictator-run country to land a government contract in the state. As the influential *Journal of Commerce* noted, "The targets are far from home, but suddenly local governments are showing they can reach around the world."[19]

Another popular purchasing restriction is one that does not broadly target all corporations in particular countries, but rather corporations engaged in a particularly objectionable practice – for instance, the practice of employing sweatshop or child labor. One such case involved the Los Angeles Monroe High School. After reading the *Life* magazine article on the Pakistani soccer-ball industry, a Monroe student, Sharon Paulson, recalled that she and her classmates "ran out during one practice and we were checking all the balls and they all said, 'Made in Pakistan.' That kind of made everything more real. Before, it was something that we read about, but then it was like, 'We won a city championship using these balls!' It gave us something to fight for." What they fought for – and won – was a commitment from the Los Angeles Board of Education to halt an order of balls made in Pakistan, and another from Los Angeles City Council "to investigate the production of soccer balls made in countries using child labor."[20] According to the Investor Responsibility Research Center, "in 1997, some 20 U.S. cities and towns... adopted 'anti-sweatshop' ordinances that require that goods purchased by those city governments – including uniforms for police, fire and public works personnel – be made without sweatshop labor."

Though selective purchasing agreements have been primarily an American phenomenon, they are beginning to pop up elsewhere. The city of St. John's, Newfoundland, passed an anti-sweatshop resolution in June 1998, and a group of kids in Fort McMurray, Alberta, succeeded in getting their city council to pass a resolution banning the use of child-made soccer balls and fireworks on public property. Free Burma resolutions, meanwhile, are reaching even farther – on March 17, 1998, the Marrickville Council in New South Wales, Australia, "voted unanimously to become the first local authority outside the United States to enact a Burma selective purchasing law."[21]

In the past four years, the Berkeley City Council has passed so many boycott resolutions – against companies doing business in Burma, Nigeria, Tibet; companies associated with the arms industry or with nuclear power – that, as Councilor Polly Armstrong joked, "Pretty soon we'll have to do our own offshore drilling."[22] It's true that between the Nigeria and Burma resolutions, and one about the *Exxon Valdez* oil spill, the council is prevented from using every single major oil company and is forced to fuel its ambulances and street-cleaning vehicles with gas from the little-known Golden Gate Petroleum Company. Berkeley banished Pepsi from its municipal vending machines because of its investment in Burma, returned to the company after it cut ties with Rangoon and then decided to boycott Coke because of its involvement in Nigeria.

It may all sound like Alice in Wonderland, but the boycotts do affect the multinationals. They may smile at a tie-dyed college town like Berkeley boycotting everything but hemp paper and Bridgehead coffee, but when rich states like Massachusetts and Vermont get in on the action, the corporate sector is not amused. In May 1999, three more states – Texas, Washington and New York – had Burma laws pending. And it was beginning to cost. For instance, before it pulled out of Burma over the Celine Dion controversy, the telecommunication firm Ericsson lost a major bid to upgrade San Francisco's 911 services because of its business ties with Burma, and Hewlett-Packard is alleged to have lost several large municipal contracts as well.

Understandably, many companies have bowed to the demands of the human-rights campaigners. Since Massachusetts adopted its Burma Law in June 1996, there has been an exodus of big-name multinationals from the dictatorship, including Eastman Kodak, Hewlett-Packard, Philips Electronics, Apple computers and Texaco. But just because these companies decided to give in does not mean they plan to accept these new local roadblocks in international commercial transactions without a fight. As Robert S. Greenberger explains in *The Wall Street Journal*, "Procurement contracts in

California alone, for example, are worth more to some U.S. companies than any business they could secure in many countries, *but they don't want to have to choose*" (italics mine).[23]

Precisely because it forces such a stark choice, many people are convinced that localized foreign policy initiatives are the most effective political tool available for wresting back some control over transnational corporations. "Selective purchasing based on the Burma model," says Danny Kennedy, coordinator of mining lobby group Project Underground, "is our greatest hope."[24]

It's the kind of statement that has come to enrage the business community, which, after being caught off guard by the sudden rise of selective purchasing laws, is determined not to make the same mistake twice. A coalition of companies, including key Burma investors such as Unocal, and Nigeria investors such as Mobil, have banded together under the National Foreign Trade Council to launch an all-out assault on local selective purchasing agreements. In April 1997, the council formed USA*Engage, claiming to represent over 670 corporations and trade associations. Its express purpose is to fight these laws collectively, allowing individual corporations to avoid putting their own practices in the firing line. Frank Kittredge, who is both president of the NFTC and vice chairman of USA*Engage, explains that "a lot of companies are not anxious to be spotlighted as supporters of countries like Iran or Burma. The way to avoid that is to band together in a coalition."[25]

The group argues that foreign policy is a federal matter, and municipal and state governments have no business wading into the fray. To these ends, USA*Engage has developed a "State and Local Sanctions Watch List" to monitor all the towns, cities and states where selective purchasing agreements have passed, as well as communities that are considering passing them and are therefore still vulnerable to outside pressure. Aggressive lobbying by USA*Engage members has already succeeded in squashing a proposed law on Nigeria that was about to be adopted by the State of Maryland (in March 1998); and Unocal (which has not managed to keep its name out of this debate) succeeded in convincing the California state legislature not to adopt a Massachusetts-style Burma law.

The attacks have also come from farther afield. Acting at the behest of multinationals based in Europe, the European Union has launched an official challenge to the Massachusetts Burma Law at the World Trade Organization. At issue is the allegation that the law violates a WTO regulation that prohibits government purchases from being made on "political" grounds.[26] There has even been talk that municipal and state governments in the U.S. could be sued by their own federal government for violating the WTO clause. Though federal legislators categorically deny that is their intention, on August 5, 1998, Congress narrowly defeated a resolution that would have barred the government from using public money for such a court challenge.

While this trade wrangling went on, the multinationals didn't wait around to see if the selective purchasing agreements would survive. In April 1998, the National Foreign Trade Council filed a lawsuit in the federal district court in Boston that challenged the Massachusetts Burma Law as unconstitutional. The NFTC argued that "the Massachusetts Burma Law directly intrudes on the exclusive power of the national government to determine foreign policy, discriminates against companies engaged in foreign commerce, and conflicts with the policies and objectives of the federal statute imposing sanctions on the Union of Myanmar."[27] Though the NFTC succeeded in winning a protective order that concealed the identities of the individual corporations funding the case, it claimed in court that thirty of its members had been affected by the law. And in November 1998, the NFTC won: the court ruled the Massachusetts Burma Law unconstitutional because it "impermissibly infringes on the federal government's power to regulate foreign affairs."[28]

The state has already lost one appeal, but both sides have said they are willing to take the case all the way to the Supreme Court. The NFTC openly acknowledges that the court challenge is an attempt to set a precedent that would effectively stamp out all municipal selective purchasing agreements, as well as campus and school-board bans. "We regard this law suit to be an important test case that will determine the very significant, perplexing and continuing issue concerning the constitutionality of state and local sanctions," said Frank Kittredge.[29]

For their part, proponents of selective purchasing argue that they are not trying to implement their own foreign policy. They say calling these laws "sanctions," as their critics invariably do, is a misnomer because selective purchasing agreements are not regulations placed on businesses, they are simply large-scale consumer pressure. Simon Billenness, the Burma campaigner who has helped draft these pieces of legislation, characterizes them colorfully as "boycotts on steroids."[30] Just as consumers have the right to personal choice in the marketplace, so too do they have the right collectively, whether as schools, town councils or state governments. He also points out that the agreements have a proven track record of meaningful human-rights victories. During the anti-apartheid movement, five U.S. states, nine cities and fifty-nine universities passed resolutions that either barred purchases from companies in South Africa outright, or compelled them to adopt the Sullivan principles. "If USA*Engage had succeeded with their tactics during the apartheid years, Nelson Mandela might still be in prison," says Simon Billenness.[31]

Perhaps most important, the assault on selective purchasing agreements has turned what were campaigns on behalf of citizens in faraway lands into battles for local rights and liberties as well. Billenness, for his part, describes the attempt to criminalize selective purchasing as "a violation of state sovereignty and local democracy."[32] It may also prove a tactical miscalculation. In taking aim at these locally based actions, the NFTC has actively reinforced the very beliefs that led to their enactment in the first place: that corporations have become more powerful than governments; that federal governments have stopped serving the interests of people; and that in the light of these two facts, citizens have no choice but to confront corporate power themselves.

The proposed Multilateral Agreement on Investment would not help matters. The MAI is stalled for the moment, but its supporters have in no way abandoned the project. According to a draft leaked in 1997, selective purchasing agreements could become instantly illegal. The agreement explicitly bans "discrimination" against corporations based on their trade relations with other countries, and states clearly that this clause would override any preexisting laws at all levels of government – including municipalities. Not only that, but multinationals would be granted the legal standing to sue governments directly for any alleged discrimination on this basis. Many now believe that these parts of the MAI will be part of the next round of World Trade Organization negotiations.

In the same way that citizens' groups from around the world mobilized against the MAI in 1998, many such groups have declared themselves ready to resist the business community's frontal assault on selective purchasing. Free Burma campaigners are vowing to "out" the corporations behind the NFTC lawsuit and target them for boycott campaigns. They also point out that local governments can easily carry out their "boycotts on steroids" with or without formal resolutions on the books. The city of Vancouver is a case in point. In 1989, at the tail end of the apartheid boycott, Vancouver passed a selective purchasing resolution that banned Shell gasoline from city-owned vehicles because of the company's controversial dealings in South Africa. Similar resolutions were passed – mostly relating to banks that issued loans to South Africa – by councils in Toronto, Ottawa and Victoria. But Shell Canada decided to take the City of Vancouver to court for discrimination. The case dragged on for nearly five years and in February 1994 the Supreme Court of Canada ruled, by a margin of five to four, in Shell's favor. Judge John Sopinka wrote that the council had indeed discriminated against Shell, and that councilors only had the jurisdiction to make procurement decisions based on concerns for Vancouver residents – not for people in South Africa. The purpose of the Shell boycott, he concluded, "is to affect matters beyond the boundaries of the City without any identifiable benefit to its inhabitants."[33]

Shell got what it wanted: the City of Vancouver's gas contract. But the company's problems were far from over. When Shell again became the subject of international approbation after the hanging of Ken Saro-Wiwa, local Sierra Club activists again began lobbying the Vancouver council to cut its ties to Shell. In light of the Supreme Court ruling, the council could not formally pass another selective purchasing resolution but, coincidentally, on July 8, 1997, it handed over a $6 million contract to fuel the fleet of ambulances and police and fire vehicles for the entire Greater Vancouver Regional District to Shell's competitor, Chevron. It is possible that the city's decision was based solely on the merits of each company's bid, but there is little doubt that the human-rights issue was also a factor. Included in the Greater Vancouver catchment area is the smaller municipality of North Vancouver; less than four months before the contract went to Chevron, North Vancouver councilors had voted unanimously to condemn Shell's behavior in Ogoniland

and to direct its staff not to buy Shell gas. "We have to take a stand on corporations, against the way Shell has raped the Ogoni people," one councilor said at the time.[34] But since the North Vancouver resolution was simply an expression of council's beliefs – with no mention of municipal contracts – Shell could not appeal legally. When the contract went to Chevron, local environmentalists, who had been keeping weekly vigils outside Shell gas stations in Vancouver for over a year, celebrated it as a victory.

But was it a victory? Less than a year later, Bola Oyinbo, a thirty-three-year-old activist who led an occupation of a Chevron oil barge in Nigeria's Ondo State, would be writing the following report:

> Just as we were preparing to leave we saw three helicopters (choppers). They came like eagles, swooping on chickens. We never expected what followed. As the choppers landed one after the other discharging soldiers, what we heard were gunshots and fire. In fact they started shooting commando style at us even before they landed. They shot everywhere. Arulika and Jolly fell. They died instantly. Larry who was near him rushed to his aid, wanting to pick him up, he was also shot. More soldiers came and more shooting followed. Some of my colleagues jumped over board into the Atlantic, others ran into the platform. There was pandemonium. They shot teargas. White men flew all the helicopters... We were defenseless, harmless.[35]

The protest had begun peacefully on May 25, 1998, and it ended three days later in a bloodbath, with two activists dead. The circumstances were eerily similar to those that had prompted Ken Saro-Wiwa's campaign against Shell five years earlier. "Go to Awoye community and see what they have done," Oyinbo writes. "Everything there is dead: mangroves, tropical forests, fish, the freshwater, wildlife etc. All killed by Chevron.... our people complain of 'dead creeks.'" According to Oyinbo, the community attempted on several occasions to negotiate with Chevron, but its executives never showed up at the meetings. The occupation of the moored barge was a last resort, they say, and the only demand was for a formal meeting with Chevron.

Oyinbo and his comrades accuse the company of hiring the soldiers who raided the barge, killing two men and injuring as many as thirty others.

Chevron says it is not responsible for the actions taken by police officers on its rig – they were simply enforcing the law against "pirates." Chevron spokesperson Mike Libbey denies that the company paid the security officers to intervene, though he admits to alerting the authorities and providing transportation to the platform. "We think it is unfortunate that people died, perhaps unnecessarily, but that doesn't change the fact that in order for Chevron to do business in ninety countries around the world, we must cooperate with governments of many kinds," he told reporters.[36] The company has further enraged the community by refusing to pay damages to the families of the deceased – only burial costs. "If they want other compensations, they should write to us and the company may decide to assist them on compassionate grounds," said Deji Haastrup, Chevron's community relations manager.[37] Perhaps fittingly, Chevron's CEO, Ken Derr, is one of the most active members of USA*Engage and its crusade against sanctions and selective purchasing.

Unlike Shell, Chevron has not yet become the subject of an international brand boycott, though there is a growing public awareness about the deaths that took place on May 28. Perhaps because Bola Oyinbo lacks Ken Saro-Wiwa's international connections, the deaths of his two colleagues were at first not even reported outside the Nigerian press. And it is sadly ironic that Chevron has undoubtedly benefited from the fact that activists have made a strategic decision to focus their criticism on Shell, rather than on the Nigerian oil industry as a whole. It points to one of the significant, at times maddening, limitations of brand-based politics.

Top: Craig Kielburger, the teenager who successfully brought child labor to the world's attention, receives an award from Reebok, a company that has been embroiled in several sweatshop scandals. *Bottom:* "Certified Organic," "Recycled" and "Dolphin Friendly." Will "No Sweat" become just another logo for the conscientious consumer?

CHAPTER EIGHTEEN

BEYOND THE BRAND

The Limits of Brand-Based Politics

In this industry, the only reason to change is because someone has got a great cattle prod that keeps jabbing you in the rear end.

— Bud Konheim, president of Nicole Miller Inc. clothing company, September 4, 1997

When Good Things Happen to Bad Brands

In *One World, Ready or Not*, William Greider writes that "focusing on the moral values of particular companies – or their immorality – invites a self-righteous response among readers that is too easy and undeserved. . . . Nike has concocted a particularly sick ideology to sell its shoes – glamorous images of superstar athletes concealing the human brutalities – but why single out Nike or Michael Jordan when the U.S. government itself is implicated in the same sickness?"[1] Greider has a valid point. The conduct of the individual multinationals is simply a by-product of a broader global economic system that has steadily been removing almost all barriers and conditions to trade, investment and outsourcing. If companies make deals with brutal dictators, sell off their factories and pay wages too low to live on, it's because there is nothing in our international trading rules to prevent them. But eliminating the inequalities at the heart of free-market globalization seems a daunting task for most of us mortals. On the other hand, focusing on a Nike or a Shell and possibly changing the behavior of one multinational can open an important door into this complicated and challenging political arena.

The all-star multinationals that have been the focus of this book are the celebrity face of global capitalism, but when they come under close public scrutiny, the entire system is hauled under the microscope as well. This is often a quite conscious strategy on the part of brand-based campaigns. The Campaign for Labor Rights, for instance, openly admits that "when we debate Nike, we are debating the new global economy."[2] Click on the Beyond McDonald's icon on the McSpotlight Web site and you'll learn that "due to its massive public prominence and indisputable arrogance [McDonald's] has simply been used as a symbol of all corporations pursuing their profits at any price." And Stephen Coats, in explaining why he chose to make Starbucks the center of a drive to improve conditions in the Guatemalan coffee industry, said simply: "You have to start somewhere. You start with one company."[3] Even the small-town battles against Wal-Mart take place at least partly on this symbolic level. John Jarvis, a historical preservationist and one of Wal-Mart's most vocal foes, explains that "the good thing about Wal-Mart was that it was big enough, nasty enough, and aggressive enough to make the problem of uncontrolled growth clear."[4]

But when one logo gets all the attention, even when it is being used tactically to illustrate broader issues, others are unquestionably let off the hook. As we have seen, Chevron has been awarded contracts that Shell lost, and Adidas has enjoyed a massive market comeback by imitating Nike's labor and marketing strategies, while sidestepping all of the controversy. The most hypocritical of all is Reebok, which has rushed to capitalize on Nike's controversies by positioning itself as the ethical shoe alternative. "Consumers are looking for what a company stands for," Jo Harlow, Reebok's vice president of marketing, says in relation to Nike's fall from grace.[5] And to make sure that consumers find what they are looking for in Reebok, the company has taken to handing out high-profile Reebok Human Rights Awards to activists who fight against child labor and repressive dictatorships. This is all rather sanctimonious, coming from a company that produces many of its shoes in the very same factories as Nike, and that has seen more than its own share of human-rights violations, though with less attendant publicity.

Gerard Greenfield, whose firsthand research into garment, shoe and toy factories in Asia has been the backbone of dozens of international campaigns, admits that he has become tired of the double standard. He points out that in March 1997, the international community was outraged by a report that a group of women at a Nike factory called Pou Chen in Vietnam were beaten by a supervisor and forced to run laps around the grounds. But, he writes, "less than a month later the same severe punishment was imposed on workers at another Taiwanese-invested shoe factory, Giant V. . . . News of this case was sent out to the groups campaigning on labor practices at Pou Chen. However, despite the similarity in these cases, it was not taken up by human-rights and labor-rights campaign groups in Europe, North America and Australia, simply because the factory does not produce Nike shoes. . . . It seems that unless the Nike connection is made, such incidents are irrelevant."[6] And so a warped hierarchy of oppression is emerging from the factories of the Third World: when it comes to seeking international solidarity, only designer injustices need apply.

Bob Ortega makes a similar point about the anti–Wal-Mart movement in his book *In Sam We Trust*.

> There is a terrific irony – if one not much appreciated by Wal-Mart executives – in the fact that hundreds of towns and suburbs across North America will fight mightily to keep the dreaded Wal-Mart at bay, even as many of these communities let in scores of other superstore retailers that try to ape Wal-Mart in every way they can. . . . To the extent that Wal-Mart's critics blast it for wiping out Main Street businesses, for homogenizing communities, for trying to crush any and all rivals, for selling goods made in sweatshops here and abroad, they are missing the forest for the biggest tree.[7]

But there is also a clear value in the big-tree approach. Ken Saro-Wiwa's brother Owens points out that although all the gas companies have skeletons in their closets, focusing on one company – Shell Oil in the case of Nigeria – can have concrete advantages. "It is important not to make people feel powerless. After all, they need to fill their cars with something. If we tell them all companies are guilty, they will feel they can do nothing. What we are trying to really do, now that we have this evidence against this one company, is to let people have the feeling that they can at least have the moral force to make one company change."[8] He also says that since Shell controls more than half of Nigeria's oil, whatever happens to Shell will serve as a lesson to all the other oil companies, including Chevron.

When Bad Things Happen to the Unbrandables

Wiwa is convinced that with continued pressure, Shell will eventually meet the campaign's demands for economic and environmental reparations in Ogoniland. The millions Shell has poured into public relations and restructuring already show how seriously even the most profitable company in the world must take its public image. But much of that has to do with the visibility and vulnerability of the Shell brand. Shell extracts a raw resource from the land and water in Nigeria, but it brands that resource with its logo and sells it at its own branded gas stations around the world. And it is at that point that consumers make a choice between Shell and Texaco or Shell and Chevron – a choice that is as arbitrary and image based as the one between Coke and Pepsi, McDonald's and Burger King. Oil is a raw resource, but it only really becomes accessible to most people as a brand.

The same cannot be said of most multinationals in the natural-resource industries. Mining, natural gas, seed and logging multinationals all trade in virtually unbranded commodities that they sell to governments and corporate clients who then transform them into consumer goods. Since resource companies don't sell directly to the public, they barely have to worry about their public image – a factor that brings up what is perhaps the most significant limitation of brand-based campaigns: they can be powerless in the face of corporations that opt out of the branding game.

So all over the world, children work in fields with toxic pesticides, in dangerous mines and in rubber and steel factories where small fingers and hands are sliced off or mangled in heavy machinery. Many of these children are producing goods for the export market: canned fish, tea, rice, rubber for tires. But their plight has never captured the world's imagination like that of the kids who make soccer balls with swooshes on them or clothing for Barbie dolls, because their exploitation is unbranded, and therefore less identifiable, less visible, in our image-obsessed world.

The Free Burma movement has felt this limitation keenly. The campaign has been astonishingly successful at shaming nearly every brand-name company out of the country, from Pepsi to Texaco. When Heineken pulled out in July 1996, CEO Karel Vuursteen minced no words in explaining his decision: "Public opinion and issues surrounding this market have changed to a degree that could have an adverse effect on our brand and corporate reputation" – another casualty of the branding boomerang.[9] Relatively speaking, however, beer, soft-drink and clothing companies were never major players on the Burmese scene. The largest foreign development – accounting for half of all foreign investment – is a $1.2 billion gas line financed by Unocal, which is U.S. based, and Total, which is French-owned. But "Unocal," as Human Rights Watch noted in its 1997 World Report, "remained indifferent to protests." CEO Roger Beach defiantly told the press: "Let me say unequivocally that the only way we will leave is if we are forced to by the enactment of law."[10] And why should Beach care what a bunch of university students and church groups have to say? In 1997, Unocal sold the last of its U.S. retail outlets and refiners. So we don't go and buy our bottles of Unocal at Wal-Mart, or slap the Unocal logo on baseball hats and T-shirts. Activists have tried to fight the gas company through the courts, but so far without luck. When brand image is the weapon, an unbranded company can get off the hook entirely.

Secondary Boycotts

There are, however, ways around this obstacle, as the Lubicon Cree discovered when the Japanese pulp-and-paper giant Daishowa Marubeni-International unveiled plans for a major logging and mill operation on land that the Cree claimed was rightfully theirs. The area in Northern Alberta has been the subject of a fierce land-claim dispute in which the Canadian government has managed to avoid negotiating a settlement for sixty-five years. In the meantime, logging and mining have caused massive damage to the ecosystem and to the Lubicon way of life. So when Daishowa refused to withdraw its $500 million logging operation until the land claim was settled, the Lubicon saw it as the final straw. If neither the government nor the company would listen, they would have to go after Daishowa directly. But how? Daishowa is hardly a household name – it cuts down trees and turns them into paper goods that it then sells in bulk to other large corporations. How can you target a company that has absolutely no interaction with the general public?

The Friends of the Lubicon, an activist support group, were struggling with that very question over pizza one night in 1989 when one member of the group looked down at the Pizza Pizza paper bag on the table and saw, printed in small lettering at the bottom, the Daishowa trademark. There it was. The campaign strategy, the Lubicon soon decided, would be to call for a "secondary boycott": they would ask Daishowa's clients – among them Pizza Pizza, the Canadian clothing retailer Roots and Woolworths to sever their ties to Daishowa or face boycotts themselves. Though Daishowa has no brand image itself, many of its clients do, and good customer relations are of central importance to them. It wasn't long before many of them started getting their paper bags elsewhere. The strategy proved so successful that in 1995, Daishowa took the Friends of the Lubicon to court, claiming the boycott was unlawful and had cost the company $14 million in lost revenue.[11] But on April 14, 1998, an Ontario court judge ruled in favor of the Friends of the Lubicon. After the ruling, the Lubicon vowed to bring back the boycott with renewed force unless Daishowa agreed to stay off the disputed land. Two weeks later, Daishowa pledged "not to harvest or purchase timber" in the entire contested area until the land claim was resolved.[12]

From the beginning of its clash with the Lubicon Cree, Daishowa insisted it was being unfairly targeted, caught in the crossfire of a dispute between the band and government. In many ways, that is true. The targeting of the multinational and its clients was an act of desperation. As Kevin Thomas, spokesperson of the Friends of the Lubicon, says, "The government was never going to settle so long as the Lubicon people were the only ones suffering – the only ones unable to carry on with business as usual."[13] By making sure that Daishowa was also unable to carry out its business, the Lubicon drew one step closer to achieving a sustainable political solution. Greider is right: individual corporations are only one piece of the puzzle. But as the Daishowa case shows, they can be a piece of the puzzle that acts as a lever to achieve broader and more lasting political change.

The Daishowa precedent serves as a powerful warning to all the other faceless, resource-based corporations that conduct their operations in relative obscurity. Investigative activists are starting to track the progress of harvested natural resources through the economy to the point that they turn into consumer goods; at that stage public pressure can be applied at the mall, superstore or grocery chain. Nickel turns into batteries, genetically modified agricultural crops into packaged foods, old-growth wood into furniture, gold into jewelry.... There is no harvested resource that does not, eventually, turn into a brand.

This strategy has already proved enormously successful in the European campaign against genetically modified foods. For years campaigners had been railing against agribusiness giant Monsanto (that most impenetrable of multinationals) and its refusal to label which foods had been modified and which had not – and, in the case of soybeans, actually mixing the two together. But when campaigners broadened their focus to include not just companies like Monsanto and Novartis, responsible for the genetic engineering, but also the supermarkets that sold their foods, the issue finally grabbed the world's attention. With shoppers shouting about "Frankenfoods" on their doorsteps and Greenpeace campaigners leading customers on "gene food tours" through their aisles, the supermarkets could not afford to share Monsanto's cloistered attitude. Eventually, several large British supermarket chains including Sainsbury, Tesco and Safeway all removed bio-engineered foods from their private-label brands. Marks & Spencer went further, banning from its stores, in March 1999, all foods containing genetically modified ingredients.[14] Chains across Western Europe followed suit, as did food giants Unilever U.K., Nestlé U.K. and Cadbury.

Environmentalists have taken a similar tack with the forestry companies logging old-growth trees in British Columbia. Rather than continuing to face off against loggers in the deep woods – as was the strategy during the Clayoquot Sound demonstrations of 1993 – Greenpeace and the Rainforest Action Network now target high-profile brands that buy products derived from old-growth timber. In response to this pressure, in December 1998, twenty Fortune 500 companies – including 3M, Kinko's, Hallmark, IBM and Nike – agreed to phase out their use of old-growth products, and they made their commitment public in a full-page ad in *The New York Times*. But Home Depot – "the world's largest retailer of ancient forest products," according to the campaign – refused to sign on, sparking a wave of protests at dozens of Home Depot outlets across North America, as well as at the company's annual general meeting in Atlanta, in May 1999. The strategy proved successful: in August 1999, Home Depot announced it would phase out old-growth wood products by 2002.

The Writing on the Wall

Despite the success of these strategies, it nonetheless seems odd that we need to go to such great lengths to reshape social and environmental injustices just so that they can be brought home to us shoppers. In a way, these campaigns help us to care about issues not because of their inherent justice or importance but because we have the accessories to go with them: Nike shoes, Pepsi, a sweater from the Gap. If we truly need the glittering presence of celebrity logos to build a sense of shared humanity and collective responsibility for the planet, then maybe brand-based activism is the ultimate achievement of branding. According to Gerard Greenfield, international political solidarity is becoming so dependent on logos that these corporate symbols now threaten to overshadow the actual injustices in question. Talk about government, talk about values, talk about rights – that's all well and good, but talk about *shopping* and you really get our attention. "If we can only talk about workers' collective rights and struggles in the context of what people choose to buy as consumers," writes Greenfield, "then it seems we face a greater challenge to building a critical, popular social consciousness than we might imagine."[15]

There is no doubt that anticorporate activism walks a precarious line between self-satisfied consumer rights and engaged political action. Campaigners can exploit the profile that brand names bring to human-rights and environmental issues, but they have to be careful that their campaigns don't degenerate into glorified ethical shopping guides: how-to's on saving the world through boycotts and personal lifestyle choices. Are your sneakers "No Sweat"? Your rugs "Rugmark"? Your soccerballs "Child Free"? Is your moisturizer "Cruelty-Free"? Your coffee "Fair Trade"? Some of these initiatives have genuine merit, but the challenges of a global labor market are too vast to be defined – or limited – by our interests as consumers.

It took almost no time, for instance, for the White House Task Force on Sweatshops, set up in response to the Kathie Lee Gifford scandal, to become just another shopping exercise. Any substantial demands for labor-law reform were immediately hijacked by a new agenda: what provisions would U.S. companies have to meet before they could sew a "No Sweat" label on their garments? The immediate priority was finding a quick and easy way to protect the right of Westerners to buy branded goods without guilt. Tellingly, Bill Clinton's "No Sweat" labeling initiative is modeled after the "Dolphin Safe" stamp on cans of tuna, which reassures buyers that the much-loved dolphin was not killed in the canning of the fish. What this proposal fails to grasp is that the rights of garment workers, unlike dolphins, cannot be assured by a symbol on a label, the equivalent of a best-before date; and that trying to do so represents nothing less than the wholesale privatization of their (and our) political rights. The whole charade reminds me of a *New Yorker* cartoon that shows a Norman Rockwell–esque family unwrapping gifts under a Christmas tree. The parents are pulling out a new pair of sneakers, as the mother asks: "How are the human rights on these?"

There is another problem with the consumer-based approach. We are living, as Susan Sontag said, in the "Age of Shopping" and any movement that is primarily rooted in making people feel guilty about going to the mall is a backlash waiting to happen. Besides, the activists who are leading this movement aren't austere Luddites who are against shopping on principle. Many of them are creative twenty-somethings designing ad jams on their Mac laptops who happen to believe that there should be *some* space left over that isn't trying to sell them something or cluttered with the debris of our consumer culture. They are young men and women in Hong Kong and Jakarta who wear Nikes and eat at McDonald's, and tell me they are too busy organizing factory workers to bother with Western lifestyle politics. And while Westerners sweat over what kinds of shoes and shirts are most ethical to buy, the people sweating in the factories line their dorm rooms with McDonald's advertisements, paint "NBA Homeboy" murals on their doors and love anything with "Meeckey." The organizers in the Cavite zone often dress for work in ersatz Disney or Tommy T-shirts – cheap knockoffs from the local market. How do they reconcile the contradiction between their clothes and their anger at these multinationals? They told me they had never really thought about it like that; politics in Cavite is about fighting for concrete improvements in workers' lives – not about what name happens to be on a T-shirt you happen to have on your back.

Corporate codes of conduct are in many ways the most controversial byproduct of brand-based activism. The moment multinational companies like Nike, Shell, Mattel and the Gap stopped denying the existence of abuses at their sites of production and resource extraction, they began drafting statements of principles, codes of ethics, memorandums of understanding and other non-legally-binding documents of good intentions. These pieces of paper espoused high standards of business ethics: nondiscrimination, respect for the environment and for the rule of law. If any busybody customer wanted to know how their products were made, the public-relations department simply mailed them a copy of the code, as if it were the list of nutritional information on the side of a box of Lean Cuisine.

When you read the codes, it's difficult not to get swept up in the starry-eyed idealism of it all. These documents stare back at their readers with a look of perfect ahistorical innocence as if to ask, Why are you surprised? We have been like this all along.... And the reader would be forgiven for wondering, at least for a moment, if perhaps it is just as the companies say: one big misunderstanding, a "communication breakdown" with a rogue contractor, something lost in the translation.

Codes of conduct are awfully slippery. Unlike laws, they are not enforceable. And unlike union contracts, they were not drafted in cooperation with factory managers in response to the demands and needs of employees. Without exception, they were drafted by public-relations departments in cities like New York and San Francisco in the immediate aftermath of an embarrassing media investigation: Wal-Mart's code arrived after reports surfaced that its supplier factories in Bangladesh were using child labor; Disney's code was born of the Haitian revelation; Levi's wrote its policy as an answer to prison labor scandals. Their original purpose was not reform but to "muzzle the offshore watchdog" groups, as Alan Rolnick, lawyer for the American Apparel Manufacturers Association, advised his clients.[16]

But the companies who rushed to adopt these codes made a serious miscalculation: they underestimated, once again, the amount of information flowing between workers and villagers in Africa, Central America and Asia and campaigns in Europe and North America – and so rather than "muzzling" anyone, the documents have only raised more questions. Why did Shell fail to translate its manifesto, *Profits and Principles*, into any language other than English and Dutch? Why, until two years ago, were Nike and the Gap's codes available only in English? Why weren't they distributed to workers in the factories? Why was there such a great discrepancy between the intentions espoused in the codes and the firsthand reports coming out of the zones and the oil fields? Who is supposed to monitor these codes through all the layers of contracting and subcontracting? Who will enforce them? What is the penalty of failure?

In short, these hybrids of advertising copy and the Communist Manifesto backfired. The offshore watchdogs kept on barking – and no wonder. Anti-corporate campaigns are fueled, at least in part, by people's deep sense of marketing overload – and for this reason, the last thing likely to mollify them is more marketing. A group of anti-Shell campaigners made that point dramatically in March 1999 after Shell launched a $32 million marketing campaign that flawlessly absorbed the rhetoric of both the Brent Spar and the Ogoni campaigns. "Exploit or Explore?" asks the glossy Shell ad.

> Every business wants to make its mark. However, in the sensitive regions of the world, like our tropical rainforests and our oceans, the scars of industrialisation are all too apparent. Our shared climate and finite natural resources concern us as never before, and there's no room for an attitude of "It's in the middle of nowhere, so who's to know?" Time and again at Shell, we're discovering the rewards of respecting the environment when doing business. If we're exploring for oil and gas reserves in sensitive areas of the world, we consult widely with the different local and global interest groups. Working together, our aim is to ensure that bio-diversity in each location is preserved. We also try to encourage these groups to monitor our progress so that we can review and improve the ways in which we work.[17]

But rather than stemming the flow of criticism, Shell's extravagant spending on public relations – with the Ogoni's grievances left unresolved and demands for outside monitoring repeatedly rejected – sparked its own kind of backlash: a backlash against "greenwash." Essential Action, the hub of the Shell boycott, launched a postcard campaign urging Shell's executives to "Spend the money cleaning up your mess, not your image!" And, in April 1999, activists in London threw green and red paint on the doors of the company's international headquarters. The green paint, said the anonymous perpetrators, was an attempt to give Shell "a taste of its own greenwash."[18]

Throwing paint is one approach. Another, and one that has become increasingly popular, is throwing the promises in the codes of conduct back in the face of the corporations who drafted them. Once again, it's the Saul Alinsky theory of political jujitsu: "No organization ... can live up to the letter of its own book. You can club them to death with their 'book' of rules and regulations."[19] Bama Athreya of the U.S.-based International Labor Rights Fund explains how this strategy can work with relation to Nike's high-minded code of conduct: "Let's face it, hypocrites are far more interesting than mere wrongdoers, and it's been much easier to sensitize press and public to Nike's failure to implement its own code of conduct than to its failure to comply with Indonesian labor law."[20]

As it became clear that these flimsy codes of conduct had failed to quiet dissent (and may have exacerbated it), several multinationals moved on to a more advanced brand of corporate code. These codes, while still not legally binding and still implemented on a voluntary basis with little monitoring, are nonetheless more substantial than a simple statement of good intentions. And by 1998, there were so many different models of these codes floating around that even the most committed anti-sweatshop campaigners had to admit that they had lost track. Some were drafted in cooperation with human-rights groups or ethical investment specialists in the West. Others, like Bill Clinton's Apparel Industry Partnership's code, were organized according to where the multinationals were headquartered. The Gap has a code that applies to one factory in El Salvador, allowing it to be monitored by local human-rights activists; a code adopted by Levi's, Mattel and Reebok refers specifically to doing business in China. A code on child labor drafted by Unicef, the International Labour Organization and an association of Pakistani manufacturers was signed by all the major soccer-ball manufacturers; it provides for outside monitoring as well as education and rehabilitation for the child laborers. Following the wave of anti-sweatshop student activism in

1998 and 1999, dozens of universities adopted their own codes, only to decide subsequently to sign on to the Clinton Apparel Industry Partnership's code *en masse* – a totally different text. Meanwhile, the Collegiate Licensing Company proposed its own anti-sweatshop code, to apply to all 160 American schools it represents – meaning that some schools were looking at three tiers of codes. And unlike the tough codes adopted by universities like Duke, the CLC code has no provision for disclosure and does not require contractors to pay a living wage, only the minimum wage.

Layered on top of these stacks of codes was one drafted by the Council on Economic Priorities, a consumer watchdog group in New York, along with several large corporations. The CEP plan would inspect factories for adherence to a set of standards covering key issues such as health, safety, overtime, child labor and the like. Under this model, brand-name multinationals like Avon and Toys 'R' Us, rather than trying to enforce their own codes around the world, simply place their orders with factories that have been found to be in compliance with the code. Then, the factories are monitored by a private auditing company, which certifies factories that meet the code as "SA8000" (SA stands for "Social Accountability"). For many multinationals, this plan was far too demanding; the American Apparel Manufacturers Association, for instance, launched its own, less stringent, voluntary code, which would also certify factories "sweatshop free."

Not surprisingly, by mid-1999 the entire sweatshop issue had degenerated into a maze of warring codes. Unions and religious groups who had been participating in the Clinton Apparel Industry Partnership walked out in protest at its weak enforcement and monitoring measures, and accused the human-rights groups that stayed in the Partnership of "selling out." The student anti-sweatshop activists launched an offensive against their own universities' participation in the Clinton Partnership, insisting that no code drafted or monitored by the corporations themselves – even at arm's length – could possibly have any merit. Monitoring had to be done by unions or by human-rights groups.

Confusing matters still further was a strange conflation of several large human-rights groups and the corporate sector. In 1999, some of the most maligned multinationals on the planet – Dow Chemical, Nestlé, Rio Tinto,

Unocal – rushed into partnership with human-rights groups and the United Nations Development Programme. Together they formed brand-new umbrella organizations with names like the Business Humanitarian Forum, Partners in Development and the Global Sustainable Development Facility, which promised to "improve communications and cooperation between global corporations and humanitarian organizations."[21] Multinationals and human-rights groups, they claimed, actually have the same goals; human rights are good for business – they are the "third bottom line."

It's tempting to take this dramatic shift in direction on the part of so many multinationals as a massive victory for the campaigners who have battled the Nikes and the Shells all these years. Maybe corporations really have seen the light, and we're all on the same page now.... Harvard business professor Debora L. Spar is among those hailing the dawn of this new age. She argues that the rise of brand-based activism has been so successful in shaming corporations, it is no longer in the financial interest of brand-name multinationals to allow abuses to occur. She calls this theory the "spotlight phenomenon." There is no need for outside regulations because "firms will cut off abusive suppliers or make them clean up because it is now in their financial interest to do so," she writes. "The spotlight does not change the morality of U.S. managers. It changes their bottom line."[22]

There is no doubt that companies like Nike have learned that labor-rights abuses can cost them. But the spotlight being shined on these companies is both roving and random: it is able to shine down on a few corners of the global production line, but darkness still shrouds the rest. Human rights, far from being protected by this process, are selectively respected: reforms seem to be implemented solely on the basis of where the spotlight's beam was last directed. There is absolutely no evidence that any of this reform activity is coalescing into a universal standard of ethical corporate behavior that will be applied around the world; and no system of universal enforcement is on the horizon.

Instead, what we have with the proliferation of voluntary codes of conduct and ethical business initiatives is a haphazard and piecemeal mess of crisis management. In mid-1999, for instance, when Nike was coming off as a savior in Indonesia for increasing wages, it was also cutting its ties with higher-waged workers in the Philippines and rushing into China, where workers' rights are least protected, monitoring is next to impossible and wages are lowest. Levi's pulled out of Burma, because its conscience simply would not allow it to stay, only to go back to China, which it had abandoned a few years earlier for the same reason. It then drafted a breakthrough code of conduct for China, but at the same moment it was laying off thousands of workers in Europe and North America. The Gap, meanwhile, was being held up as a model of openness and reform in El Salvador, while protestors outside its stores in New York and San Francisco shouted about the abhorrent conditions at its factories in Saipan and Russia. In addition, there were wildly diverging reports about whether even the toughest codes were actually being implemented in the factories, and whether the vast majority of workers around the world had even heard of them. And, of course, there is still no monitoring system in place to get an accurate picture of what's really happening in the factories. Without question, some imaginative and effective initiatives have come out of these PR scrambles, but the fact remains that this patchwork approach is no way to draft a sustainable labor or environmental policy for the global economy.

If the way multinationals like Nike and Shell have handled their respective scandals seems uncharacteristically chaotic for such streamlined operators, that chaos could well be deliberate. Even when the codes fail to stamp out abuses, what they do manage to do, rather effectively, is obscure the fact that multinationals and citizens do not actually share the same goals when it comes to deciding how to regulate against labor and environmental abuses. Even when there is genuine agreement on the need to address a problem (child labor, for instance), beneath the talk of ethics and partnerships the two parties are still engaged in a classic power struggle.

Ever since key multinationals stopped denying the existence of any human-rights abuses in their global production operations, the struggle has not been over whether controls need to be put in place, but rather over who will get to place those controls. Will it be the people and their democratically elected representatives? Or will it be the global corporations themselves? It's clear from the privatized codes which direction the corporations want to go. The question is, What will citizens do in response?

The subtext of the codes of conduct is a settled hostility toward the idea that citizens can – through unions, laws and international treaties – take control of their own labor conditions and of the ecological impact of industrialization. In the twenties and thirties, when the crises of sweatshops, child labor and workers' health were at the forefront of the political agenda in the West, these problems were tackled with mass unionization, direct bargaining between workers and employers and governments enacting tough new laws. That type of response could be marshaled again, only this time on a global scale, through the enforcement of existing International Labor Organization treaties, if compliance with those treaties were observed with the same commitment that the World Trade Organization now shows in its enforcement of the rules of global trade.

The United Nations Declaration of Human Rights already recognizes the right to freedom of association. If respecting that right became a condition of trade and investment, it would transform the free-trade zones overnight. If workers in the zones had the freedom to bargain for their rights without living in fear of either a government crackdown or immediate factory flight, the need for private codes and independent monitors would virtually disappear. In countries like the Philippines and Indonesia, governments would enforce these standards, and their own laws, or fear the economic repercussions. But then, this type of rigid regulation is exactly what the corporate sector has so aggressively fought since free trade was introduced – by yanking the teeth out of the UN's declarations and treaties, and by steadfastly opposing all proposals to link trade deals to enforceable labor and environmental codes. In fact, it is precisely this kind of regulation that multinationals are now trying so frantically to circumvent by drafting their own voluntary codes.

And so, after Nike and dozens of universities had joined the White House Partnership, Charles Kernaghan saw clearly that the anti-sweatshop movement he helped launch had become an entirely new ball game. Gone were the days when the most pressing task was convincing companies that they even had a problem. "Nike hopes to co-opt our movement," he writes. "In this view, what we are seeing is no less than a struggle over who will control the agenda for eradicating sweatshop abuses. Nike's implicit message is: 'Leave it to us. We have voluntary codes of conduct. We have a task force. We'll take care of it from here. Go home and forget about sweatshops.'"[23]

There is something Orwellian about the idea of turning the enforcement of basic human rights into a multinational industry, as the private codes would do, to be checked like any other quality control. Global labor and environmental standards should be regulated by laws and governments – not by a consortium of transnational corporations and their accountants, all following the advice of their PR firms. The bottom line is that corporate codes of conduct – whether drafted by individual companies or by groups of them, whether independently monitored mechanisms or useless pieces of paper – are not democratically controlled laws. Not even the toughest self-imposed code can put the multinationals in the position of submitting to collective outside authority. On the contrary, it gives them unprecedented power of another sort: the power to draft their own privatized legal systems, to investigate and police themselves, as quasi nation-states.

So this is a power struggle, make no mistake. In an editorial in *The Journal of Commerce*, codes of conduct are explicitly presented to employers as a less threatening alternative to externally imposed regulation. "The voluntary code helps diffuse a contentious issue in international trade negotiations: whether to make labor standards part of trade agreements. If... the sweatshop problem is solved outside the trade context, labor standards will no longer be tools in the hands of protectionists."[24]

Such warnings hint that despite the ineffectiveness of governments and the rhetoric of corporate triumphalism, there are still some mechanisms left with which to regulate multinationals. As we have seen, there are trade agreements and local selective purchasing laws, as well as ethical investment drives – but conditions can also be attached to government loans and insurance offered to foreign investors, as well as to involvement in government trade missions. It may seem unrealistic to suggest that multinationals would ever accept such restrictions on their global mobility. Then again, the past four years have seen the world's most powerful and profitable brand-name multinationals forced to continually raise the bar of their own public relations. If the public will is there, the bar can be raised further still, taking these issues away from corporate control and forcing them into the public domain.

Top: French farmers protest cuts to farm subsidies by throwing bags of corn gluten and chicken feed into the Seine, Paris, November 1992. *Bottom:* G-8 leaders pose for official family photo, Cologne, June 1999.

CONCLUSION

CONSUMERISM VERSUS CITIZENSHIP

The Fight for the Global Commons

The beers at my hotel bar in Rosario were blissfully cold, and the gang from the Workers' Assistance Center were all getting a little drunk. We were arguing, once again, about whether codes of conduct have any merit whatsoever. Zernan Toledo (who personally favors armed revolution – it's just a question of when) pounded the table. "These documents are written by the transnational corporations, so they will only serve the transnational corporations – haven't you read Marx?"

"It's different now," I countered. "With globalization, there need to be some common standards – and the governments certainly aren't setting them."

"Globalization is nothing new. We have always had globalization," said Arnel Salvador, another of the WAC organizers. His eyes were fixed not on me, but on something across the bar. Since the hotel where I stayed is the only one near the Cavite Export Processing Zone, it was, as usual, packed with visiting factory owners, contractors and buyers who were here to stay up all night singing karaoke and cutting deals for cheap clothes and electronics. I followed Arnel's eyes to a young man slouched in his chair, his feet up on the table across from him, his knees splayed apart as if he owned the world. Trendy and modern, he looked like a character from one of the many cell-phone commercials on Asian TV. "You can always tell the foreigners," Arnel said slowly, his usually warm voice icy. "No Filipino would sit like that."

The foreign investors who sing karaoke at the Mountain and Sea Hotel in Rosario are part of a long and bitter history of colonialists in the Philippines: first the Spaniards came and conquered, then the Americans arrived, setting up army bases and turning teenage prostitution into one of the country's largest industries. Now colonialism is dead, the U.S. military has receded and the new imperialists are the Taiwanese and Korean contractors in the export processing zones, sexually harassing the eighteen-year-old Filipinas on the assembly lines. Several of the free-trade zones in the Philippines (though not Cavite) are actually built on land that only a few years back housed U.S. military bases, and all over the country workers are shuttled to and from the zones in U.S. army jeeps converted into mini-buses. To Arnel Salvador and Zernan Toledo, the much-vaunted joys of economic globalization amount to pretty much more of the same: the boss has just traded in his military uniform for an Italian suit and an Ericsson cell phone.

The day after our night of drinking, I sat with Nida Barcenas in the backyard of the Workers' Assistance Center and asked her what motivates her, night after night, to trudge out to the dorms at 11 p.m. to meet with garment workers when they finally get off work. My question took Nida by surprise. "Because I want to help the workers. I really want to help them," she said. Then the tough composure that helps her stand up to zone bosses and petty local bureaucrats disappeared and fat tears rolled down her smooth cheeks. All she managed to say was "It's like Arnel said – it's just been so long." What has been so long is not the fight for rights for her fellow factory workers, although she means that as well. What has been so long is the fight against feudal landlords, against military dictators and now against foreign factory owners. I turned the tape recorder off and we sat in silence until her colleague, Cecille Tuico, quietly brought us mugs of syrupy-sweet vanilla ice cream that turned to soup in the hot sun.

Because the Workers' Assistance Center's chief mission is to empower workers to stand up for their rights, WAC organizers don't much like the idea of Westerners swooping into the zone brandishing codes of conduct, with teams of well-meaning monitors trailing behind. "The more significant way to resolve those problems," says Nida Barcenas, "lies with the workers themselves, inside the factory." And codes of conduct, she says, have little hope of helping because the workers have no hand in drafting them. As for third-party independent monitoring, Zernan Toledo believes that no matter who performs it, it's just that: third party. All it will do is reinforce the idea that somebody else is looking after the workers' destiny, not the workers themselves. To some this flat-out rejection sounds stubborn and ungrateful, an unfair dismissal of all the well-meaning work being conducted in boardrooms in Washington, London and Toronto. But the right to sit down and bargain – even when you don't get the perfect deal – is the fundamental right for which the international trade union movement has struggled from its inception; it has always been about self-determination. Zernan Toledo invokes an old and familiar aphorism to explain the distinction: "If you give a man a fish, he will eat for one day. But if you teach him how to fish he will eat forever." And so, every evening at the Workers' Assistance Center, Zernan, Arnel, Cecille and Nida give the workers their fishing lessons. A little blackboard stands in the backyard with the chickens, and the organizers take turns leading seminars. Sometimes fifty workers show up, other times just one. Though this route will no doubt take longer than the ready-made codes and monitoring, the WAC organizers say they are willing to wait. As Nida says, it has already "been so long," they may as well get it right.

It's a message that applies not just in Cavite, but to all those concerned about corporate abuses around the world. When we start looking to corporations to draft our collective labor and human rights codes for us, we have already lost the most basic principle of citizenship: that people should govern themselves. As we have seen, Nike, Shell, Wal-Mart, Microsoft and McDonald's have become metaphors for a global economic system gone awry, largely because, unlike the back-door wheeling and dealing at NAFTA, GATT, APEC, WTO, MAI, the EU, the IMF, the G-8 and the OECD, the methods and objectives of these companies are plain to see: workers and foreign observers alike understand very well what they are up to. They have become the planet's best and biggest popular education tools, providing some much-needed clarity inside the global market's maze of acronyms and centralized, secretive dealings.

By attempting to enclose our shared culture in sanitized and controlled brand cocoons, these corporations have themselves created the surge of opposition described in this book. By thirstily absorbing social critiques and political movements as sources of brand "meaning," they have radicalized that opposition still further. By abandoning their traditional role as direct, secure employers to pursue their branding dreams, they have lost the loyalty that once protected them from citizen rage. And by pounding the message of self-sufficiency into a generation of workers, they have inadvertently empowered their critics to express that rage without fear.

But the fact that the brands have led us into this maze does not mean we should look to them to lead us out. Nike and Shell are shiny new doorways opening onto the much more complicated and less glamorous world of international law. And though it won't be easy and it won't come quickly, we will find our way out as citizens, on our own. We may feel a little like Theseus, clutching his thread as he entered the Minotaur's labyrinth, but nothing else will do. Political solutions – accountable to people and enforceable by their elected representatives – deserve another shot before we throw in the towel and settle for corporate codes, independent monitors and the privatization of our collective rights as citizens.

It is a daunting task but it does have an upside. The claustrophobic sense of despair that has so often accompanied the colonization of public space and the loss of secure work begins to lift when one starts to think about the possibilities for a truly globally minded society, one that would include not just economics and capital, but global citizens, global rights and global responsibilities as well. It has taken many of us a while to find our footing in this new international arena, but thanks in large part to the crash course provided by the brands, we are closer than ever before.

The first step has been an astonishingly successful network of popular-education projects. In 1995, the International Forum on Globalization held its first Global Teach-In in New York, which brought together leading scientists, activists and researchers to examine the impacts of a single, unfettered world market on democracy, human rights, labor and the natural environment. There were seminars on NAFTA, APEC, the IMF, the World Bank, Structural Adjustment of the North and every other global body or trade agreement you never understood but were afraid to ask. The New York Conference attracted several hundred people, but at the second conference in Berkeley, California, two thousand people showed up (with zero pre-publicity and no media coverage – just some posters and E-mail lists). A conference a few months later in Toronto attracted even more people and there have been similar gatherings on university campuses around the world.

And world leaders can't have lunch together these days without somebody organizing a counter-summit – gatherings that bring together everyone from sweatshop workers trying to unionize the zones to teachers fighting the corporate takeover of education. At these events – in Geneva, Cologne and Birmingham – alternative models of globalization spill onto the streets during the day, and the Reclaim the Streets parties go on all night.

It is sometimes difficult to tell whether these trends are the start of something genuinely new or the last gasps of something very old. Are they, as the engineering professor and peace activist Ursula Franklin asked me, simply "wind blocks," creating temporary shelter from the corporate storm, or are they the foundation stones of some as yet unimagined, free-standing edifice? When I started this book, I honestly didn't know whether I was covering marginal atomized scenes of resistance or the birth of a potentially broad-based movement. But as time went on, what I clearly saw was a movement forming before my eyes.

Three years ago, when I attended the Berkeley teach-in on globalization, I was frustrated that the speakers were all over fifty and that the links with the college-age culture jammers and anticorporate campaigners had yet to be made. A year later, these generations of activists and theorists were already enmeshed on several fronts, lending urgency and depth of analysis to each other's actions. During this same time, campaigns focusing on a single corporation in a single place – Shell in Nigeria, say, or Nike in Indonesia – had also found each other and were starting a process of intellectual cross-pollination, often at the click of a hotlink, thanks to the Net.

This emerging movement even has a major victory under its belt: getting the Multilateral Agreement on Investment taken off the agenda of the Organization for Economic Co-operation and Development in April 1998. As the *Financial Times* noted with some bewilderment at the time, "The opponents' decisive weapon is the Internet. Operating from around the world via web sites, they have condemned the proposed agreement as a secret conspiracy to ensure global domination by multinational companies, and mobilized an international movement of grassroots resistance." The article went on to quote a World Trade Organization official who said, "The NGOs have tasted blood. They'll be back for more."[1] Indeed they will.

On June 18, 1999, these virtual connections were made real when a coalition of groups including Reclaim the Streets and People's Global Action held the second Global Street Party, this time to coincide with the G-8 meeting in Cologne, Germany. The event, billed as a "global carnival against capital," took aim squarely at corporate power. All around the world, parties and protests were held in financial districts, outside stock exchanges, superstores, banks and multinational headquarters. With simultaneous action in seventy different cities, the day was the coming-out party for this new global political player: it displayed all of the movement's promise and creativity – and showed more forcefully than ever before just how much anticorporate rage is brewing.

Though they were organized locally, a common theme ran through all the events. In Bangladesh, women garment workers held a protest against sweatshop conditions; in San Francisco, they protested those same conditions outside Gap stores. In Montevideo, Uruguay, activists turned the main square of the city's financial district into a "fair trade" show, with exhibits on every corporate abuse from child labor to the arms trade; in Madrid, the entrance to the stock exchange was blocked. And in Cologne, site of the G-8 summit, European activists held a counter-summit and demanded debt forgiveness for Third World countries. The event was joined by five hundred Indian farmers who were traveling across Western Europe in an "intercontinental caravan." Along the way, they stopped off at the corporate headquarters of agribusinesses such as Cargill and Monsanto whose seed patents and genetic engineering of crops have burdened many Indian farmers with massive debts.

On the same day that the Indian farmers were peacefully protesting in Cologne, London's financial district turned into a war zone – the city hasn't seen anything like it since the 1990 poll-tax riots. The 10,000-strong gathering started as a classic RTS surreal political party. The streets were cleared by a Critical Mass bike ride and were flooded by activists dressed in second-hand suits with slogans painted on the backs. They danced in the doorsteps of office towers, formed a human chain around the Treasury and held peaceful sit-ins at several banks. The bankers and investment brokers, meanwhile, came to work disguised in casual sports clothes, having been advised by police to "blend in" with the activists so as not to catch a flying pie. But as the day wore on, the crowd splintered into smaller groups and became gradually more violent. One group stormed the Futures Exchange, breaking all the glass in the lobby, disrupting automated stock trading and forcing an evacuation of the building. In other parts of London, a McDonald's outlet, a bank and a Mercedes Benz dealership were trashed, a protestor was run over by a police van and several police were injured. There was also mob violence in Eugene, Oregon: windows were broken at banks and fast-food restaurants, cars were stormed, protestors attacked cops with rocks and cops attacked protestors with pepper spray. In both cities, the political messages about widening economic disparities and the brutalities of free-market globalization were drowned out by the sound of shattering glass.

In Geneva, the message was clear as day: rather than throwing rocks through windows, activists arrived with sponges, soap and squeegees to wash the façades of the big downtown banks. The organizers explained to the press that they only wanted to help these fine institutions clean up the stains left behind by crippling Third World debt and Nazi gold. In Port Harcourt, Nigeria, the mood at the "Carnival of the Oppressed" was militant but celebratory. A crowd of 10,000 welcomed Ken Saro-Wiwa's brother back to his homeland after years in exile. After listening to a speech by Owens Wiwa, the crowd proceeded to the gates of the city's Shell Oil headquarters and blocked entry for several hours. The next stop was a street named after the late Nigerian dictator, General Sani Abacha, where members of the crowd lowered the street sign and temporarily renamed the road after one of the men whose lives he stole: Ken Saro-Wiwa. According to the organizers, "There was dancing and singing in the streets, bringing Port Harcourt, Nigeria's petroleum capital, to a standstill."

And all of this happened on a single day.

When this resistance began taking shape in the mid-nineties, it seemed to be a collection of protectionists getting together out of necessity to fight everything and anything global. But as connections have formed across national lines, a different agenda has taken hold, one that embraces globalization but seeks to wrest it from the grasp of the multinationals. Ethical shareholders, culture jammers, street reclaimers, McUnion organizers, human-rights hacktivists, school-logo fighters and Internet corporate watchdogs are at the early stages of demanding a citizen-centered alternative to the international rule of the brands. That demand, still sometimes in some areas of the world whispered for fear of a jinx, is to build a resistance – both high-tech and grassroots, both focused and fragmented – that is as global, and as capable of coordinated action, as the multinational corporations it seeks to subvert.

AFTERWORD

TWO YEARS ON THE STREETS

Moving Through the Symbols

The first edition of *No Logo* ends with an image of activists speaking in hushed tones about their plan to build a global anti-corporate movement. Then, when the book was at the printers, something happened that changed everything: on November 30, 1999, the streets of Seattle exploded in protests against the World Trade Organization. Overnight, that hushed whisper turned into a shout, one heard around the world. This movement was no longer a secret, a rumour, a hunch. It was a fact.

Seattle took the political campaigns described in this book to a much more prominent place in the political discourse. As the mass demonstrations spread to Washington D.C., Quebec City, New Delhi, Melbourne, Genoa, Buenos Aires and elsewhere, debates raged in the press about police and protester violence, as well as what alternatives there are – if any – to what the French call "wild capitalism" (*capitalisme sauvage*). The issues behind the protests changed too. In very short order, college-age activists who started off concerned with the unethical behaviour of a single corporation began questioning the logic of capitalism itself, and the effectiveness of trickle down economics. Church groups who had previously demanded only the "forgiveness" of Third World debt were now talking about the failure of the "neo-liberal economic model," which holds that capital must be freed of all encumbrances to facilitate future development. Instead of reform, many were calling for the outright abolition of the World Bank and the International Monetary Fund. And ad-busters were longer satisfied with jamming a single billboard, but were busy creating new and exciting networks of participatory media like the Independent Media Centres, now in dozens of cities around the world.

Meanwhile, the institutions that have been the primary enforcers and defenders of global neo-liberal policies have been going through their own metamorphoses. The World Bank, the IMF, the WTO and the World Economic Forum have stopped denying that their model of globalization is failing to deliver the promised results, and have begun preoccupying themselves – at least in public statements – with the paradoxes of debt slavery, the AIDS pandemic, and the billions left out of the global market.

It became clear to me that I would need to update *No Logo*. The problem was that despite all these changes – or more accurately, because of them – I could never quite get around to doing the updating. Like so many other activists and theorists in this field, ever since Seattle exploded onto the world stage I have been swept up in the unstoppable momentum of the globalization battles: speaking, debating, organizing, and travelling way too much. We've been doing, in other words, what movements should do – we've been moving. Often so fast that it has seemed impossible to keep up with the latest twists and turns, let alone to step back and reflect about where all this motion is leading us.

It was only after the September 11 attacks that, at least in North America, this context began to change. All of a sudden, everyone seemed to be talking about the gap between the global haves and have nots, as well as the absence of democracy in so many parts of the world. But though the North American public was more aware of the failings of the global economy – failures glossed over in the press by the euphoria of boom-time prosperity – it was suddenly much harder to transform that awareness into political action. Rather than pushing governments to change clearly faulty policies, a fearful population was instead handing their politicians stacks of blank cheques, freeing them to barrel ahead with more of the same: new tax cuts for wealthy corporations, new trade deals, new privatization plans. To engage in dissent in this climate was cast as unpatriotic.

There are other challenges that North American activists have faced since September 11. As this book has argued, activists began targeting corporations in the mid-nineties as a response to the fact that so much that is powerful today is virtual: currency trades, stock prices, intellectual property, brands and arcane trade agreements. By latching onto symbols, whether a famous brand like Nike or a prominent meeting of world leaders, the intangible was made temporarily actual, the vastness of the global market more human-scale. Yet the dominant iconography of this movement – the culture-jammed logos, the guerrilla-warfare stylings, the choices of brand name and political targets – look distinctly different to eyes changed by the horrors of September 11. Today, campaigns that rely even on a peaceful subversion of powerful capitalist symbols find themselves in an utterly transformed semiotic landscape.

This struck me recently, looking at a slide show I had been pulling together just before the attacks. It is about how anti-corporate imagery is increasingly being absorbed by corporate marketing. One slide shows a group of activists spray-painting the window of a Gap outlet during the protests in Seattle. The next shows Gap's recent window displays featuring its own prefab graffiti - the word "Independence" sprayed in black. And the next is a frame from Sony PlayStation's State Of Emergency game featuring cool-haired anarchists throwing rocks at sinister riot cops protecting the fictitious American Trade Organisation. When I first looked at these images beside each other, I was amazed by the speed of corporate co-optation. But looking at them after September 11, the images had all been instantly overshadowed, blown away by the terrorist attacks like so many toy cars and action figures on a disaster movie set.

It could hardly have been other otherwise. The attacks on the World Trade Centre and Pentagon were acts of real and horrifying terror, but they were also acts of symbolic warfare, and instantly understood as such. As many commentators have put it, the towers were not just tall buildings; they were "symbols of American capitalism." Predictably, many political opponents of the anti-corporate position have begun using the symbolism of the attacks to argue that these acts of terrorism represent an extreme expression of the ideas held by the protesters. Some have argued that the attacks were only the far end of a continuum of anti-American and anti-corporate violence: first the Starbucks window in Seattle, then the WTC.

Others have gone even farther, arguing that free-market polices are the economic front of the war on terrorism. Supporting "free trade" has been rebranded, like shopping and baseball, as a patriotic duty. United States

Trade Representative Robert Zoellick has explained that trade "promotes the values at the heart of this protracted struggle," so the U.S., he says, needs a new campaign to "fight terror with trade." In an essay in the *New York Times Magazine*, Michael Lewis makes a similar conflation between freedom fighting and free trading when he explains that the traders who died were targeted as "not merely symbols but also practitioners of liberty . . . They work hard, if unintentionally, to free others from constraints. This makes them, almost by default, the spiritual antithesis of the religious fundamentalist, whose business depends on a denial of personal liberty in the name of some putatively higher power."

The new battle lines have been drawn, crude as they are: to criticize the U.S. government is to be on the side of the terrorists, to stand in the way of market-driven globalization is to further the terrorists' evil goals.

There is, of course, a glaring problem with this logic: the idea that the market can, on its own, supply solutions to all of our social problems has been profoundly discredited by the experience of September 11 itself. From the privatized airport security officers who failed to detect the hijackers' weapons, to the private charities that have so badly bungled aid to the victims, to the corporate bailouts that have failed to stimulate the economy, market-driven policies are not helping to win the war on terrorism. They are liabilities. So while criticizing politicians may be temporarily out of favor, "People Before Profit," the street slogan from the globalization protests, has become a self-evident and viscerally felt truth for many more people in the U.S. since the attacks.

The most dramatic manifestation of this shift is the American public's changing relationship to its public sector. Many of the institutions and services that have been underfunded, vilified, deregulated and privatized during the past two decades – airports, post-offices, hospitals, mass transit systems, water and food inspection – were forced to take center stage after the attacks, and they weren't ready for their close-up. Americans found out fast what it meant to have a public health care system so overburdened it cannot handle a routine flu season, let alone an anthrax outbreak. There were severe drug shortages, and private labs failed to come up with enough anthrax vaccines for U.S. soldiers, let alone civilians. Despite a decade of pledges to safeguard the U.S. water supply from bioterrorist attack, scandalously little had been done by the overburdened U.S. Environmental Protection Agency. The food supply proved to be even more vulnerable, with inspectors managing to check about 1 per cent of food imports - hardly a safeguard against rising fears of "agroterrorism."

And most wrenchingly of all, it was the fire-fighters who rushed in to save the lives of the bond traders and other employees in the towers, demonstrating that there is indeed still a role for a public sector after all. So it seems fitting that on the streets of New York City the hottest-selling T-shirts and baseball hats are no longer the ones displaying contraband Nike and Prada logos, but the logo of the Fire Department of New York.

The importance of a strong public realm is not only being rediscovered in rich countries like the U.S., but also in poor countries, where fundamentalism has been spreading so rapidly. It is in countries where the public infrastructure has been ravaged by debt and war that fanatical sugar daddies like Osama bin Laden are able to swoop in and start providing basic services that are usually in the public domain: roads, schools, health clinics, even basic sanitation. And the extreme Islamic seminaries in Pakistan that indoctrinated so many Taliban leaders thrive precisely because they fill a huge social welfare gap. In a country that spends 90 per cent of its budget on its military and debt - and a pittance on education – the madrassas offer not only free classrooms but also food and shelter for poor children.

In understanding the mechanics of terrorism - north and south – one theme is recurring: we pay a high price when we put the short-term demands of business (for lower taxes, less "red tape," more investment opportunities) ahead of the needs of people. Post-September 11, clinging to *laissez faire* free-market solutions, despite overwhelming evidence of their failings, looks a lot like blind faith, as irrational as any belief system clung to by religious fanatics fighting a suicidal jihad.

For activists, there are many connections to be made between the September 11 attacks and the many other arenas in which human needs must take precedence over corporate profits, from AIDS treatment in Africa to homelessness in our own cities. There is also an important role to be played in arguing for more reciprocal international relations. Terrorism is indeed an international threat, and it did not begin with the attacks in the U.S. As Bush invites the world to join America's war, sidelining the UN and the international courts, globalization activists need to become passionate defenders of true multilateralism, rejecting once and for all the label "anti-globalization." From the start, it was clear that President Bush's coalition did not represent a genuinely global response to terrorism but the internationalization of one country's foreign policy objectives - the trademark of U.S. conduct on the world stage, from the WTO negotiating table to the abandonment of the Kyoto Protocol on climate change. These arguments can be made not in a spirit of anti-Americanism, but in a spirit of true internationalism.

By far the most important role for those concerned with the explosion of corporate power is to act not only as voices of opposition but also as beacons – beacons of other ways to organize a society, ways that exist outside of the raging battles between "good" and "evil." In the current context, this is no small task. The attacks on the U.S. and the U.S. attacks on Afghanistan have ushered in an era of ideological polarization not seen since the Cold War. On the one hand, there is George W Bush claiming "You are either with us, or you are with the terrorists"; on the other, there is bin Laden, asserting that, "These events have divided the world into two camps, the camp of the faithful and the camp of the infidels." Anti-corporate and pro-democracy activists should demonstrate the absurdity of this duality and insist that there are more than two choices available. We can spread rumors about the existence of routes not taken, choices not made, alternatives not built. As Indian novelist and activist Arundhati Roy wrote after September 11, "the people of the world do not have to choose between the Taliban and the U.S. government. All the beauty of human civilization–our art, our music, our literature–lies beyond these two fundamentalist, ideological poles." Confronted with a deadly multiple-choice exam, the answer should be, "None of the above."

Well before September 11, there was a growing awareness in movement circles that attention needed to shift from "summit-hopping" to articulating and building these alternatives. For more than a year, the largely symbolic attacks on individual corporations and trade summits were being vocally challenged by many who feared that globalization battles – with their smashed McDonald's windows and running fights with police – were beginning to look like theater, cut off from the issues that affect people's day to day lives. And there is much that is unsatisfying about fighting a war of symbols: the glass shatters in the storefront, the meetings are driven to ever more remote locations - but so what? It's still only symbols, facades, representations.

In response, a new mood of impatience was already taking hold, an insistence on putting forward social and economic alternatives that address the roots of injustice, from land reform in the developing world to slavery reparations in the U.S., to participatory democracy at the municipal level in cities around the world. Rather than summit-hopping, the focus was moving to forms of direct action that attempt to meet people's immediate needs for housing, food, water, life-saving drugs, and electricity. This is being expressed in countless unique ways around the world.

In India, it means defiantly producing generic AIDS drugs for the rest of the developing world. In Italy, it means taking over dozens of abandoned buildings and turning them into affordable housing and lively community centres. You can see the same spirit coursing through the actions of the Landless Peasants' Movement of Brazil, which seizes tracts of unused farmland and uses them for sustainable agriculture, markets and schools under the slogan '*Ocupar, Resistir, Producir*' (Occupy, Resist, Produce).

It is in South Africa where this spirit of direct action may be spreading most rapidly. Since a sweeping privatization program was instituted in 1993, half a million jobs have been lost, wages for the poorest 40 per cent have dropped by 21 per cent – poor areas have seen their water costs go up 55 per cent, and electricity as much as 400 per cent. Many have resorted to drinking polluted water, leading to a cholera outbreak that infected 100,000 people. In Soweto, 20,000 homes have their electricity cut off each month. In the face of this system of "economic apartheid," as privatization is called by many South African activists, unemployed workers in Soweto have been reconnecting their neighbours' cut off water and the Soweto Electricity Crisis Committee has illegally reconnected power in thousands of homes.

No matter where it takes place, the theory behind this defiant wave of direct action is the same: activism can no longer be about registering symbolic dissent. It must be about taking action to make people's lives better – where they live, right away.

The question now facing this movement is how to transform these small, often fleeting, initiatives into broader, more sustainable social structures. There are many attempts to answer this question but by far the most ambitious is the annual World Social Forum, launched in January 2001 in Porto Alegre, Brazil. The WSF's optimistic slogan is "Another World Is Possible" and it was conceived as an opportunity for an emerging movement to stop screaming about what it is against and start articulating what it is for. In its first year, more than 10,000 people attended a week of more than 60 speeches, dozens of concerts and 450 workshops. The particular site was chosen because Brazil's Workers Party (Partido dos Trabalhadores, the PT) is in power in the city of Porto Alegre, as well as in the state of Rio Grande do Sul, and has become known world-wide for its innovations in participatory democracy.

But the World Social Forum is not a political convention: there are no policy directives made, no official motions passed, and no attempts to organize the parts of this movement into a political party, with subordinate cells and locals. And that fact, in a way, is what makes this wave of activism unlike anything that has come before it. Thanks to the Net, mobilizations are able to unfold with sparse bureaucracy and minimal hierarchy; forced consensus and labored manifestos are fading into the background, replaced instead by a culture of constant, loosely structured and sometimes compulsive information-swapping. While individual intellectuals and key organizers may help shape the ideas of the people on the streets, they most emphatically do not have the power or even the mechanisms to lead them in any one direction. It isn't even, if truth be told, a movement. It is thousands of movements, intricately linked to one another, much as 'hotlinks' connect their websites on the Internet. And while this network is wildly ambitious in its scope and reach, its goals are anything but imperial. This network unrelentingly challenges the most powerful institutions and individuals of our time, but does not seek to seize power for itself. Instead it seeks to disperse power, as widely and evenly as possible.

The best example of this new revolutionary thinking is the Zapatistas uprising in Chiapas, Mexico. When the Zapatistas rose up against the Mexican military in January 1994, their goal was not to win control over the Mexican state but to seize and build autonomous spaces where "democracy, liberty and justice" could thrive. For the Zapatistas, these free spaces, created from reclaimed land, communal agriculture and resistance to privatization, are an attempt to create counter-powers to the state, not a bid to overthrow it and replace it with an alternate, centralized regime.

It's fitting that the figure that comes closest to a bona fide movement 'leader' is Subcomandante Marcos, the Zapatista spokesperson who hides his real identity and covers his face with a mask. Marcos, the quintessential anti-leader, insists that his black mask is a mirror, so that "Marcos is gay in San Francisco, black in South Africa, an Asian in Europe, a Chicano in San Ysidro, an anarchist in Spain, a Palestinian in Israel, a Mayan Indian in the streets of San Cristobal, a Jew in Germany, a Gypsy in Poland, a Mohawk in Quebec, a pacifist in Bosnia, a single woman on the Metro at 10 pm, a peasant without land, a gang member in the slums, an unemployed worker, an unhappy student and, of course, a Zapatista in the mountains." In other words, he says, he is us: we are the leader we've been looking for.

Marcos's own story is of a man who came to his leadership not through swaggering certainty, but by coming to terms with political doubt, by learning to follow. The most repeated legend that clings to him goes like this: an urban Marxist intellectual and activist, Marcos was wanted by the state and was no longer safe in the cities. Filled with revolutionary rhetoric and certainty, he fled to the mountains of Chiapas in southeast Mexico to convert the poor indigenous masses to the cause of armed proletarian revolution against the bourgeoisie. He said the workers of the world must unite, and the Mayans just stared at him. They said they weren't workers but people, and, besides, land wasn't property but the heart of their communities. Having failed as a Marxist missionary, Marcos immersed himself in Mayan culture. But the more he learned, the less he realised he knew.

Out of this process, a new kind of army emerged – the EZLN is not controlled by an elite of guerrilla commanders but by the communities themselves, through clandestine councils and open assemblies. Marcos isn't a commander barking orders, but a subcomandante, a conduit for the will of the councils. His first words in his new persona were: "through me speaks the will of the Zapatista National Liberation Army."

The Zapatista struggle has become a powerful beacon for other movements around the world precisely because it is organized according to principles that are the mirror opposite of the way states, corporations and religions tend to be organized. It responds to concentration with a maze of fragmentation, to centralization with localization, to power consolidation with radical power dispersal. The question is: could this be a microcosm for a global strategy to reclaim the commons from the forces of privatization?

Many of today's activists have already concluded that globalization is not simply a good idea that has been grabbed by the wrong hands. Nor do they believe that the situation could be righted if only international institutions like the WTO were made democratic and accountable. Rather, they are arguing that alienation from global institutions is only the symptom of a much broader crisis in representative democracy, one that has seen power and decision-making delegated to points further and further away from the places where the effects of those decisions are felt. As one-size-fits-all logic sets in, it leads at once to a homogenization of political and cultural choices, and to widespread civic paralysis and disengagement.

If centralization of power and distant decision making are emerging as the common enemies, there is also an emerging consensus that participatory democracy at the local level – whether through unions, neighborhoods, city governments, farms, villages, or aboriginal self-government – is the place to start building alternatives. The common theme is an overarching commitment to self-determination and diversity: cultural diversity, ecological diversity, even political diversity. The Zapatistas speak of building a movement of 'one "no" and many "yeses"', a description that defies the characterization that this is one movement at all, and challenges the assumption that it should be. What seems to be emerging organically is not a movement for a single global government but a vision for an increasingly connected international network of very local initiatives, each built on reclaimed public spaces, and, through participatory forms of democracy, made more accountable than either corporate or state institutions. If this movement has an ideology it is democracy, not only at the ballot box but woven into every aspect of our lives.

All of this makes it terribly ironic when critics attempt to make ideological links between anti-corporate protesters and religious fundamentalists like bin Laden, as British secretary of state for international development Clare Short did in November 2001. "Since September 11, we haven't heard from the protesters," she observed. "I'm sure they are reflecting on what their demands were because their demands turned out to be very similar to those of Bin Laden's network." She couldn't have been more mistaken. Bin Laden and his followers are not driven not by a critique of centralized power but by a rage that more power is not centralized in their own hands. They are furious not at the homogenization of choices, by that the world is not organized according to their own homogenous and imperialist belief system.

In other words, this is a classic power struggle over which great, all-knowing system will govern the day; where the battle lines were once Communism versus Capitalism, they are now the God of the Market squaring off against the God of Islam. For bin Laden and his followers, much of the allure of this battle is clearly the idea that they are living again in mythic times, when men were god-like, battles were epic and history was spelled with a capital H. "Screw you, Francis Fukuyama," they seem to be saying, "History hasn't ended yet. We are still making it."

It's an idea we've heard from both sides since September 11, a return of the great narrative: chosen men, evil empires, master plans, and great battles. All are ferociously back in style. This grand redemption narrative is our most persistent myth, and it has a dangerous flip side. When a few men decide to live their myths, to be larger than life, it can't help but have an impact on all the lives that unfold in regular sizes. People suddenly look insignificant by comparison, easy to sacrifice by the thousands in the name of some greater purpose.

Thankfully, anti-corporate and pro-democracy activists are engaged in no such fire and brimstone crusades. They are instead challenging systems of centralized power *on principle*, as critical of left-wing, one-size-fits-all state solutions as of right-wing market ones. It is often said disparagingly that this movement lacks ideology, an overarching message, a master plan. This is absolutely true, and we should be extraordinarily thankful. At the moment, the anti-corporate street activists are ringed by would-be leaders, anxious for the opportunity to enlist them as foot soldiers. It is to this young movement's credit that it has as yet fended off all of these agendas and has rejected everyone's generously donated manifesto, holding out for an acceptably democratic, representative process to take its resistance to the next stage. Will it be a ten-point plan? A new political doctrine?

Perhaps it will be something altogether new. Not another ready-made ideology to do gladiatorial combat with free-market fundamentalism and Islamic fundamentalism, but a plan to protect the possibility and development of many worlds – a world, as the Zapatistas say, with many worlds in it. Maybe instead of meeting the proponents of neoliberalism head on, this movement of movements will surround them from all directions.

This movement is not, as one newspaper headline recently claimed, "so yesterday." It is only changing, moving, yet again, to a deeper stage, one that is less focused on acts of symbolic resistance and theatrical protests and more on "living our alternatives into being," to borrow a phrase from a recent direct action summit in New York City. Shortly after *No Logo* was published, I visited the University of Oregon to do a story on anti-sweatshop activism at the campus that is nicknamed Nike U. There I met student activist Sarah Jacobson. Nike, she told me, was not the target of her activism, but a tool, a way to access a vast and often amorphous economic system. "It's a gateway drug," she said cheerfully.

For years, we in this movement have fed off our opponents' symbols - their brands, their office towers, their photo-opportunity summits. We have used them as rallying cries, as focal points, as popular education tools. But these symbols were never the real targets; they were the levers, the handles. The symbols were only ever doorways. It's time to walk through them.

Naomi Klein, *January 2002*

NOTES

INTRODUCTION: A WEB OF BRANDS

1. Industry Canada, "Canadian Imports – Top 25 Products. Origin: Indonesia."

2. Levi Strauss Web site, 1996.

CHAPTER ONE: NEW BRANDED WORLD

1. "Government Spending Is No Substitute for the Exercise of Capitalist Imagination," *Fortune*, September 1938, 63–64.

2. Ellen Lupton and J. Abbott Miller, *Design Writing Research: Writing on Graphic Design* (New York: Kiosk, 1996), 177.

3. Roland Marchand, "The Corporation Nobody Knew: Bruce Barton, Alfred Sloan, and the Founding of the General Motors 'Family,'" *Business History Review*, 22 December 1991, 825.

4. Randall Rothberg, *Where the Suckers Moon* (New York: Vintage, 1995), 137.

5. Stats are from McCann-Erikson's ad spending forecast appearing in *Advertising Age* and the United Nations Human Development Report 1998. Most industry watchers estimate that U.S. spending from the global brands represents 40 percent of the total ad spending in the rest of the world. Canadian ad spending, which is less rigorously tracked by the industry, follows the same rate of growth, but with smaller figures. Between 1978 and 1994, for instance, it grew from a $2.7 billion industry to a $9.2 billion industry (source: "A Report Card on Advertising Revenues in Canada," 1995).

6. Yumiko Ono, "Marketers Seek the 'Naked' Truth in Consumer Psyches," *Wall Street Journal*, 30 May 1997, B1.

7. *Daily Mail* (London), 17 November 1997.

8. *Wall Street Journal*, 14 April 1998.

9. *Boston Globe*, 21 July 1993.

10. *Marketing Management*, Spring 1994.

11. *Economist*, 10 April 1993.

12. U.S. stats from "100 Leading National Advertisers," *Advertising Age*, 29 September 1993. In Canada, overall ad spending also dropped in 1991 by 2.95% and dipped again 0.3% in 1993. (Source: "A Report Card on Advertising Revenues in Canada," 1995.)

13. Jack Myers, *Adbashing: Surviving the Attacks on Advertising* (Parsippany, N.J.: American Media Council, 1993), 277.

14. *Guardian*, 12 June 1993.

15. Shelly Reese, "Nibbling at Brand Loyalty," *Cincinnati Enquirer*, 11 July 1993, G1.

16. Scott Bedbury (as vice president of marketing with Starbucks, speaking to the Association of National Advertisers) as quoted in *The New York Times*, 20 October 1997.

17. Howard Shultz, *Pour Your Heart into It* (New York: Hyperion, 1997), 5.

18. Tom Peters, "What Great Brands Do," *Fast Company*, August/September 1997, 96.

19. Geraldine E. Willigan, "High-Performance Marketing: An Interview with Nike's Phil Knight," *Harvard Business Review*, July 1992, 92.

20. Tom Peters, *The Circle of Innovation* (New York: Alfred A. Knopf, 1997), 16.

21. Jennifer Steinhauer, "That's Not a Skim Latte, It's a Way of Life," *New York Times*, 21 March 1999.

22. Association of National Advertisers.

23. *Wall Street Journal*, 1 April 1998, from "Trends in Corporate Advertising, a joint project of the Association of National Advertisers and Corporate Branding Partnership, in association with the Wall Street Journal."

24. Donald Katz, *Just Do It: The Nike Spirit in the Corporate World* (Holbrook: Adams Media Corporation, 1994), 25.

25. "In the Super Bowl of Sport Stuff, the Winning Score is $2 Billion," *New York Times*, February 11, 1996, Section 8, 9.

26. John Heilemann, "All Europeans Are Not Alike," *New Yorker*, 28 April & 5 May 1997, 175.

27. "Variations: A Cover Story," *New York Times Magazine*, 13 December 1998, 124.

28. *Report on Business Magazine, World in 1997*.

29. *Business Week*, 22 December 1997.

30. Sam I. Hill, Jack McGrath and Sandeep Dayal, "How to Brand Sand," *Strategy & Business*, Second Quarter 1998.

31. Peters, *The Circle of Innovation*, 337.

CHAPTER TWO: THE BRAND EXPANDS

1. *Business Week*, 24 May 1999, and *Wall Street Journal*, 12 February 1999.

2. Matthew P. McAllister, *The Commercialization of American Culture* (Thousand Oaks: Sage, 1996), 177.

3. Ibid., 221.

4. *Wall Street Journal*, 12 February 1999.

5. Lesa Ukman, "Assertions," *IEG Sponsorship Report*, 22 December 1997, 2.

6. *Advertising Age*, 28 September 1998.

7. "Old-fashioned Town Sours on Candymaker's New Pitch," *Wall Street Journal*, 6 October 1997, A1.

8. Gloria Steinem, "Sex, Lies & Advertising," *Ms.*, July/August 1990.

9. "Chrysler Drops Its Demand for Early Look at Magazines," *Wall Street Journal*, 15 October 1997.

10. *Independent*, 5 January 1996, 1, and *Evening Standard*, 5 January 1996, 12; and Andrew Blake, "Listen to Britain," in *Buy This Book*, edited by Mica Nava, Andrew Blake, Iain MacRury and Barry Richards (London: Routledge, 1997), 224.

11. *Saturday Night*, July/August 1997, 43–51.

12. "MTV Man Warns about Branding," *Globe and Mail*, 19 June 1998, B21.

13. "Sing a Song of Selling," *Business Week*, 24 May 1999.

14. Michael J. Wolf, *The Entertainment Economy* (New York: Times Books, 1999), 66.

15. Interview aired on Citytv's *New Music* "Smokes and Booze" special on 22 February 1997.

16. Interview aired on Citytv's *New Music* on 9 September 1995.

17. Kyle Stone, "Promotion Commotion," *Report on Business Magazine*, December 1997, 102.

18. Ann Powers, "Everything and the Girl," *Spin*, November 1997, 74.

19. Wolf, *The Entertainment Economy*, 29.

20. "And the Brand Played On," *Elm Street*, April 1999.

21. "Star Power, Star Brands," *Forbes*, 22 March 1999.

22. Willigan, "High-Performance Marketing," 94.

23. Katz, *Just Do It*, 8.

24. *New York Times*, 20 December 1997, A1.

25. Katz, *Just Do It*, 284.

26. Ibid., 34, 231.

27. Ibid., 30–31.

28. Ibid., 36, 119.

29. Ibid., 233.

30. Ibid., 24.

31. Ibid., 24.

32. "Michael Jordan's Full Corporate Press," *Business Week*, 7 April 1997, 44.

33. Katz, *Just Do It*, 35.

34. "Space Jam Turning Point for Warner Bros., Jordan," *Advertising Age*, 28 October 1996, 16.

35. "Merchandise Upstages Box Office," *Wall Street Journal*, 24 September 1996.

36. "Armchair Adventures," *Globe and Mail*, 11 January 1999, C12.

37. Katz, *Just Do It*, 82.

38. Roy F. Fox, "Manipulated Kids: Teens Tell How Ads Influence Them," *Educational Leadership*, September 1995, 77.

CHAPTER THREE: ALT. EVERYTHING

1. Mean Fiddler promotional material obtained by author.

2. "Woodstock at 25" (editorial), *San Francisco Chronicle*, 14 August 1994, 1.

3. "Hits replace jingles on TV Commercials," *Globe and Mail*, 29 November 1997.

4. Robert Goldman and Stephen Papson, *Sign Wars: The Cluttered Landscape of Advertising* (New York: Guilford Press, 1996), 43.

5. *Greater Baton Rouge Business Report*, 28 June 1994, 30. Decoteau is co-owner of the store Serape in Baton Rouge.

6. Eric Ransell, "IBM's Grassroots Revival," *Fast Company*, October/November 1997, 184.

7. *Campaign*, 30 May 1997.

8. *USA Today*, 4 September 1996.

9. "Levi's Blues," *New York Times Magazine*, 21 March 1998.

10. "Job Titles of the Future," *Fast Company*, October/November 1997, 54.

11. Marc Gunther, "This Gang Controls Your Kids' Brains," *Fortune*, 27 October 1997.

12. Ibid.

13. Robert Sullivan, "Style Stalker," *Vogue*, November 1997, 182, 187–88.

14. Janine Lopiano-Misdom and Joanne De Luca, *Street Trends: How Today's Alternative Youth Cultures are Creating Tomorrow's Mainstream Markets* (New York: HarperCollins Business, 1997), 11.

15. Norman Mailer, "The Faith of Graffiti," *Esquire*, May 1974, 77.

16. "Off the Street...," *Vogue*, April 1994, 337.

17. Lopiano-Misdom and De Luca, *Street Trends*, 37.

18. Erica Lowe, "Good Rap? Bad Rap? Run-DMC Pushes Rhyme, Not Crime," *San Diego Union-Tribune*, 18 June 1987, E-13.

19. Christopher Vaughn, "Simmons' Rush for Profits," *Black Enterprise*, December 1992, 67.

20. Lisa Williams, "Smaller Athletic Firms Pleased at Super Show; Shoe Industry Trade Show," *Footwear News*, 16 February 1987, 2.

21. *Advertising Age*, 28 October 1996.

22. Josh Feit, "The Nike Psyche," *Willamette Week*, 28 May 1997.

23. *Tommy Hilfiger 1997 Annual Report*.

24. Paul Smith, "Tommy Hilfiger in the Age of Mass Customization," in *No Sweat: Fashion, Free Trade, and the Rights of Garment Workers*, edited by Andrew Ross (New York: Verso, 1997), 253.

25. Nina Munk, "Girl Power," *Fortune*, 8 December 1997, 137.

26. "Old Navy Anchors Micro-Radio Billboard," *Chicago Sun-Times*, 28 July 1998.

27. Editorial, *Hermenaut #10: Popular Culture*, 1995.

28. Nick Compton, "Who Are the Plastic Palace People?" *Face*, June 1996, 114–15.

29. Lopiano-Misdom and De Luca, *Street Trends*, 8–9.

30. Ibid., 110.

31. James Hibberd, "Bar Hopping with the Bud Girls," *Salon*, 1 February 1999.

32. *Business Week*, 12 April 1996.

33. *Hype!*, Doug Pray, 1996.

34. Susan Sontag, "Notes on Camp," in *Against Interpretation*, edited by Susan Sontag (New York: Anchor Books, 1986), 275.

35. Ibid., 283.

36. Ibid., 288.

37. *Globe and Mail*, 22 November 1997.

38. *Women's Wear Daily*, 7 November 1997.

CHAPTER FOUR: THE BRANDING OF LEARNING

1. Myers, *Adbashing*, 151.

2. *Wall Street Journal*, 24 November 1998.

3. "A La Carte Service in the School Lunch Program," fact sheet prepared for Subway by Giuffrida Associates, Washington, D.C.

4. *Wall Street Journal*, 15 September 1997.

5. The Center for Commercial-Free Public Education, Oakland, California, 9 October 1997 release.

6. *Extra! The magazine of Fairness and Accuracy in Reporting*, May/June 1997 10, no. 3.

7. *Wall Street Journal*, 24 November 1997, B1.

8. "Captive Kids: Commercial Pressures on Kids at School," Consumers Union paper, 1995.

9. Josh Feit, "Nike in the Classroom: Nike's effort to teach kids about treading lightly on Mother Nature meet with skepticism from educators and consumer watchdogs," *Willamette Week*, 15 April 1998.

10. "ZapMe! Sparks Battle Over Ads," Associated Press, 6 December 1998.

11. "Schools Profit from Offering Pupils for Market Research," *New York Times*, 5 April 1999.

12. *Advertising Age*, 14 August 1995.

13. Kim Bolan, *Vancouver Sun*, 20 June 1998, B5.

14. Ibid.

15. Associated Press, 25 March 1998.

16. Stuart Ewen, *Captains of Consciousness* (New York: McGraw-Hill, 1976), 90.

17. *Wisconsin State Journal*, 21 May 1996.

18. *Kentucky Gazette*, 17 June 1997.

19. Associated Press, 13 April 1996.

20. Both quotations come from personal interviews with participants in the Kent State incident.

21. Mark Edmundson, *Harper's*, September 1997.

22. *Science*, vol. 273, 26 July 1996, and *Science*, vol. 276, 25 April 1997.

23. "A Cautionary Tale," *Science*, vol. 273.

24. Michael Valpy, "Science Friction," *Elm Street*, December 1998.

25. *Science*, vol. 276, 25 April 1997.

26. W. Cohen, R. Florida, W.R. Goe, "University-Industry Research Centers in the United States" (Pittsburgh: Carnegie Mellon University Press, 1994).

27. *Business Week*, 22 December 1997.

28. Julianne Basinger, "Increase in Number of Chairs Endowed by Corporations Prompt New Concerns," *Chronicle of Higher Education*, 24 April 1998, A51.

29. "ZapMe! Invites Ralph Nader Back to School," PR Newswire, 10 December 1998.

30. Janice Newson, "Technical Fixes and Other Priorities of the Corporate-Linked University: The Humanists' Challenge," paper presented to the Humanities Research Group of the University of Windsor, October 1995.

CHAPTER FIVE: PATRIARCHY GETS FUNKY

1. Jeanie Russell Kasindorf, "Lesbian Chic," *New York*, 10 May 1993, 35.
2. Dinesh D'Souza, "Illiberal Education," *Atlantic Monthly*, March 1991, 51.
3. John Taylor, "Are You Politically Correct?" *New York*, 21 January 1991.
4. J. Walker Smith and Ann Clurman, *Rocking the Ages* (New York: HarperCollins, 1997), 88.
5. *Vogue*, November 1997.
6. "Starbucks Is Ground Zero in Today's Coffee Culture," *Advertising Age*, 9 December 1996.
7. Jared Mitchell, "Out and About," *Report on Business Magazine*, December 1996, 90.
8. Powers, "Everything and the Girl," 74.
9. Gary Remafedi, Simone French, Mary Story, Michael D. Resnick and Robert Blum, "The Relationship between Suicide Risk and Sexual Orientation: Results of a Population-Based Study," *American Journal of Public Health*, January 1998, 88, no. 1, 57–60.
10. Goldman and Papson, *Sign Wars*, v.
11. Richard Goldstein, "The Culture War Is Over! We Won! (For Now)," *Village Voice*, 19 November 1996.
12. Theodore Levitt, "The Globalization of Markets," *Harvard Business Review*, May–June 1983.
13. Sumner Redstone made the remark at the Drexel Burnham Lambert annual media conference in January 1990. He made a similar comment in an October 1994 interview with *Forbes*: "MTV is associated with the forces of freedom and democracy around the world. When the Berlin Wall came down, there were East German guards holding MTV umbrellas. MTV is on the cutting-edge. It's irreverent. It's antiestablishment."
14. Scripps Howard News Service, 19 July 1997.
15. *Times* (London), 2 September 1993, A5, an edited version of a speech Rupert Murdoch delivered on 1 September 1993.
16. From a speech made by UN Secretary General Kofi Annan on 17 October 1997.
17. The 1997 United Nations Human Development Report, Table 2.1.
18. Ibid., Figure 2. 2b.
19. "Western Companies Compete to Win Business of Chinese Babies," *Wall Street Journal*, May 15, 1998. The quotation comes from Robert Lipson, president of U.S.-China Industrial Exchange Inc., which is opening joint-venture pediatric hospitals in China.
20. Bernard Wysocki, "In Developing Nations Many Youths Splurge, Mainly on U.S. Goods," *Wall Street Journal*, June 26, 1997.
21. Chip Walker, "Can TV Save the Planet?" *American Demographics*, May 1996, 42.
22. Cyndee Miller, "Teens Seen as the First Truly Global Consumers," *Marketing News*, 27 March 1995.
23. Renzo Rosso, *FoRty*, self-published.
24. *Wall Street Journal*, 26 June 1997.

25. Dirk Smillie, "Tuning in First Global TV Generation," *Christian Science Monitor*, 4 June 1997.

26. Walker, "Can TV Save the Planet?", 42.

27. Tim Brennan, "'PC' and the Decline of the American Empire," *Social Policy*, Summer 1991, 16.

28. Ibid.

29. *New York Times*, 8 December 1997, D12.

30. Sarah Eisenstein, *Give Us Bread But Give Us Roses* (New York: Routledge, 1983), 32.

31. Dorothy Inglis, *Bread and Roses* (New York: Killick Press, 1996), 1.

CHAPTER SIX: BRAND BOMBING

1. James Howard Kunstler, *The Geography of Nowhere* (New York: Touchstone, 1993).

2. Bob Ortega, *In Sam We Trust* (New York: Times Books, 1998), 75.

3. Sam Walton with John Huey, *Sam Walton, Made in America: My Story* (New York: Doubleday, 1992), 110.

4. *Globe and Mail*, 9 February 1998, B1.

5. Ortega, *In Sam We Trust*, 293. According to Ortega, in New England, "Six of the first 30 stores Wal-Mart proposed in the region had sparked heated fights."

6. "Judge Rules That Toys 'R' Us Illegally Limited Supplier Sales," *Wall Street Journal*, 30 September 1997 (on-line).

7. Starbucks 1995 Annual Report.

8. Ibid.

9. John Barber, "Something Bitter Brewing Over Annex Cafe," *Globe and Mail*, 9 November 1996.

10. Nina Munk, "Gap Gets It," *Fortune*, 3 August 1998.

11. Scott Bedbury has stated that "challenges come when we're compared to McDonald's. All of us take offense at that." *Advertising Age*, 9 December 1996, 49.

12. Nicole Nolan, "Starbucked!," *In This Times*, 11 November 1996.

13. *Globe and Mail*, 7 February 1998, A1.

CHAPTER SEVEN: MERGERS AND SYNERGY

1. Alexis de Tocqueville, *Democracy in America*, vol. II, translated by Henry Reeve (New York: Schocken Books, 1961), 94. First edition published 1835–40.

2. *New York Times*, 7 January 1998, D4; article quotes EPM Communications' *Licensing Letter*.

3. Wolf, *The Entertainment Economy*, 224.

4. Ibid, xvii.

5. "Nelvana Acquires Leading Children's Publisher Kids Can Press," Canadian NewsWire, 19 August 1998.

6. *Forbes*, 17 October 1994.

7. "Diesel's Guide to Working, Living Large," *Ad Age International*, May 1997.

8. "Why Foot Locker Is in a Sweat," *Business Week*, 27 October 1997, 52.

9. Frances Anderton, "Hawking the Hustler Sensibility," *New York Times*, 21 March 1999.

10. "Mass Marketers Invade the Land of Chic," *Wall Street Journal*, 4 October 1996, B1.

11. Kelly Barron, "Theme Players," *Fortune*, 22 March 1999, 53.

12. Wolf, *The Entertainment Economy*, 70.

13. Ibid., 68.

14. Geoff Pevere, *Team Spirit: A Field Guide to Roots Culture* (Toronto: Doubleday, 1998), 50.

15. Ibid., 47.

16. Ibid., 137.

17. From "Atopias: The Challenge of Imagineering," a lecture given at Toronto's Design Exchange.

18. Wolf, *The Entertainment Economy*, 11.

19. The American Booksellers Association and twenty-six independent bookstores filed a suit in March 1998 against Borders and Barnes & Noble charging that the two largest players in U.S. bookselling used their size to extract "secret and illegal" deals from publishers, making it difficult for independent book retailers to compete. At the time of writing, the case was still pending in California court. (The ABA is the source of bookstore numbers by year.)

20. *Globe and Mail*, 21 November 1997.

21. *Vancouver Sun*, 10 December 1996, C7.

22. *Forbes*, 17 October 1994.

23. Myers, *Adbashing*, 253.

24. Thomas Ferraro, UPI, 10 November 1983.

25. Ibid.

26. Peter Szekely, Reuters, 12 July 1985.

27. *New York Times*, 14 November 1993, 21.

28. Andrew L. Shapiro, "Memo to Chairman Bill," *Nation*, 10 November 1997.

CHAPTER EIGHT: CORPORATE CENSORSHIP

1. *Wall Street Journal*, 22 October 1997, A1.

2. *New York Times*, 12 November 1996.

3. *Billboard*, 2 October 1993.

4. *Globe and Mail*, 7 January 1998, C2.

5. "Guardian Angels," *New Yorker*, 25 November 1996, 47.

6. *Wall Street Journal*, 22 October 1997, A1.

7. *Sacramento Bee*, 10 December 1997, E1.

8. Gail Shister, Knight Ridder Newspapers, 20 October 1998.

9. "Fresh Air," National Public Radio, 29 September 1998.

10. Lawrie Mifflin, "ABC News Reporter Discovers the Limits of Investigating Disney," *New York Times*, 19 October 1998.

11. Jennet Conant, "Don't Mess with Steve Brill," *Vanity Fair*, August 1997, 62–74.

12. "'Controls Eased' Over Journalists and Artists; Deng Provides New Freedoms for Media," *South China Morning Post*, 30 September 1992, 1.

13. *Wall Street Journal*, 5 March 1998. The statements were made on 20 January 1998 at a gathering of Freedom Forum, a media foundation.

14. Japan Economic Newswire, 22 October 1993.

15. Seth Faison, "Dalai Lama Movie Imperils Disney's Future in China," *New York Times*, 26 November 1996.

16. "Gere's 'Corner' on Saving Tibet," *San Francisco Chronicle*, 26 October 1997.

17. *Wall Street Journal*, 3 November 1997.

18. Constance L. Hays, "Math Book Salted with Brand Names Raises New Alarm," *New York Times*, 21 March 1999.

19. Grant McCracken, *Culture and Consumption: New Approaches to the Symbolic Character of Consumer Goods and Activities* (Bloomington: Indiana University Press, 1988).

20. Susan Fournier, "The Consumer and the Brand: An Understanding within the Framework of Personal Relationships," Harvard Business School, Division of Research, working paper, September 1996, 64.

21. *Los Angeles Times*, 17 September 1997, E2.

22. *Sydney Morning Herald*, 21 March 1998.

23. David Gans, "The Man Who Stole Michael Jackson's Face," *Wired*, February 1995.

24. Eileen Fitzpatrick, "Lawsuit Doesn't Sink Aqua 'Barbie Girl' Driving Album Sales," *Billboard*, 27 September 1997.

25. Joan H. Murphy, "Mattel – Where Security Isn't Child's Play," *Security Management*, January 1990, 39.

26. Chuck Taylor, "Danish Breakout Group Aqua Toys with U.S. Pop Success with Its 'Barbie Girl,'" *Billboard*, 30 August 1997.

27. *Vancouver Sun*, 10 December 1996, C7.

28. Barnes & Noble Booksellers, background paper. Supplied by company.

29. Michael Moore, "Banned by Borders," *Nation*, 20 November 1996.

30. Amy Harmon, "As America Online Grows, Charges That Big Brother Is Watching," *New York Times*, 31 January 1999, A1.

31. Ibid.

32. Ibid.

33. Noam Chomsky, "Market Democracy in a Neoliberal Order," Davie Lecture, University of Cape Town, May 1997, reprinted in *Z Magazine*, September 1997, 40–46.

34. Excerpts from "Corporatism and Plutocracy," speech given at Harvard University, date unknown.

35. James Christie, "Bailey Satellites Do Damage Control," *Globe and Mail*, 17 July 1996.

36. Michael Walker, "Scally? Not Me, says Fowler," *Guardian*, 19 April 1997.

37. Nick Harris, "Footballer Falls Foul of the Rules," *Independent*, 22 March 1997.

38. Associated Press, 23 April 1997, quotation comes from Jill Krutick, an entertainment analyst at Smith Barney.

39. John Lippman, "Godzilla Opening Weekend Receipts Disappoint Despite Big Ad Campaign," *Wall Street Journal*, 26 May 1998.

40. Peters, *The Circle of Innovation*, 349.

41. "MTV Man Warns about Branding," *Globe and Mail*, 19 June 1998.

42. "Nike's Problems Don't Seem to Be Short Term to Investors," *New York Times*, 26 February 1998.

43. *Globe and Mail*, 8 May 1999.

CHAPTER NINE: THE DISCARDED FACTORY

1. Landor Web site.

2. "People Buy Products Not Brands," by Peter Schweitzer (J. Walter Thompson White Papers series, undated).

3. "Big Brand Firms Know the Name Is Everything," *Irish Times*, 27 February 1998.

4. Ortega, *In Sam We Trust*, 342.

5. "Trade and Development Report, 1997," United Nations Conference on Trade and Economic Development.

6. Katz, *Just Do It*, 204.

7. Cathy Majtenyi, "Were Disney Dogs Treated Better Than Workers?" *Catholic Register*, 23–30 December 1996, 9.

8. "Extreme Spreadsheet Dude," *Baffler* no. 9, 79, and *Wall Street Journal*, 16 April 1998 (on-line).

9. John Gilardi, "Adidas Share Offer Set to Win Gold Medal," Reuters, 26 October 1995.

10. *Globe and Mail*, 26 September 1997.

11. Charles Kernaghan, "Behind the Label: 'Made in China,'" prepared for the National Labor Committee, March 1998.

12. *Los Angeles Times*, 16 September 1997, D5. Furthermore, Sara Lee's investors had been getting a solid return on their investment but the stock "had gained 25 per cent over the prior 12 months, lagging the 35 per cent increase of the benchmark Standard & Poor's 500-stock index."

13. Peters, *The Circle of Innovation*, 16.

14. David Leonhardt, "Sara Lee: Playing with the Recipe," *Business Week*, 27 April 1998, 114.

15. Ibid.

16. Jennifer Waters, "After Euphoria, Can Sara Lee Be Like Nike?" *Crain's Chicago Business*, 22 September 1997, 3.

17. Nina Munk, "How Levi's Trashed a Great American Brand," *Fortune*, 12 April 1999, 83.

18. "Levi Strauss & Co. to Close 11 of Its North American Plants," Business Wire, 22 February 1999, B1.

19. *Wall Street Journal*, 4 November 1997, B1.

20. Joanna Ramey, "Levi's Will Resume Production in China After 5-Year Absence," *Women's Wear Daily*, 9 April 1998, 1.

21. "Anti-Sweatshop Activists Score in Campaign Targeting Athletic Retailers," *Boston Globe*, 18 April 1999.

22. Richard S. Thoman, *Free Ports and Foreign Trade Zones* (Cambridge: Cornell Maritime Press, 1956).

23. These are International Labor Organization figures as of May 1998 but in "Behind the Label: 'Made in China,'" by Charles Kernaghan, March 1998, the figures on China's zone are much higher. Kernaghan estimates that there are 30 million inside the zones, and that there are 400 – as opposed to 124 – special economic zones inside China.

24. The International Labor Organization's Special Action Program on Export Processing Zones. Source: Auret Van Heerden.

25. This estimate was provided by Michael Finger at the World Trade Organization in a personal correspondence. No official figures are available.

26. Figures for 1985 and 1995 provided by the WTO. Figures for 1997 supplied by the Maquila Solidarity Network/Labor Behind the Label Coalition, Toronto.

27. *World Accounting Report*, July 1992.

28. Saskia Sassen, *Losing Control? Sovereignty in an Age of Globalization* (New York: Columbia University Press, 1996), 8–9.

29. "Castro Dampens WTO Party," *Globe and Mail*, 20 May 1998.

30. Martin Cottingham, "Cut to the Bone," *New Statesman & Society*, 12 March 1993, 12.

31. Personal interview, 2 September 1997.

32. The Workers' Assistance Center, Rosario.

33. "Globalization Changes the Face of Textile, Clothing and Footwear Industries," International Labor Organization press release, 28 October 1996.

34. "Working Conditions in Sports Shoe Factories in China Making Shoes for Nike and Reebok," by Asia Monitor Resource Centre and Hong Kong Christian Industrial Committee, September 1997.

35. Personal interview, 1 September 1997.

36. "Behind the Wire: Anti-Union Repression in the Export Processing Zones," a report by the International Confederation of Free Trade Unions.

37. Steven Greenhouse, *New York Times*, 29 February 1999.

38. Suzanne Goldenberg, "Colombo Stitch-Up," *Guardian*, 7 November 1997.

39. Cottingham, "Cut to the Bone," 12.

40. Goldenberg, "Colombo Stitch-Up."

41. "The Globe-Trotting Sports Shoe" by Peter Madden and Bethan Books, published by Christian Aid.

42. Kernaghan, "Behind the Label: 'Made in China.'"

43. From a panel discussion at "International Relocation" conference held in Brussels on 19–20 September 1996.

44. "Globalization Changes the Face of Textile . . . ," ILO.

45. "Submission Concerning Pregnancy-Based Discrimination in Mexico's Maquiladora Sector to the United States National Administrative Office," submitted by Human Rights

Watch Women's Rights Project, Human Rights Watch/Americas, Internationals Labor Rights Fund, and Asociacion National de Abogados Democraticos, 15 May 1997.

46. "No Guarantees: Sex Discrimination in Mexico's Maquiladora Sector," Human Rights Watch Women's Rights Project, August 1996.

47. Laura Eggertson, "Abuse Part of Jobs at Mexican Firms," *Globe and Mail*, 14 October 1997.

48. Cottingham, "Cut to the Bone."

49. "General Motors Corporation's Response to June 28, 1996 Letter from Human Rights Watch." The statement was attached to a letter dated 14 August 1996 signed by Gregory E. Lau, Executive Director, Worldwide Executive Compensation and Corporate Governance.

50. *Wall Street Journal*, 21 November 1997 (on-line).

51. Kate Bronfenbrenner, "We'll Close! Plant Closings, Plant-Closing Threats, Unions Organizing and NAFTA," *Multinational Monitor*, 18, no. 3, March 1997.

52. David Fischer, "Global Hopscotch," *U.S. News and World Report*, 5 June 1995.

53. Henny Sander, "Sprinting to the Forefront," *Far Eastern Economic Review*, 1 August 1996, 50.

54. Personal interview, 3 September 1997.

55. Ortega, *In Sam We Trust*, 250.

56. "South Korea Will Leave Indonesia if Strikes Continue," *Straits Times* (Singapore), 30 April 1997, 18. The article reported that Reebok's Indonesian executive Scott Thomas had met with South Korean officials, saying that if the worker strikes continued in Indonesia, the company might relocate again, saying Reebok "could place its orders easily with other countries if the situation persisted."

57. *Jakarta Post*, 30 April 1997.

58. "Nike in China" (abridged), Harvard Business School, 9-390-092, 12 August 1993.

59. "Nike Joins President Clinton's Fair Labor Coalition," PR NewsWire, 2 August 1996.

60. Christopher Reed, "Sweatshop Jobs Don't Put Food On Table," *Globe and Mail*, 9 May 1997.

61. Allen R. Myerson, "In Principle, a Case for More 'Sweatshops,'" *New York Times*, 22 June 1997, 4–5.

62. Ibid.

63. "Labour-Women Say Nike Supports Women in Ads, But Not in Factories," Inter Press Service, 29 October 1997.

64. "Raising Wages a Penny an Hour," National Labor Committee press release, 29 March 1999. Wages fell from 27 cents an hour to 15 cents an hour, even after Nike announced a 6 percent raise.

65. "High Unemployment, Higher Prices and Lower Wages," Ibon press release, 15 March 1999.

66. "Two Shoe Firms Close RP Shops," *Philippine Daily Enquirer*, 22 February 1999. The two factories were P.K. Export, which laid off 300 workers in 1998 and employed another 767

when the closure was announced, and Lotus Footwear, which employed 438 workers when it filed a notice of factory closure.

67. Aaron Bernstein, "Outsourced – And Out of Luck," *Business Week*, 17 July 1995, 60–61.

CHAPTER TEN: THREATS AND TEMPS

1. "A Conversation with Charles Handy," *Organizational Dynamics*, Summer 1996, 15–26.

2. For instance, in Canada, "between 1976 and 1997, the proportion of Canadians working in goods-producing industries shrank to 27 percent from 36 percent, according to Statistics Canada. Meanwhile, the proportion of the population working in the service industries rose to 73 per cent from 65 per cent." *Report on Business Magazine*, April 1998, 74.

3. Donna Smith and Carole Lusby, "Analysis of Educational Needs Assessment of Retail Employees," Ryerson Polytechnic University, 14 February 1997.

4. Personal interview, 7 October 1997.

5. Personal interview, 7 October 1997.

6. Personal interview, 24 November 1997.

7. In the U.S., the average hourly wage for a retail worker was $8.26.

8. Ortega, *In Sam We Trust*, 361. In Canada, Wal-Mart employees earn Can$8 an hour and have an average annual income of around $12,000.

9. *San Francisco Chronicle*, 3 October 1997, A19.

10. Dan Gallin is general secretary of the International Union of Food, Agricultural, Hotel, Restaurant, Catering, Tobacco and Allied Workers Association (IUF) based in Geneva. He offered the definition in an interview on the McSpotlight Web site. A good illustration of the place where trademark law interferes with public discourse about the reality of the corporate political landscape is in the McDonald's Corporation's threat to sue the Oxford Dictionary (among several other parties) over the word "McJobs." Not only has McDonald's, which employs over 1 million people worldwide, played a huge role in pioneering the low standards now equated with the word "McJobs," but it has also decided to restrict our ability to have a public discussion about the impact of the McJobs phenomenon.

11. Verdict delivered 19 June 1997.

12. *Good Morning America*, 16 April 1998, interviewer Kevin Newman with guests Bryan Drapp and Dominic Tocco.

13. Personal interview, 7 October 1997.

14. Letter addressed to "Borders Booksellers, Musicsellers, and Cafe Staff," from Richard L. Flanagan, President, Borders Stores, 30 May 1997.

15. Personal interview, 24 November 1997.

16. "Why Store 21 Tried to Unionize," Borders Books & Union Stuff Web site.

17. Source: Dan Gallin, general secretary of the International Union of Food, Agricultural, Hotel, Restaurant, Catering, Tobacco and Allied Workers Association (IUF), McSpotlight Web site.

18. *Globe and Mail*, 13 June 1998.

19. Ontario Labour Relations Board, File No. 0387-96-R. Decision of Janice Johnston, vice chair, and board member H. Peacock, 10 February 1997.

20. The number of part-time workers in the U.S. in 1997 was 23.2 million. *Handbook of U.S. Labor Statistics*, Bureau of Labor Statistics, 1998. According to Harry Pold, labor force researcher at Statistics Canada, between 1975 and 1997, part-time employment in Canada increased by 4.2 percent and full-time employment increased by a rate of only 1.5 percent ("Employment & Job Growth," Labour & Household Surveys Analysis Division, 1998).

21. Andrew Jackson, "Creating More and Better Jobs Through Reduction of Working Time," policy paper for Canadian Labour Conference, February 1998.

22. Personal interview, 24 November 1997.

23. Ortega, *In Sam We Trust*, 351.

24. *USA Today*, 5 August 1997, B1.

25. Ibid.

26. Ortega, *In Sam We Trust*, xiii.

27. Jim Frederick, "Internment Camp: The Intern Economy and the Culture Trust," *Baffler*, no. 9, 51–58.

28. Ibid.

29. U.S. Department of Labor.

30. "Staffing Services Annual Update," National Association of Temporary and Staffing Services, 1999.

31. In fact, Manpower, which employs over 800,000 workers, is a larger employer than Wal-Mart, which employs 720,000, but since Manpower's workers aren't out working every day, on any given day Wal-Mart has more workers on the payroll than Manpower.

32. *USA Today*, 5 August 1997, B1.

33. Helen Cooper and Thomas Kamm, "Europe Firms Lift Unemployment by Laying Off Unneeded Workers," *Wall Street Journal*, 3 June 1998.

34. Ibid.

35. Ibid.

36. Cooper and Kamm, "Europe Firms Lift Unemployment."

37. United States Bureau of Labor Statistics.

38. Bruce Steinberg, "Temporary Help Annual Update for 1997," *Contemporary Times*, Spring 1998.

39. Bernstein, "Outsourced – And Out of Luck."

40. Ibid.

41. Peters, *The Circle of Innovation*, 240.

42. Steinberg, "Temporary Help Annual Update for 1997."

43. Chris Benner, "Shock Absorbers in the Flexible Economy: The Rise of Contingent Employment in Silicon Valley," May 1996. Published by Working Partnerships USA.

44. Leslie Helm, "Microsoft Testing Limits on Temp Worker Use," *Los Angeles Times*, 7 December 1997, D1.

45. Ibid.

46. Microsoft won't divulge how many temps it uses but the 5,750 figure comes from the National Writers Union, which came to it by counting the number of E-mail addresses at

Microsoft that begin with the "a-" prefix. The "a" stands for "agency" and is on all the temps' accounts.

47. Helm, "Microsoft Testing Limits."

48. TechWire, 26 July 1997.

49. *Business Insurance*, 9 December 1996, 3.

50. Kevin Ervin, "Microsoft Clarifies Relationship with Temporary Workers," Knight Ridder Tribune Business News, 24 June 1998.

51. Alex Fryer, "Temporary Fix at Microsoft?" *Seattle Times*, 16 December 1997, A1.

52. Remarks by Bob Herbold, Seattle, Washington, 24 July 1997. From transcript.

53. Helm, "Microsoft Testing Limits."

54. Jonathan D. Miller, "Microsoft cutting back? In one sense it has, official says," *Eastside Journal* (Bellevue WA), 17 July 1997.

55. Remarks by Bob Herbold, Executive Vice President and Chief Operating Officer, Microsoft Corporation Annual Shareholders' Meeting, 14 November 1997, Seattle, Washington.

56. Peters, *The Circle of Innovation*, 184–85.

57. Daniel H. Pink, "Free Agent Nation," *Fast Company*, December 1997/January 1998.

58. "Opportunity Rocks!" *Details*, June 1997, 103.

59. Ron Lieber, "Don't Believe the Hype," *Details*, June 1997, 113.

60. "How We Work Now," *Newsweek*, 1 February 1999.

61. "Nonstandard Work, Substandard Jobs: Flexible Work Arrangements in the U.S.," Economic Policy Institute, Washington, DC.

62. Benner, "Shock Absorbers in the Flexible Economy."

63. *Employment and Unemployment in 1997: The Continuing Jobs Crisis*, Canadian Labour Congress.

64. Clive Thompson, "The Temp," *This Magazine*, February 1998, 32.

65. *San Francisco Examiner*, 27 April 1998, D27.

66. *Wall Street Journal*, 22 May 1998 (on-line).

67. Pink, "Free Agent Nation."

68. *Wall Street Journal*, 23 February 1998, A22.

69. "Runaway CEO Pay," on AFL-CIO's Executive PayWatch Web site.

70. "Executive Excess '98: Fifth Annual Executive Compensation Survey" (Boston: United for a Fair Economy), 23 April 1998.

71. *Globe and Mail*, "Report On Business," 21 April 1998.

72. Jennifer Reingold, "Executive Pay," *Business Week*, 20 April 1998, 64–70.

73. From "Corporate Success, Social Failure, Corporate Credibility," a speech given to the Canadian Club of Toronto, 23 February 1998.

CHAPTER ELEVEN: BREEDING DISLOYALTY

1. Keffo, editorial, *Temp Slave*, Issue 11.

2. "A Wake-up Call for Business," *Business Week*, 1 September 1997, 26–27.

3. World Development Movement, "Corporate Giants: Their grip on the world's economy."

The 5 percent of world employment relates to both direct and indirect employment (73 million, or two-thirds, are directly employed). This figure comes from the United Nations Research Institute for Social Development (UNRISD), occasional paper no. 5, written by Eric Kolodner. Figure on percentage of world's productive assets comes from the UNCTAD's 1994 World Investment Report.

4. "Global 500," *Fortune*, 29 July 1991 and 3 August 1998. Companies are ranked by revenue.

5. Challenger, Gray & Christmas and the U.S. Bureau of Labor Statistics, 1999.

6. "UNCTAD Sounds Warning on Globalization," UNCTAD press release, 11 September 1997.

7. "Poverty Amid Consumer Affluence," UN Human Development Report 1998 press release, 9 September 1998.

8. Aaron Bernstein, "The Wage Squeeze," *Business Week*, 17 July 1995, 54–62.

9. Bertrand Russell, *Ideas and Beliefs of the Victorians: An historic revaluation of the Victorian Age* (London: Sylvan Press, 1949), 20.

10. When Alan Greenspan commented that the rate of growth could not continue without, eventually, a corresponding increase in wages, *The Wall Street Journal* responded: "At the moment, the share of total gross domestic product that businesses keep is at a three-decade high of about 10 per cent, while the portion going to workers has slipped somewhat in recent years to 58 per cent. Changing that balance could be healthy for the economy, putting more money in the hands of consumers and dampening the possibility of social unrest."

11. J. Walker Smith and Ann Clurman, *Rocking the Ages: The Yankelovich Report on Generational Marketing* (New York: HarperCollins, 1997), 102.

12. *American Demographics*, May 1996.

13. *Business Week*, 3 November 1997.

14. Debbie Goad, "Hello, My Name Is Temp 378," *Temp Slave*, Issue 10, 6.

15. Steven Greenhouse, New York Times Service, printed in *International Herald Tribune*, 31 March 1998, 1.

16. Helm, "Microsoft Testing Limits..."

17. Greenhouse, New York Times Service. The worker quoted is Rebecca Hughes, who edited CD-ROMs at Microsoft as a permatemp for three years.

18. Charles Handy, *The Hungry Spirit* (London: Hutchinson, 1997), 70–71.

19. Hal Niedzviecki, "Stupid Jobs Are Good to Relax With," *This Magazine*, January/February 1998, 16–19.

CHAPTER TWELVE: CULTURE JAMMING

1. Personal interview.

2. Saul D. Alinsky, *Rules for Radicals: A Pragmatic Primer for Realistic Radicals* (Random House: New York, 1971), 152.

3. Personal interview. Many adbusters I interviewed chose to remain anonymous.

4. Personal interview.

5. Mary Kuntz, "Is Nothing Sacred," *Business Week*, 18 May 1998, 130–37.

6. Ibid.

7. Personal interview.

8. Personal interview.

9. Katz, *Just Do It*, 39.

10. *New York Times*, 4 April 1990, B1. DeWitt F. Helm Jr., president of the Association of National Advertisers, called the whitewashing over cigarette and alcohol ads by church groups "vigilante censorship."

11. Alison Fahey, "Outdoor Feels the Drought," *Advertising Age*, 6 August 1990, 3.

12. "The Greatest Taste Around," *Dispepsi*, Negativland, 1997.

13. "Soda Pop," *Entertainment Weekly*, 26 September 1997.

14. Susan J. Douglas, *Where the Girls Are* (Times Books: New York, 1994), 227.

15. Personal interview.

16. Stephanie Strom, "Billboard Owners Switching, Not Fighting," *New York Times*, 4 April 1990, B1.

17. Steinem, "Sex, Lies & Advertising."

18. Personal interview.

19. John Seabrook, "The Big Sellout," *New Yorker*, 20 & 27 October 1997, 182–95.

20. Bob Paquin, "E-Guerrillas in the Mist," *Ottawa Citizen*, 26 October 1998.

21. Personal interview.

22. Manifesto produced by Earth First! in Brighton, England.

23. *Guerrilla Shots* 1, no. 1.

24. Carrie McLaren, "Advertising the Uncommercial," *Escandola*, published by Matador Records, November 1995.

25. Jim Boothroyd, "ABC Opens the Door," *Adbusters*, Winter 1998, 53–54.

26. Personal interview.

27. Mitchel Raphael, "Corporate Perversion," *Toronto Star*, 7 February 1998, M1.

28. Doug Saunders, "One Person's Audio Debris Is Another's Musical Treasure," *Globe and Mail*, 25 September 1997, C5.

29. Barnaby Marshall, "Negativland: Mark Hosler on the Ad Assault," *Shift* on-line, 22.

30. *Time*, 17 November 1997.

31. *Advertising Age*, 18 November 1996.

32. Lopiano-Misdom and De Luca, *Street Trends*, 27–28.

33. "Anarchy in the U.K.," *Times* (London), 16 May 1998.

34. Martin Espada, *Zapata's Disciple* (Boston: South End Press, 1998).

35. "Nader Nixes Nike $25K Run," *Washington Post*, 13 May 1999.

36. Wilson Bryan Key, *Subliminal Seduction* (New York: Penguin, 1973), 7.

37. James Twitchell, *Adcult USA: The Triumph of Advertising in American Culture* (New York: Columbia University Press, 1996), 12.

38. The term "creative psychiatry" comes from a speech made by Columbia journalism professor Walter B. Pitkin to the 1933 convention of the Association of National Advertisers.

39. Rorty, *Our Master's Voice*, 382–83.
40. Ibid.
41. C.B. Larrabee, "Mr. Schlink," *Printer's Ink*, 11 January 1934, 10.

CHAPTER THIRTEEN: RECLAIM THE STREETS

1. John Jordan, "The Art of Necessity: The Subversive Imagination of Anti-Road Protest and Reclaim the Streets," in *DiY Culture: Party & Protest in Nineties Britain*, edited by George McKay (London: Verso, 1998).
2. Ibid.
3. Ibid.
4. Ibid.
5. RTS agitprop.
6. Personal interview.
7. *Mixmag*, no. 73, June 1997, 101.
8. A Liverpool dock worker quoted in *Do or Die*, no. 6, 9.
9. *Daily Telegraph*, 14 April 1997.
10. *Express*, 13 April 1997.
11. Personal interview.
12. RTS Sydney official report.
13. Both quotations come from the *SchNEWS* report of the event.
14. *San Francisco Weekly*, 27 May 1998.
15. This list of cities represents everyone who tried to plan a street party. Not everyone was successful and there were a few last-minute cancellations.

CHAPTER FOURTEEN: BAD MOOD RISING

1. From an interview conducted by Oxblood Ruffin of the hacker group Cult of the Dead Cow. Authenticity of the interview confirmed with source.
2. "Derr Pied!" Biotic Baking Brigade press release, 10 March 1999.
3. "Entarteurs Take Note: Custard Wins Test of Best Pies for Throwing," *Wall Street Journal*, 26 May 1999.
4. Kitty Krupat, "From War Zone to Free Trade Zone," in *No Sweat*, 56.
5. "Toy Story," *Dateline*, NBC, 17 December 1996.
6. Sydney H. Schamberg, "Six Cents an Hour," *Life*, June 1996.
7. "Suu Kyi Calls for Halt to Investment in Burma," *Australian Herald*, 4 September 1995.
8. "Disney Labor Abuses in China," report produced by the Hong Kong Christian Industrial Committee.
9. Ortega, *In Sam We Trust*, 236.
10. Philip S. Foner, *Women and the American Labor Movement* (New York: The Free Press, 1979), 358.
11. William Greider, *One World, Ready or Not* (New York: Simon & Schuster, 1997), 338.

12. The libel suit was against "a Swiss group who translated War on Want's publication *The Baby Killer...*" Source: "Baby Milk: Destruction of a World Resource" (London: Catholic Institute for International Relations, 1993), 3.

13. Fred Pearce, "Legacy of a Nightmare," *Guardian*, 8 August 1998. These numbers represent conservative estimates. Satinath Sarangi, a researcher based in Bhopal, puts the death toll at 16,000.

14. Myriam Vander Stichele and Peter Pennartz, *Making It Our Business: European NGO Campaigns on Transnational Corporations* (London: Catholic Institute for International Relations, 1996).

15. Personal interview.

16. "Corporations and Human Rights," *Human Rights Watch 1997 World Report*.

17. Julie Light, "Repression, Inc.: The Assault on Human Rights," *Corporate Watch*, 4 February 1999.

18. "The 'Enron Project' in Maharashtra: Protests Suppressed in the Name of Development," Amnesty International, 17 July 1997, 2.

19. Doonesbury, *Toronto Star*, 25 and 27 June 1997.

20. Pico Iyer, "India's Night of Death," *Time*, 17 December 1984.

21. Personal interview.

22. Personal interview.

CHAPTER FIFTEEN: THE BRAND BOOMERANG

1. Stuart Ewen, *Captains of Consciousness* (New York: McGraw-Hill, 1976), 80.

2. Personal interview.

3. Lorraine Dusky, "What Jogging Has to Do with Jogjakarta," *USA Today*, 21 May 1998.

4. Neal Stevenson, *Snow Crash* (New York: Bantam Books, 1992), 7.

5. "Testimony of Wendy Diaz before the Committee on International Relations," Congressional Testimony by Federal Document Clearing House, 11 June 1996.

6. Joyce Barrett and Joanna Ramey, "Sweatshop-buster Charles Kernaghan: Fashion Hits Its Nader; Ralph Nader," *Women's Wear Daily*, 6 June 1996, 1.

7. Steven Greenhouse, "Anti-Sweatshop Crusader Makes Celebrities, Big Business Tremble," *New York Times*, 4 July 1996.

8. Kernaghan, "Behind the Label: 'Made in China.'"

9. Various sources. Confirmed with NLC.

10. Cathy Majtenyi, "Were Disney Dogs Treated Better Than Workers?" *Catholic Register*, 23–30 December 1996, 9.

11. "An Appeal to Walt Disney" in *No Sweat*, edited by Andrew Ross, 101.

12. Cynthia Enloe, "We Are What We Wear," in *Of Common Cloth*, edited by Wendy Chapkis and Cynthia Enloe (Amsterdam: Transnational Institute, 1983), 119.

13. Personal interview.

14. Ortega, *In Sam We Trust*, 228.

15. "What's Wrong with McDonald's Factsheet," first published by London Greenpeace in 1986.

16. "Honduran Child Labor Described," *Boston Globe*, 30 May 1996.
17. Andrew Ross, "Introduction," in *No Sweat*, 27.
18. "Tarnished Rings?" *Wall Street Journal*, 6 January 1999.
19. "Hootie & the Blowfish Criticizes Suzuki for Ties to Burmese Military Junta," press release from Hootie & the Blowfish, 24 May 1999.
20. Josh Feit, "Stepping on Nike's Toes," *Now*, 27 November–3 December 1997, reprinted from *Willamette Week*.
21. "Pond Crowd Pummels Eisner's Mighty Bucks," *Variety*, 2 March 1997.
22. *Hollywood Reporter*, 26 February 1997 and *Variety*, 3 March 1997.
23. *The Sweatshop Quandary: Corporate Responsibility on the Global Frontier*, edited by Pamela Valery (Washington, D.C.: Investor Responsibility Research Center, 1998), 19.

CHAPTER SIXTEEN: A TALE OF THREE LOGOS

1. Memo, 4 May 1998, from the Maquila Solidarity Network, "Nike Day of Action Canada Report & Task Force Update."
2. "Nike protest update," *Labour Alerts*, 18 October 1997.
3. "Nike Mobilization: Local Reports," *Labor Alerts*, Campaign for Labor Rights, 26 October 1998.
4. Mark L. Zusman, "Editor's Notebook," *Willamette Week*, 12 June 1996.
5. *Oregonian*, 16 June 1996.
6. Campaign for Labor Rights Web site, regional reports.
7. Nick Alexander, "Sweatshop Activism: Missing Pieces," *Z Magazine*, September 1997, 14–17.
8. Personal interview, 6 October 1997.
9. Katz, *Just Do It*, 271.
10. Letter dated 24 October 1997.
11. Personal interview.
12. David Gonzalez, "Youthful Foes Go Toe to Toe with Nike," *New York Times*, 27 September 1997, B1.
13. Personal interview.
14. Minutes from 10 September meeting between Nike executives and the Edenwald-Gun Hill Neighborhood Center.
15. Personal interview.
16. "Wages and Living Expense for Nike Workers in Indonesia," report released by Global Exchange, 23 September 1998.
17. "Nike Raises Wages for Indonesian Workers," *Oregonian*, 16 October 1998.
18. "Nike to Improve Minimum Monthly Wage Package for Indonesian Workers," Nike press release, 19 March 1999.
19. Steven Greenhouse, "Nike Critic Praises Gains in Air Quality at Vietnam Factory," *New York Times*, 12 March 1999.
20. Shanthi Kalathil, "Being Tied to Nike Affects Share Price of Yue Yuen," *Wall Street Journal*, 25 March 1998.

21. "Third quarter brings 70 percent increase in net income for sneaker giant," Associated Press, 19 March 1999.

22. "Cole Haan Joins Ranks of Shoe Companies Leaving Maine," Associated Press, 23 April 1999.

23. Zusman, "Editor's Notebook."

24. Robin Grove-White, "Brent Spar Rewrote the Rules," *New Statesman*, 20 June 1997, 17–19.

25. "Battle of Giants, Big and Small," *Guardian*, 22 June 1995, 4.

26. "Giant Outsmarted: How Greenpeace Sank Shell's Plan to Dump Big Oil Rig in Atlantic," *Wall Street Journal*, 7 July 1995.

27. Grove-White, "Brent Spar Rewrote the Rules."

28. "Battle of Giants," *Guardian*.

29. Suzanne Moore, "Sea Changes in Political Talk," *Guardian*, 22 June 1995.

30. Megan Tresidder, "Slick Answers in Oily Waters," *Guardian*, 24 June 1995, 27.

31. "Giant Outsmarted," *Wall Street Journal*.

32. Memo written by Major Paul Okuntimo, dated 5 May 1994, reprinted in *Harper's*, June 1996.

33. Andrew Rowell and Stephen Kretzmann, "The Ogoni Struggle," a report by Project Underground, Berkeley, 1996.

34. Nadine Gordimer, "In Nigeria, the Price for Oil Is Blood," *New York Times*, 25 May 1997.

35. Letter posted on the shell-nigeria-action list dated 2 June 1998.

36. From a personal letter from R.B. Blakely to the author, 6 June 1997.

37. Janet Guyon, "Why Is the Most Profitable Company Turning Itself Inside Out?" *Fortune*, 4 August 1997, 120.

38. Jonathan Schorr, "Board Holds Off on Shell Decision," *Oakland Tribune*, 8 August 1997.

39. Victor Dania, a Shell spokesman said "We're working on healing the wounds," quoted in "Shell Cleaning Up Act for Ogoniland Return," *Times* (London), 11 July 1997.

40. Matthew Tostevin, "Attacks Cut Nigerian Oil Output by One Fifth," Reuters, 7 October 1998.

41. "Nigerian Protesters Seize Shell Helicopters," BBC, 8 October 1999, and "Nigerians Seize Shell Oil Stations," Associated Press, 7 October 1998.

42. Franny Armstrong, "Why Won't British TV Show a Film about McLibel?" *Guardian*, 19 June 1998.

43. Ibid.

44. "McLibel in London," *Fortune*, 20 March 1995.

45. "Anti-McDonald's Activists Take Message Online," Associated Press, 27 March 1996.

46. "Activists Win Partial Victory in Appeal Over McDonald's Libel Case," Associated Press, 31 March 1999.

47. "Few Nuggets and Very Small Fries," *Guardian*, 20 June 1997, 22.

48. "Guess Who's Still in Trouble?" Campaign for Labor Rights, Newsletter #9, October 1997, 4.

49. John Vidal, *McLibel* (London: Macmillan, 1997), 314–15; quotation taken from the afterword written by Steel and Morris.

50. "Asian Workers Sue Retailers in U.S., Apparel Firms," *Wall Street Journal*, 14 January 1999.

51. David A. Love, "A New Leaf for Nigeria?" *Washington Post*, 22 August 1998, A17. Jenny Green quote: "Shell Faces Saro-Wiwa Legal Action," *Independent*, 21 May 1999.

52. "Sites for Sore Consumers," *Washington Post*, 29 March 1998.

53. Eveline Lubbers, "Counterstrategies Against Online Activism," *Telepolis*, 22 September 1998.

54. That figure, provided by McSpotlight, is a tally of every "hit" and therefore includes repeat visitors.

55. Personal interview.

56. Lubbers, "Counterstrategies Against Online Activism."

57. John Vidal, "Modem Warfare," *Guardian*, 13 January 1999.

CHAPTER SEVENTEEN: LOCAL FOREIGN POLICY

1. Personal interview.

2. G. Kramer, "Suu Kyi Urges U.S. Boycott," Associated Press, 27 January 1997.

3. Farhan Haq, "Burma-Finance: Oil company digs in heels despite Rangoon's record," Inter Press Service, 4 February 1997.

4. "Pepsi, Burma, Take 2: Pepsi Responds to Aims of Target Audience," Dow Jones News Service, 27 January 1997.

5. Free Burma Coalition Web site.

6. "NUS Withdraws from McDonald's 'Privilege Card' Scheme," McLibel Support Campaign press release, 14 April 1998.

7. "Nike Campaign Strategy, Part 1," *Labor Alerts*, 14 January 1998.

8. "Reports on Nike Demos," *Labor Alerts*, 21 April 1998.

9. Verena Dobnik, "Anti-Sweatshop Protesters March up Fifth Avenue," Associated Press, 6 March 1999.

10. Feit, "Stepping on Nike's Toes."

11. "Was Your School's Cap Made in This Sweatshop? A UNITE Report on Campus Caps Made by BJ&B in the Dominican Republic," released 13 April 1998.

12. "Dominican Republic Workers Urge University of Illinois to Demand Humane Factory Conditions," *Daily Illini*, 24 April 1998.

13. Steven Greenhouse, "Duke to Adopt a Code to Prevent Apparel from Being Made in Sweatshops," *New York Times*, 8 March 1998.

14. Steven Greenhouse, "Activism Surges at Campuses Nationwide, and Labor Is at Issue," *New York Times*, 29 March 1999.

15. Ibid.

16. "An Open Letter to the Students," by Charles Kernaghan, undated.

17. William Cahn, *Lawrence 1912: The Bread & Roses Strike* (New York: The Pilgrim Press, 1977), 174.

18. Reagan administration trade representative Clayton Yeutter said, "Resort to unilateral sanctions has become almost a fad." Quoted by Harry Dunphy, "States Dictate Own Foreign Policy," Associated Press, 13 April 1998.

19. "Struggle over States," *Journal of Commerce*, 4 August 1998.

20. Lucille Renwick, "Teens' Efforts Give Soccer Balls the Boot," *Los Angeles Times*, 23 December 1996.

21. Simon Billenness, "Investing for a Better World," 15 April 1998, published by Franklin Research & Development Corporation.

22. Elaine Herscher, "Berkeley Running Out of Gas," *San Francisco Chronicle*, A1.

23. Robert Greenberger, "State and Local Sanctions Trouble U.S. Trade Partners," *Wall Street Journal*, 1 April 1998.

24. Craig Forcese, *Putting Conscience into Commerce: Strategies for Making Human Rights Business as Usual* (Montreal: International Centre for Human Rights and Democratic Development, 1997), 75.

25. Ken Silverstein, "So You Want to Trade with a Dictator," *Mother Jones*, May 1998.

26. "E.U. Raps U.S. on Trade Barriers," *European Report*, 11, 28 December 1998.

27. "Test Case Filed Contesting Validity of State and Local Sanctions Laws," 30 April 1998. ICFTU press release.

28. "Judge Says Law Is Unconstitutional," Associated Press, 5 November 1998.

29. "Test Case Filed Contesting Validity of State and Local Sanctions Laws," 30 April 1998. ICFTU press release.

30. Charles Oliver, "What Do You Do When a City Enacts Its Own Foreign Policy?" *Investor's Business Daily*, 19 August 1997.

31. Silverstein, "So You Want to Trade with a Dictator."

32. Statement made by Simon Billenness in 10 July 1998 "Massachusetts Burma Law Case Update."

33. *Shell Canada Products v. Vancouver (City)* [1994] 1 S.C.R. 231, 110 D.L.R. (4th) 1, 163 N.R. 81.

34. Comments made by councilor Barbara Perrault. "NV City takes a swing at Shell," *North Shore News*, 21 March 1997, 3.

35. "ERA's Environmental Testimonies #5," published by Environmental Rights Action/Friends of the Earth Nigeria, 10 July 1998.

36. Danielle Knight, "Oil Giant Had Role in Killing," International Press Service, 2 October 1998, and personal interview with Mike Libbey, 4 June 1999.

37. "Chevron, oil communities fail to agree on compensation," *Punch* (Lagos), 16 July 1998, 9.

NOTES

CHAPTER EIGHTEEN: BEYOND THE BRAND

1. Greider, *One World, Ready or Not*, 497.

2. "Nike and Free Trade Failures: An Analysis by Campaign for Labor Rights," *Labor Alerts*, 14 July 1998.

3. "Starbucks Criticized on Coffee-Workers Promise," *Seattle Post-Intelligencer*, 6 March 1997.

4. Ortega, *In Sam We Trust*, 317.

5. William J. Holstein, "Casting Nike as the Bad Guy," *U.S. News and World Report*, 22 September 1997, 49.

6. Gerard Greenfield, "The Impact of TNC Subcontracting on Workers in Asia: Strategy Report – Part 2." Unpublished.

7. Ortega, *In Sam We Trust*, xv.

8. Personal interview.

9. "Heineken Bows to Pressure and Withdraws from Burma," Reuters, 10 July 1996.

10. "U.S. Oil Company Vows to Remain in Thai-Burmese Pipeline Project," Deutsche Press-Agentur, 17 June 1997.

11. Gordon Laird, "Speak No Evil," *This Magazine*, March/April 1998, 18–25.

12. Letter from Tokiro Kawamura, president, Daishowa-Marubeni International, to Bernard Ominayak, Chief of the Lubicon Lake Indian Nation, 20 May 1998.

13. Personal interview.

14. "M&S Bows to Shoppers' Fears and Orders Ban on Frankenfoods," *Daily Mail*, 16 March 1999.

15. Greenfield, unpublished strategy report.

16. Alan L. Rolnick, "Muzzling the Offshore Watchdogs," *Bobbin*, February 1997, 71.

17. Shell advertisement, *Business Week*, 5 April 1999.

18. "Profit, Profit, Profit: An Act of Commitment," statement released by UK Oil Overthrow Association, 21 April 1999.

19. Alinsky, *Rules for Radicals*, 152.

20. "A China Business Code," *Labor Alerts*, 5 June 1999.

21. "Business-Humanitarian Forum Holds First Meeting," Business Humanitarian Forum press release, 27 January 1999.

22. Debora L. Spar, "The Spotlight on the Bottom Line," *Foreign Affairs*, 13 March 1998.

23. "Nike, Reebok Compete to Set Labor Rights Pace," *Labor Alerts*, 25 March 1999.

24. *Journal of Commerce*, 17 April 1997.

CONCLUSION: CONSUMERISM VERSUS CITIZENSHIP

1. Guy de Jobquieres, "Network guerrillas," *Financial Times*, 30 April 1998.

APPENDIX

Table 1.3. Absolut Vodka Ad Spending, 1989–97

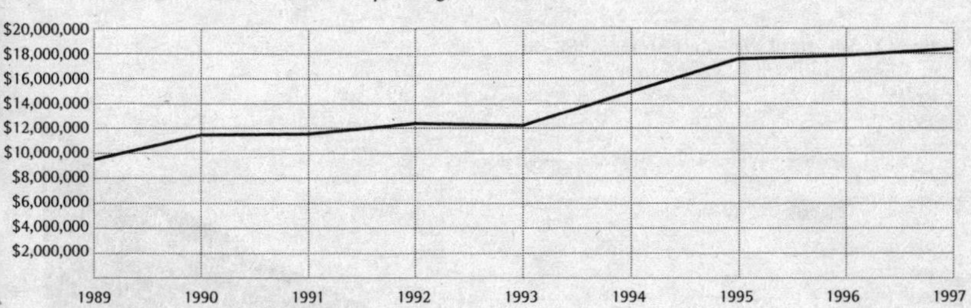

Source: Annual "Media Spending Guide" in *Food & Beverage Marketing* (August 1991, August 1993, August 1995, July 1996, August 1998).

Table 1.4. Ad Spending Patterns of the Superbrands, 1981–97

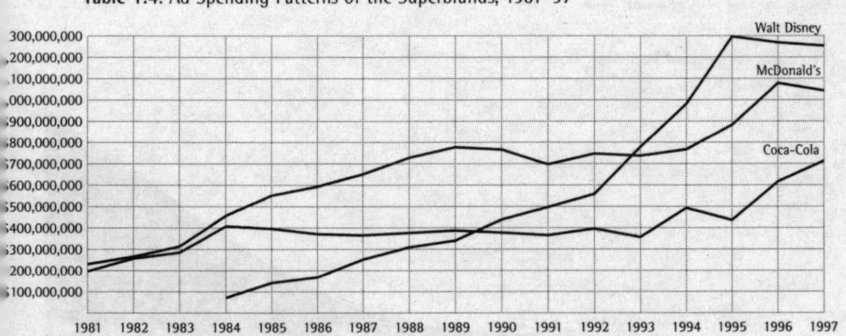

Source: "100 Leading National Advertisers," *Advertising Age*, 1982–98.

APPENDIX

Table 2.1a. Corporate Tax as a Percentage of Total Federal Revenue in Canada, 1955, 1983 and 1998

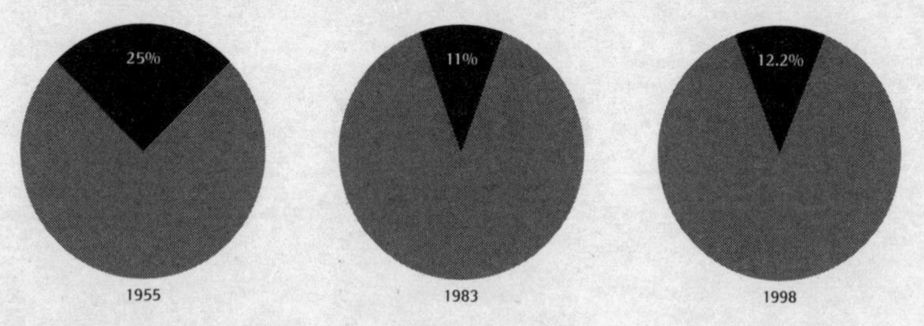

Source: Department of Finance, Canadian Economic Observer and Statistics Canada.

Table 6.1. Growth of Wal-Mart Stores, 1968–98

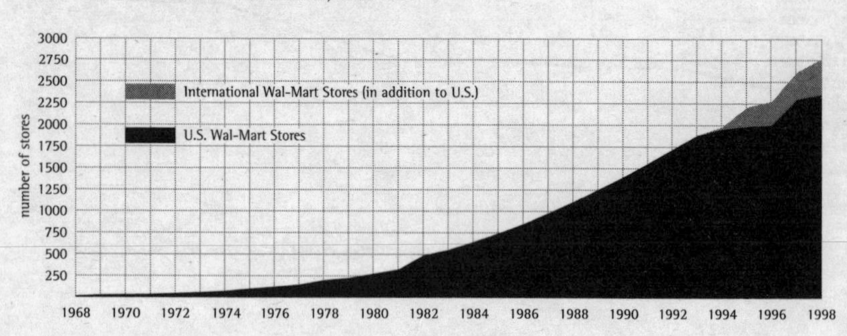

Source: Wal-Mart provided growth figures.

Table 6.2. Growth of Wal-Mart Supercenters, 1988–98

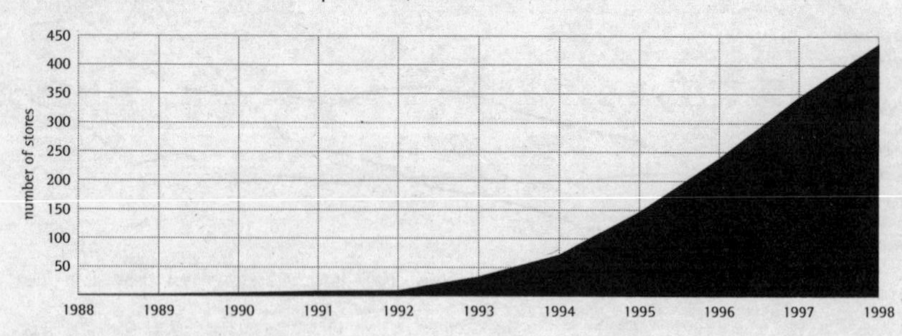

Source: Wal-Mart provided growth figures.

Table 6.3. Starbucks Percent Change in Sales per Store, 1993–98

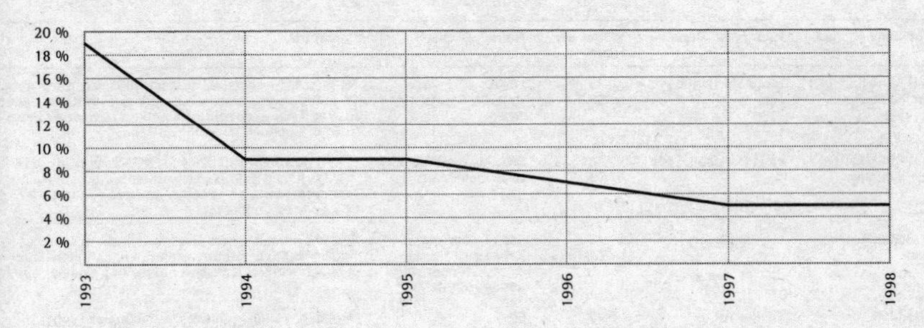

Source: Starbucks Annual Reports 1997 and 1998.

Table 6.4. Starbucks Total Net Revenues, 1993–98

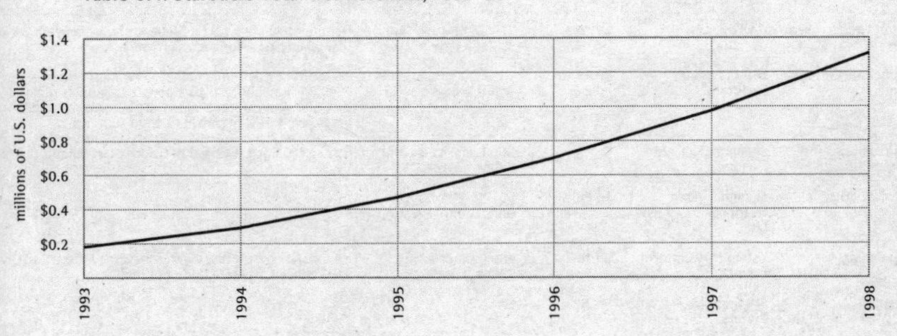

Source: Starbucks Annual Reports 1997 and 1998.

Table 9.1. Adidas Profits, 1993–97

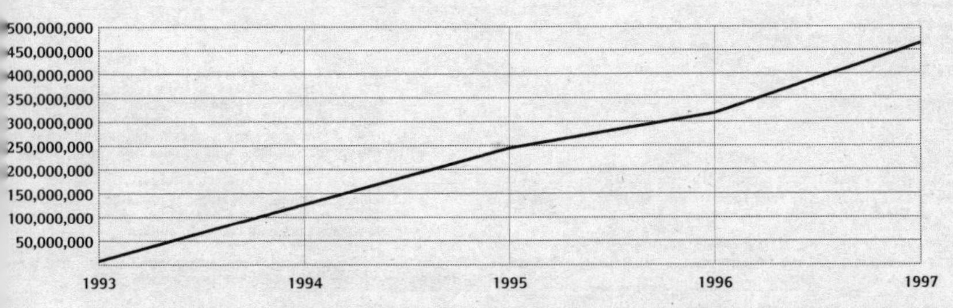

Sources: AFX News, 11 April 1995; AFX News, 7 March 1996; Reuters Financial Service, 11 March 1997; AFX News, 5 March 1998. As of September 1999, one German mark was worth US$0.53.

APPENDIX

Table 9.3. Sweatshop Profiles

Company / Label	Factory in China	Wages per Hour	Hours per Week	Conditions
Wal-Mart/Kathie Lee handbags	Liang Shi Handbag Factory	$0.13 to $0.23	60–70; 10-hour shifts; 6–7 days a week	No factory fire exits; dirty, cramped dorms, 10 to a roor 70 hours a week, warehouse workers earn $3.44; no be fits; no legal work contract; workers have never heard c Code of Conduct
Wal-Mart/Kathie Lee handbags	Ya Li Handbag, Ltd.	$0.18 to $0.28	60; plus overtime up to 16-hour shifts	Forced overtime—stiff fines for refusal; overtime premiu 2 1/2 cents an hour; some workers not paid 3–4 montl to a dorm room; no benefits, no work contract; never l of a Code of Conduct
Wal-Mart/Kathie Lee	Li Wen Factory	$0.20 to $0.35	84; 12-hour shifts; 7 days a week; mandatory 24-hour shifts during rush times	Forced overtime, severe fines for refusal to comply; no fits, no overtime rate; no fire exits in dormitories; no w contract; workers have never heard of a Code of Condu
Wal-Mart	Tianjin Yuhua Garment Factory	$0.23	60	Wal-Mart is pulling out of this factory and other large licly owned plants in the north to relocate its work to u ulated lower-wage privately owned sweatshops in the s of China
Ann Taylor and Preview	Kang Yi Fashion Manufacturers	$0.14	96; 7 days a week; 7 a.m. to midnight	Workers have never heard of a Code of Conduct; 6 to10 workers in dorm rooms
Ralph Lauren, Ellen Tracy/Linda Allard	Iris Fashions	$0.20	72–80; 12- to 15- hour shifts; 6 days a week	No union; workers paid a $0.06 an hour premium for o time; paid $0.02 for each shirt collar sewn
Esprit Label (Esprit Group)	You Li Fashion Factory	$0.13	93; 7:30 a.m. to midnight; 7 days a week	No overtime pay; no benefits; sometimes employees ne work 24-hour shifts; 6 to 8 people to a dorm room; do dark and dirty; workers afraid; under constant surveillar never heard of a corporate Code of Conduct
Liz Claiborne and Bugle Boy	Shanghai Shirt 2d Factory	$0.25	66; 8 a.m. to 8 p.m.; 6 days a week	Employees fined if they don't work overtime; no union
Liz Claiborne	Shanghai Jiang District Silk Fashions Ltd.	$0.28	60–70; 11.5-hour shift; 6 days a week	
J.C. Penney	Zhong Mei Garment Factory	$0.18	78; 11-hour shifts; 7 days a week	No union; no benefits; workers have never heard of J.C. Penney Corporate Code of Conduct
Kmart	Shanghai No.4 Shirt Factory	$0.28	70	
Cherokee Jeans	Meiming Garment Factory	$0.24	60–70	No benefits; workers have never heard of monitoring; 8 dorm room
Sears	Tianjin Beifang Garment Factory	$0.28	60	Sears is pulling out to relocate its production in lower- unregulated sweatshops in the south
Structure/ The Limited	Aoda Garment Factory	$0.32	70	No union; 6 workers to a dorm room
Nike Athletic Shoes	Wellco Factory	$0.16	77–84; 11- to12-hour shifts; 7 days a week	Workers fined if they refuse to work overtime; overtime not paid; most had no legal work contract; humiliation, screaming, some corporal punishment; arbitrary fining c pregnant women and older (25 years old and up) wome fines if talking at work; approximately 10 children in th sewing section; most workers have never heard of Nike's Code of Conduct
Nike and Adidas Athletic Shoes	Yue Yuen Factory	$0.19	60–84	Forced overtime, no overtime premium paid; excessive n pollution, fumes in the factory; no worker had heard of or Adidas Corporate Code of Conduct
Adidas Garments	Tung Tat Garment Factory	$0.22	75–87.5; 12.5-hour shifts; 6 or 7 days a week	Employees fined if late/resting/found talking; forced mc ing calisthenics; 8 workers to a dorm room

Source: "Company Profiles/Working Conditions: Factories in China Producing Goods for Export to the U.S.," "Made in China: Behind the Label," Charles Kernaghan of the National Labor Committee, March 1998. Wages are in U.S. dollars.

Table 9.2. Percentage Changes in Employment in the Textile, Clothing, Leather and Footwear Industries, 1980–93

Country	% Change		% Change
Finland	-71.7	Mauritius	344.6
Sweden	-65.4	Indonesia	177.4
Norway	-64.9	Morocco	166.5
Austria	-51.5	Jordan	160.8
Poland	-51.0	Jamaica	101.7
Syria	-50.0	Malaysia	101.2
France	-45.4	Mexico	85.5
Hungary	-43.1	China	57.3
Netherlands	-41.7	Islamic Republic of Iran	34.0
United Kingdom	-41.5	Turkey	33.7
New Zealand	-40.9	Philippines	31.8
Germany	-40.2	Honduras	30.5
Spain	-35.3	Chile	27.2
Australia	-34.7	Kenya	16.1
Argentina	-32.9	Israel	13.4
United States	-30.1	Venezuela	7.9

Source: International Labour Office

Table 10.1. 1997 Average Hourly Earnings, Retail Trade vs. Overall Average in the U.S., Canada and the U.K.

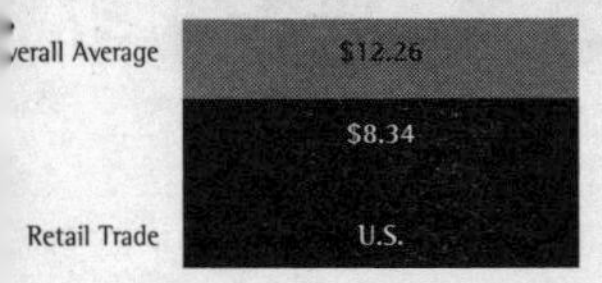

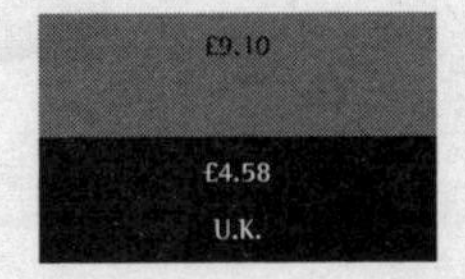

Source for U.S. figures: Bureau of Labor Statistics. Source for Canadian figures: Annual Estimates on Employment, Earnings and Hours, Statistics Canada. Source for U.K. figures: New Earnings Survey, Office for National Statistics.

Table 10.3. Aging Fast-Food Workers in Canada, 1987–97

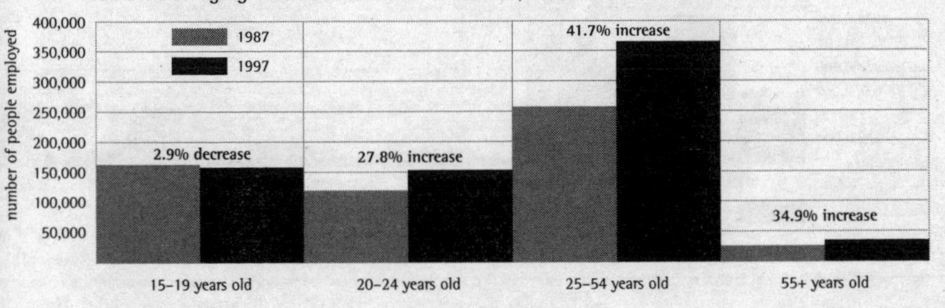

Source: Statistics Canada.

Table 10.4. Percentage of Employees Working "Full-time" versus "Part-time" in Selected Service-Sector Chains

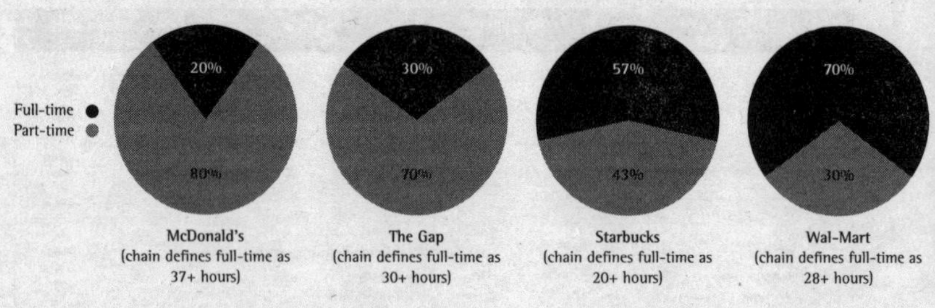

Source: Companies supplied information upon request.

Table 10.5. Part-time Employment as a Percentage of Total Employment in Canada and the U.K.

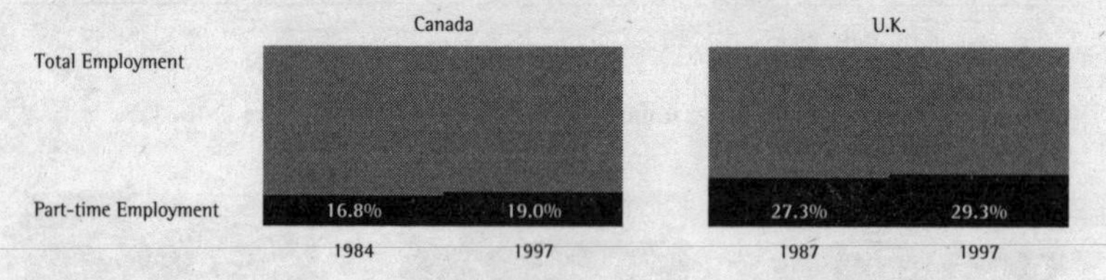

Source for Canadian figures: Statistics Canada, Labour Force Database. Source for U.K. figures: Office for National Statistics.

Table 10.6. Part-time Employment in the U.S., Canada and the U.K. by Sex, 1984/87 and 1997

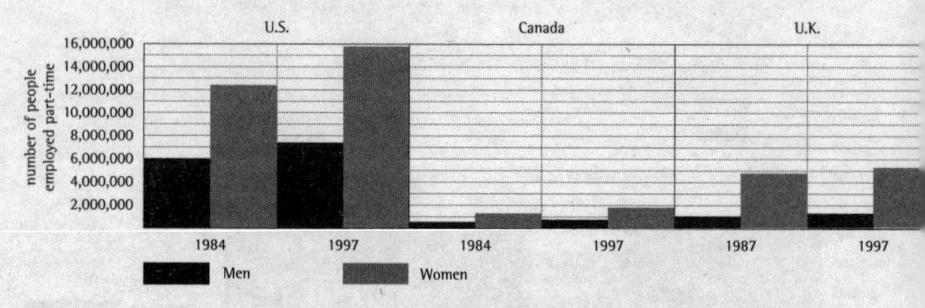

Source for U.S. figures: Bureau of Labor Statistics. Source for Canadian figures: Statistics Canada, Labour Force Database. Source for U.K. figures: Office of National Statistics; 1987 figures are estimates. No figures for part-time employment by age were available.

Table 10.7. Percentage of People Employed in Standard and Nonstandard Work Arrangements Who Receive Benefits in the U.S.

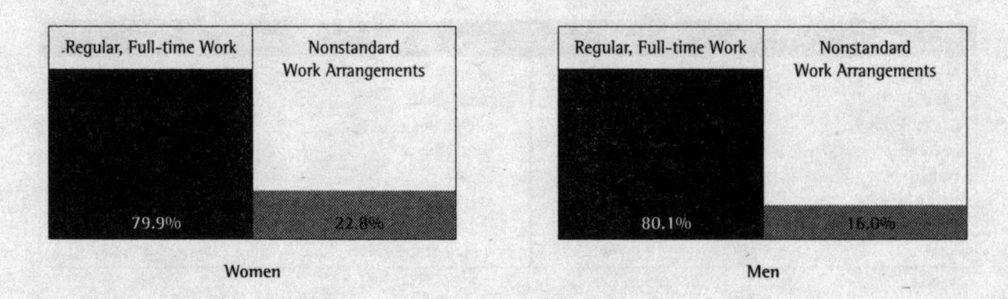

Source: Economic Policy Institute and Women's Research and Education Institute.

Table 11.5. Unemployment Rates for Selected Countries, 1970 and 1998

Country	1970	1998	Country	1970	1998
Australia	1.6 %	8.0 %	Netherlands [a]	1.1 %	5.0 %
Canada	5.9 %	8.3 %	Philippines [c]	5.2 %	7.9 %
Finland	1.9 %	11.4 %	South Korea [d]	4.5 %	6.8 %
France [a]	2.4 %	12.0 %	Spain	1.1 %	18.8 %
Germany [b]	0.7 %	12.3 %	Sweden	1.5 %	6.5 %
Italy [a]	5.4 %	12.0 %	United Kingdom	2.6 %	4.7 %
Japan [d]	1.1 %	4.1 %	United States	4.9 %	4.5 %

Source: *Yearbook of Labour Statistics*, from 1980 to 1999, International Labour Office (individual country sources: labour force sample surveys; social insurance statistics; employment office statistics; official estimates; and administrative records).
Note: Because of changes in methodology, not all data is strictly comparable.

[a] The 1998 data for France, Italy and the Netherlands is a forecast from the *World Economic and Social Survey 1998: Trends and Policies in the World Economy*, United Nations, based on data from the Organization for Economic Cooperation and Development.

[b] Germany's 1970 figure represents the Federal Republic of Germany.

[c] Philippines' figures compare 1971 and 1997.

[d] In February 1999, Japan's unemployment rate had risen to 4.7%. By March 1999, South Korea's had risen to 8.1%.

APPENDIX

Table 11.7. Youth Unemployment as a Percentage of Total Unemployment in Selected Countries, 1997

Country	1997	Country	1997
Australia	38.6 %	Japan	24.9 %
Canada	28.7 %	Netherlands	29.1 %
Czech Republic	30.5 %	Philippines	45.4 %
Finland	21.5 %	South Korea	34.4 %
France	20.3 %	Spain	30.8 %
Germany	12.2 %	Sweden	20.2 %
Indonesia*	72.5 %	United Kingdom	29.9 %
Italy	20.2 %	United States	35.9 %

Source: *Yearbook of Labour Statistics*, from 1980 to 1997, International Labour Office (individual country sources: labor-force sample surveys; social insurance statistics; employment office statistics; official estimates; and administrative records). Italy's figure from Istituto Nazionale di Statistica. All age brackets are 15–24, except Spain, Sweden, the U.K. and the U.S. which are 16–24, Germany which is 14–24, and France which is 18–24.

*Indonesia's figure is from 1996.

READING LIST

Aaker, David A. *Building Strong Brands*. New York: The Free Press, 1996.

Barlow, Maude and Heather-jane Robertson. *Class Warfare: The Assault on Canada's Schools*. Toronto: Key Porter Books, 1994.

Barnet, Richard J., and John Cavanagh. *Global Dreams: Imperial Corporations and the New World Order*. New York: Simon & Schuster, 1994.

Bey, Hakim. *T.A.Z.: Temporary Autonomous Zone, Ontological Anarchy, Poetic Terrorism*. Brooklyn: Autonomedia, 1985.

Cahn, William. *Lawrence, 1912: The Bread & Roses Strike*. New York: The Pilgrim Press, 1977.

Chapkis, Wendy, and Cynthia Enloe, eds. *Of Common Cloth: Women in the Global Textile Industry*. Amsterdam: Transnational Institute, 1983.

Clarke, Tony. *Silent Coup: Confronting the Big Business Takeover of Canada*. Ottawa: The Canadian Centre for Policy Alternatives and James Lorimer & Company, 1997.

Danaher, Kevin, ed. *Corporations Are Gonna Get Your Mama: Globalization and the Downsizing of the American Dream*. Monroe: Common Courage Press, 1996.

Debord, Guy. *The Society of the Spectacle*. Translated by Donald Nicholson-Smith. New York: Zone Books, 1994.

Dobbin, Murray. *The Myth of the Good Corporate Citizen: Democracy Under the Rule of Big Business*. Toronto: Stoddart, 1999.

Ewen, Stuart. *Captains of Consciousness: Advertising and the Social Roots of the Consumer Culture*. New York: McGraw-Hill, 1976.

Frank, Thomas. *The Conquest of Cool*. Chicago: The University of Chicago Press, 1997.

Goldman, Robert and Stephen Papson. *Sign Wars: The Cluttered Landscape of Advertising*. New York: The Guilford Press, 1996.

Greider, William. *One World, Ready or Not: The Manic Logic of Global Capitalism*. New York: Simon & Schuster, 1997.

Hargrove, Buzz. *Labour of Love: The Fight to Create a More Humane Canada*. Toronto: Macfarlane Walter & Ross, 1998.

Herman, Edward S., and Robert W. McChesney. *The Global Media: The New Missionaries of Global Capitalism*. London: Cassell, 1997.

Herman, Edward S. *Triumph of the Market: Essays on Economics, Politics, and the Media.* Boston: South End Press, 1995.

Karliner, Joshua. *The Corporate Planet: Ecology and Politics in the Age of Globalization.* San Francisco: Sierra Club Books, 1997.

Katz, Donald. *Just Do It: The Nike Spirit in the Corporate World.* Holbrook: Adams Media Corporation, 1994.

Korten, David C. *When Corporations Rule the World.* West Hartford: Kumarian Press and Berrett-Koehler Publishers, 1995.

Kunstler, James Howard. *The Geography of Nowhere: The Rise and Decline of America's Man-Made Landscape.* New York: Simon & Schuster, 1993.

Kuttner, Robert. *Everything for Sale: The Virtues and Limits of Markets.* New York: Alfred A. Knopf, 1997.

Mander, Jerry, and Edward Goldsmith, eds. *The Case Against the Global Economy.* San Francisco: Sierra Club Books, 1996.

McKay, George, ed. *DiY Culture: Party & Protest in Nineties Britain.* London: Verso, 1998.

Miller, Mark Crispin. *Boxed In: The Culture of TV.* Evanston: Northwestern University Press, 1988.

Moody, Kim. *Workers in a Lean World.* London: Verso, 1997.

Moore, Michael. *Downsize This! Random Threats from an Unarmed American.* New York: Crown Publishers, 1996.

Nava, Mica, Andrew Blake, Iain MacRury and Barry Richards, eds. *Buy This Book: Studies in Advertising and Consumption.* London: Routledge, 1997.

Ortega, Bob. *In Sam We Trust: The Untold Story of Sam Walton and How Wal-Mart Is Devouring America.* New York: Random House, 1998.

Peters, Tom. *The Circle of Innovation.* New York: Alfred A. Knopf, 1997.

Rail, Genevieve, ed. *Sport and Postmodern Times.* Albany: State University of New York Press, 1998.

Ritzer, George. *The McDonaldization of Society: An Investigation into the Changing Character of Contemporary Social Life.* Thousand Oaks: Pine Forge Press, 1996.

Rodrik, Dani. *Has Globalization Gone Too Far?* Washington: Institute for International Economics, 1997.

Rorty, James. *Our Master's Voice.* New York: The John Day Company, 1934.

Ross, Andrew, ed. *No Sweat: Fashion, Free Trade, and the Rights of Garment Workers.* London: Verso, 1997.

Rothenberg, Randall. *Where the Suckers Moon: The Life and Death of an Advertising Campaign.* New York: Alfred A. Knopf, 1994.

Sassen, Saskia. *Losing Control? Sovereignty in an Age of Globalization.* New York: Columbia University Press, 1996.

Saul, John Ralston. *The Unconscious Civilization.* Concord: Anansi, 1995.

Savan, Leslie. *The Sponsored Life: Ads, TV, and American Culture*. Philadelphia: Temple University Press, 1994.

Schiller, Herbert I. *Culture Inc.: The Corporate Takeover of Public Expression*. New York: Oxford University Press, 1989.

Shorris, Earl. *A Nation of Salesmen: The Tyranny of the Market and the Subversion of Culture*. New York: Avon Books, 1994.

Smoodin, Eric, ed. *Disney Discourse: Producing the Magic Kingdom*. New York: Routledge, 1994.

Sontag, Susan. *Against Interpretation*. New York: Anchor Books, 1986.

Twitchell, James B. *Adcult USA: The Triumph of Advertising in American Culture*. New York: Columbia University Press, 1996.

Vanderbilt, Tom. *The Sneaker Book: Anatomy of an Industry and an Icon*. New York: The New Press, 1998.

Vidal, John. *McLibel: Burger Culture on Trial*. London: Macmillan, 1997.

Wernick, Andrew. *Promotional Culture: Advertising, Ideology and Symbolic Expression*. London: Sage Publications, 1991.

Wilson, William Julius. *When Work Disappears: The World of the New Urban Poor*. New York: Alfred A. Knopf, 1996.

Wolf, Michael J. *The Entertainment Economy: How Mega-Media Forces Are Transforming Our Lives*. New York: Random House, 1999.

PHOTO CREDITS

page xii: Felix Wedgwood
page 2: (top) Quaker Oats (detail, red tint added)
page 10: Jayce Salloum, *At the End of the "Nation State"* (detail), Beirut, 1992
page 18: (top) CP Picture Archive/Elise Amendola; (bottom) CP Picture Archive/Bernd Kammerer
page 32: Absolut Vodka/Keith Haring, *Absolut Haring* (detail), 1992
page 62: (top) CP Picture Archive/Richard Drew; (bottom) Barr Gilmore
page 86: (top) Apple Computers Inc.; (bottom) Bedford Park Public School, Sesquicentennial Museum and Archives, Toronto District School Board (Channel One logo added)
page 106: (top) CP Picture Archive/Jose Goita; (bottom) Diesel ad, spring 1995
page 128: (top) CP Picture Archive/Todd Korol; (bottom) Lawrence Emerson/*The Fauquier Citizen*
page 194: (top) Robin Romano; (bottom) CP Picture Archive/Malcolm Dunlop
page 230: (top) Steve Payne/(bottom) *Adbusters*
page 234: Chris Woods, *McDonald's Nation* (detail), 1997, courtesy the artist and Diane Farris Gallery, Vancouver
page 258: (bottom) Whispered Media, *The Pie's the Limit*
page 272: Barr Gilmore
page 278: (top) *The Ballyhoo*, June 1932; (bottom) Ron English
page 310: (top and bottom) Nick Cobbing/David Hoffman Photo Library
page 322: Reclaim the Streets poster, London, 1995
page 324: (top) CP Picture Archive/Michael Schmelling; (bottom) courtesy *Justice: Do It Nike!*
page 344: (top) *San Francisco Examiner*/Darryl Bush; (bottom) Jubal Brown, *Ad Death*, 1996, performance/vandalism, courtesy the artist (red tint added)
page 364: Greenpeace/Melanie Kemper (detail, red tint added)
page 420: (top) CP Picture Archive/Susan Walsh
page 438: (top) CP Picture Archive/Lionel Cirronneau; (bottom) CP Picture Archive/Tom Hanson
page 446: courtesy McSpotlight Web site

INDEX

INDEX

INDEX

INDEX